ECOLOGY

Key Concepts in Critical Theory

Series Editor
Roger S. Gottlieb

Key Concepts
in
Critical Theory

ECOLOGY

SECOND EDITION

Carolyn Merchant

Humanity
Books

an imprint of Prometheus Books
59 John Glenn Drive, Amherst, New York 14228-2119

Published 2008 by Humanity Books, an imprint of Prometheus Books

Ecology, Second Edition. Copyright © 2008 by Carolyn Merchant. All rights reserved. No part of this publication may be reproduced, stored in a retrieval system, or transmitted in any form or by any means, digital, electronic, mechanical, photocopying, recording, or otherwise or conveyed via the Internet or a Web site without prior written permission of the publisher, except in the case of brief quotations embodied in critical articles and reviews.

Inquiries should be addressed to
Humanity Books
59 John Glenn Drive
Amherst, New York 14228–2119
VOICE: 716–691–0133, ext. 210
FAX: 716–691–0137
WWW.PROMETHEUSBOOKS.COM

12 11 10 09 08   5 4 3 2 1

Library of Congress Cataloging-in-Publication Data

Merchant, Carolyn.
    Ecology / by Carolyn Merchant. — 2nd ed.
        p. cm. — (Key concepts in critical theory)
    Includes bibliographical references and index.
    ISBN 978–1–59102–578–8
    1. Human ecology—Philosophy. 2. Social ecology—Philosophy. 3. Ecology.
I. Title.

GF21.E25 2008
304.2—dc22   2008000592

Printed in the United States of America on acid-free paper

CONTENTS

PART IV. ECOFEMINISM

PART V. ENVIRONMENTAL JUSTICE

PART VI. SPIRITUAL ECOLOGY

PART VII. POSTMODERN SCIENCE

CONCLUSION

INDEX   411

SERIES EDITOR'S PREFACE

The vision of a rational, just, and fulfilling social life, present in Western thought from the time of the Judaic prophets and Plato's *Republic*, has since the French Revolution been embodied in systematic critical theories whose adherents seek a fundamental political, economic, and cultural transformation of society.

These critical theories—varieties of Marxism, socialism, anarchism, feminism, gay/lesbian liberation, ecological perspectives, discourses by antiracist, anti-imperialist, and national liberation movements, and utopian/critical strains of religious communities—have a common bond that separates them from liberal and conservative thought. They are joined by the goal of sweeping social change; the rejection of existing patterns of authority, power, and privilege; and a desire to include within the realms of recognition and respect the previously marginalized and oppressed.

Yet each tradition of Critical Theory also has its distinct features: specific concerns, programs, and locations within a geometry of difference and critique. Because of their intellectual specificity and the conflicts among the different social groups they represent, these theories have often been at odds with one another, differing over basic questions concerning the ultimate cause and best response to injustice, the dynamics of social change, the optimum structure of a liberated society, the identity of the social agent who will direct the revolutionary change, and in whose interests the revolutionary change will be made.

In struggling against what is to some extent a common enemy, in overlapping and (at times) allying in the pursuit of radical social change, critical theories to a great extent share a common conceptual vocabulary. It is the purpose of this series to explore that vocabulary, revealing what is common and what is distinct, in the broad spectrum of radical perspectives.

For instance, although both Marxists and feminists may use the word "exploitation," it is not clear that they really are describing the same phenomenon. In the Marxist paradigm the concept identifies the surplus labor appropriated by the capitalist as a result of the wage-labor relation. Feminists have used the same term to refer as well to the unequal amounts of housework, emotional nurturance, and child raising performed by women in the nuclear family. We see some similarity in the notion of group inequality (capitalists/workers, husbands/wives) and of unequal exchange. But we also see critical differences: a previously "public" concept extended to the private realm; one first centered in the economy of goods now moved into the life of emotional relations. Or, for another example, when deep ecologists speak of "alienation" they may be exposing the contradictory and destructive relations of humans to nature. For socialists and anarchists, by contrast, "alienation" basically refers only to relations among human beings. Here we find a profound contrast between what is and is not included in the basic arena of politically significant relationships.

What can we learn from exploring the various ways different radical perspectives utilize the same terminology?

Most important, we see that these key concepts have histories and that the theories of which they are a part and the social movements whose spirit they embody take shape through a process of political struggle as well as of intellectual reflection. As a corollary, we can note that the creative tension and dissonance among the different uses of these concepts stem not only from the endless play of textual interpretation (the different understandings of classic texts, attempts to refute counterexamples or remove inconsistencies, rereadings of history, reactions to new theories), but also from the continual movement of social groups. Oppression, domination, resistance, passion, and hope are crystallized here. The feminist expansion of the concept of exploitation could only grow out of the women's movement. The rejection of a purely anthropocentric (human-centered, solely humanistic) interpretation of alienation is a fruit of people's resistance to civilization's lethal treatment of the biosphere.

Finally, in my own view at least, surveys of the differing applications of these key concepts of Critical Theory provide compelling reasons to see how complementary, rather than exclusive, the many radical perspectives are. Shaped by history and embodying the spirit of the radical movements that created them, these varying applications each have in them some of the truth we need in order to face the darkness of the current social world and the ominous threats to the earth.

Roger S. Gottlieb

ACKNOWLEDGMENTS

The following publishers and authors have kindly granted permission to reprint or quote from their articles.

Karl Marx and Friedrich Engels, "Marx and Engels on Ecology," from *Marx and Engels on Ecology*, ed. Howard Parsons (Westport, CT: Greenwood Press, 1977 [an imprint of Greenwood Publishing Group, Inc., Westport, CT], 1977), pp. 129–85, selections. Copyright © 1977 by Howard L. Parsons. Reproduced with permission of Greenwood Publishing Group, Inc., Westport, CT.

Max Horkheimer and Theodor Adorno, "The Concept of Enlightenment," from *Dialectic of Enlightenment* by Max Horkheimer and Theodor Adorno (1944), trans. John Cumming (New York: Herder & Herder, 1972), pp. 3–28, 42, excerpts. English translation copyright © by Herder & Herder, Inc. Reprinted by permission of The Continuum Publishing Company.

Herbert Marcuse. "Ecology and Revolution," from *Liberation* (September 1972): 10–12. Reprinted with permission of the Literary Estate of Herbert Marcuse, Peter Marcuse, Executor. Supplementary material from previously unpublished work of Herbert Marcuse, much now in the Archives of the Goethe University in Frankfurt/Main, has been and will be published by Routledge Publishers, England, in a six-volume series edited by Douglas Kellner and in a German series edited by Peter-Erwin Jansen published by zu Klampen Verlag, Germany. All rights to further publication are retained by the Estate.

William Leiss, "The Domination of Nature," from William Leiss, "Technology and Domination," in *The Domination of Nature* (New York: George Braziller, 1972), pp. 145–55, 161–65. Copyright © 1972 by William Leiss. Reprinted by permission of George Braziller, Inc., McGill-Queen's University Press, and William Leiss.

Robyn Eckersley, "The Failed Promise of Critical Theory." Reprinted from *Environmentalism and Political Theory: Toward an Ecocentric Approach* by Robyn Eckersley (Albany: State University of New York Press, 1992), pp. 97–106, by permission of

11

the State University of New York Press. Copyright © by the State University of New York. All rights reserved.

Rosemary Radford Ruether, "Corporate Globalization," from *Integrating Ecofeminism, Globalization and World Religions* by Rosemary Radford Ruether (Lanham, MD: Rowman & Littlefield, 2005), pp. 1–8.

Joel Kovel, "Global Capitalism and the End of Nature," from *The Enemy of Nature: The End of Capitalism or the End of the World?* by Joel Kovel (London: Zed Books, 2002), pp. 20–25, 115–16.

Brian Tokar, "Global Ecological Movements" from *Earth for Sale: Reclaiming Ecology in the Age of Corporate Greenwash* (Boston: South End Press, 1997), pp. 159–64, 168–74.

Barry Commoner, "Population and Poverty," from *Making Peace with the Planet* (New York: Pantheon, 1990), pp. 141–43, 155–60, 161–64, 168. Copyright © Barry Commoner. Reprinted by permission of Barry Commoner.

Paul Hawken, Amory Lovins, and L. Hunter Lovins, "Natural Capitalism," pp. 150–59, from *Natural Capitalism* by Paul Hawken. Copyright © 1999 by Paul Hawken, Amory Lovins, and L. Hunter Lovins. By permission of Little, Brown and Co.

Arne Naess, "Deep Ecology," reprinted from "The Shallow and the Deep, Long-Range Ecology Movement: A Summary," by Arne Naess from *Inquiry*, www.tandf.no/inquiry, 1973, 15: 95–100, by permission of Taylor & Francis AS.

Bill Devall, "The Deep Ecology Movement," from *Natural Resources Journal* 20 (April 1980): 299–313.

George Sessions, "Ecocentrism and the Anthropocentric Detour," from *ReVision* 13, no. 3 (Winter 1991): 109–15. Reprinted with permission of the Helen Dwight Reid Educational Foundation. Published by Heldref Publications, 1319 Eighteenth Street, NW Washington, DC 20036-1802. Copyright © 1991.

Murray Bookchin, "The Concept of Social Ecology," from *CoEvolution Quarterly* (Winter 1981): 15–22. Copyright © 1981 by Murray Bookchin. Reprinted with the permission of the Murray Bookchin Estate.

James O'Connor, "Socialism and Ecology," from *Capitalism, Nature, Socialism* 2, no. 3 (1991): 1–12. "Socialism and Ecology" has been reprinted in *Our Generation* (Canada), *Il Manifesto*, (Italy), *Ecologia Politica*, (Spain), *Nature and Society* (Greece), *Making Sense* (Ireland), and *Salud y Cambio* (Chile). A revised version appears in James O'Connor, *Capitalism and Nature: Essays in the Political Economy and Politics of Ecology*, published by Guilford Publications in association with *Capitalism, Nature, Socialism*.

Françoise d'Eaubonne, "The Time for Ecofeminism," trans. Ruth Hottell from *Le Féminisme ou la Mort* (Paris: Pierre Horay, 1974), pp. 213–52, excerpts. Translation reprinted by permission of Ruth Hottell.

Ariel Kay Salleh, "The Ecofeminist Connection," from "Deeper than Deep Ecology: The Ecofeminist Connection," *Environmental Ethics* 6, no. 4 (Winter 1984): 339–45. Reprinted by permission of Ariel Kay Salleh and *Environmental Ethics*.

Val Plumwood, "Ecosocial Feminism as a General Theory of Oppression," from *Ecopolitics V Proceedings*, ed. Ronnie Harding (Kensington, NSW, Australia: Centre for Liberal & General Studies, University of New South Wales, 1992), pp. 63–72. Reprinted by permission of Ronnie Harding.

Noël Sturgeon, "Ecofeminist Movements," from *Ecofeminist Natures: Race, Gender, Feminist Theory and Political Action* (New York: Routledge, 1997), pp. 23–30.

Mary Mellor, "Towards a Feminist Green Socialism," from *Breaking the Boundaries: Towards a Feminist Green Socialism* (London: Virago Press, 1992), pp. 249–51, 276–81.

Peter Wenz, "The Importance of Environmental Justice," from *Environmental Justice* by Peter Wenz (Albany: State University of New York Press, 1988), pp. xi–xii, 6–21, excerpts, by permission of the State University of New York Press. Copyright © 1988 by the State University of New York. All rights reserved.

Robert Bullard, "Confronting Environmental Racism," from *Confronting Environmental Racism: Voices from the Grassroots*, ed. Robert Bullard (Boston: South End Press, 1993), pp. 15–24, 38–39.

Luke Cole and Sheila Foster, "The Environmental Justice Movement," from *From the Ground Up: Environmental Racism and the Rise of the Environmental Justice Movement* (New York: New York University Press, 2001), pp. 19–33. Copyright © 2001 by New York University. All rights reserved. Reprinted by permission of New York University Press and by permission of Luke Cole and Sheila Foster.

Vandana Shiva, "Development, Ecology, and Women," from *Staying Alive: Women, Ecology, and Development* by Vandana Shiva (London: Zed Books, 1988), pp. 1–9, 13.

Mark Dowie, "Conservation Refugees," from "Conservation Refugees: When Protecting Nature Means Kicking People Out," *Orion Magazine* (November/December 2005): 16–27. This article first appeared in the November/December 2005 issue of *Orion*, 187 Main St., Great Barrington, MA 01230, www.orionmagazine.org. Reprinted by permission of *Orion Magazine* and Mark Dowie.

Carolyn Merchant, "Reinventing Eden," from *Reinventing Eden: The Fate of Nature in Western Culture* (New York: Routledge, 2003), pp. 11–20, 26–38.

Joanna Macy, "Toward a Healing of Self and World," from Joanna Macy, "Deep Ecology Work: Toward the Healing of Self and World," *Human Potential Magazine* 17, no. 1 (Spring 1992): 10–13, 29–31. Reprinted by permission of Joanna Macy, author of *Coming Back to Life*.

Charlene Spretnak, "The Spiritual Dimension of Green Politics," from *The Spiritual Dimension of Green Politics* by Charlene Spretnak (Santa Fe, NM: Bear & Co., 1986), pp. 52–69. Copyright © 1986, Bear & Co., Inc., PO Box 2860, Santa Fe, NM 87504. Reprinted by permission of Charlene Spretnak.

John Cobb Jr., "Ecology and Process Theology," from "Process Theology and an Ecological Model," *Pacific Theological Review* 15, no. 2 (Winter 1982): 24–7, 28. Reprinted by permission of John Cobb Jr.

Winona LaDuke, "Recovering the Sacred," from *Recovering the Sacred: The Power of Naming and Claiming* (Cambridge, MA: South End Press), pp. 11–15, 241–43, 251–53.

Fritjof Capra, "Systems Theory and the New Paradigm," reprinted from "Physics and the Current Change of Paradigms," in *The World View of Contemporary Physics: Does It Need a New Metaphysics?* ed. Richard F. Kitchener (Albany: State University of New York Press, 1988), pp. 144–54, by permission of the State University of New York Press. Copyright © 1988 by the State University of New York. All rights reserved.

Donald Worster, "Ecology of Order and Chaos," from *Environmental History Review* 14, nos. 1–2 (Spring/Summer, 1990): 4–16.

David Bohm, "Postmodern Science and a Postmodern World," from *The Reenchantment of Science: Postmodern Proposals*, ed. David Ray Griffin (Albany: State University of New York Press, 1988), pp. 57–58, 60–66, 68, by permission of the State University of New York Press. Copyright © by the State University of New York. All rights reserved.

Edward Lorenz, "Predictability: Does the Flap of a Butterfly's Wings in Brazil Set Off a Tornado in Texas?" from *The Essence of Chaos* by Edward Lorenz (Seattle: University of Washington Press, 1993), pp. 181–84, copyright © 1993 University of Washington Press. Reprinted with permission of the University of Washington Press.

Ilya Prigogine, "Science in a World of Limited Predictability," from "The Rediscovery of Time: Science in a World of Limited Predictability," paper presented to the International Congress on Spirit and Nature, Hanover, Germany, May 21–27, 1988, excerpts. Copyright © Stiftung Niedersachsen, Hanover, Germany. Reprinted by permission of Maryna Prigogine.

The First National People of Color Environmental Leadership Summit, "Principles of Environmental Justice," Washington, DC, October 24–27, 1991, adopted October 27, 1991. Public Domain.

The research for this edition was supported by the Committee on Research and the Agricultural Experiment Station (project H CA-B*-SOC-6299-H) at the University of California, Berkeley. I thank Rob Weinberg for assistance in preparing notes and obtaining permissions and Celeste Newbrough for preparing the index to the first and second editions. —C. M.

INTRODUCTION

CAROLYN MERCHANT

Domination has been one of humanity's most fruitful concepts for understanding human-human and human-nature relationships. The theme of domination and its reversal through liberation unites critical theorists and environmental philosophers whose work offers hope for the twenty-first century. When the domination of nonhuman nature is integrated with the domination of human beings and the call for environmental justice, critical theory instills the environmental movement with ethical fervor. This book brings together the Frankfurt school's analysis of domination with the insights of today's deep, social, and socialist ecologists, ecofeminists, people of color, spiritual ecologists, and postmodern scientists. But the project of analyzing and overcoming domination is not dealt with uniformly by all parties. Disagreement occurs over why and how nature and humans are linked and how to change those linkages.

The problem of domination was explored in depth by Max Horkheimer and Theodor Adorno, in exile from Germany during World War II, in New York and in Los Angeles, California. Together they wrote *Dialectic of Enlightenment*, published in 1945, and Horkheimer followed with *The Eclipse of Reason* in 1947. From the 1920s, the Institute for Social Research in Frankfurt had attempted to develop a multidisciplinary theory of society and culture that would bring together critiques and alternatives to mainstream social theory, science, and technology, and that would address social problems. They expanded on Marxism by extending its analysis of political economy to the interconnections between and among the spheres of nature, economy, society, politics, psychology, and culture. They characterized their approach as a comprehensive, totalizing theory of modernization and its alternatives and emphasized Hegel's theory of dialectical interactions among the various aspects of society.

They were deeply concerned about the problems they associated with modernity—the period from the Renaissance and Reformation to the era of state capitalism in the twentieth century—and the concept of enlightenment that epitomized the ideology of the modern world. Horkheimer and Adorno exposed the Greco-Roman roots of individualism, science, and the domination of nature that reached a crescendo in the eighteenth century's Age of Enlightenment. Rather than seeing the progressive aspects of modernity in which science, technology, and capitalism increasingly improve on the human condition, they emphasized modernity's dehumanizing tendencies, its destruction of the environment, its potential for totalitarian politics, and its inability to control technology.[1]

Critical theory drew its initial inspiration from Marxism. Karl Marx and Friedrich Engels's nineteenth-century critique of the inequalities created by industrialization's separation of capital from labor, entrepreneurs from working people, mind from body, and humanity from nature went to the heart of the problem of social justice. Under capitalism, justice and equality for every person could never be achieved because of the structural constraints built into capital's need to expand its sphere continually, by using nature and humans as resources. Nevertheless, Marx and Engels viewed capitalism as only a stage in the progression to an equitable socialist society in which the basic needs of all people for food, clothing, shelter, and energy would be fulfilled. But both capitalism and socialism would achieve these human gains over the "necessities of nature," through the domination of nonhuman nature by science and technology. While Marx and Engels displayed an extraordinary understanding of and sensitivity toward the "ecological" costs of capitalism, as revealed in the selection of their work edited by Howard Parsons, they nevertheless bought into the enlightenment myth of progress via the domination of nature.[2] It was this myth that Horkheimer and Adorno sought to expose.

Horkheimer and Adorno employed Hegelian dialectics to analyze society as a totality, drawing on the humanism of Marx's early work. They saw culture as continually changing and developing in an open-ended transforming process. They emphasized the relative autonomy of the cultural superstructure rather than the later Marx's determinism of the economic base on the legal-political superstructure. Horkheimer had set out the agenda when he became the director of the Institute for Social Research in 1931. He saw the institute's task as delineating the interconnections between economic life, psychological development of individuals, and culture, including science, technology, spiritual life, ethics, law, and even entertainment and sports. In addition to Horkheimer and Adorno, various members of the institute, such as Herbert Marcuse, Karl Wittfogel, Erich Fromm, and Leo Lowenthal, carried out critiques of philosophy, bureaucratic administration, ideology, sociology, psychology, literature, and popular music, as they contributed to the dominant ideology. They saw social theories as reproducing the dominant practices of capitalist society. Scientific theories and cultural concepts were representations of the material world rather than absolute truths. In opposition to the absolutes of idealism, they defended a materialist approach to nature and reality, emphasizing the material conditions of human needs. Material life created the human subject, while the subjects in turn transformed their own specifically historical and material conditions.

To the Frankfurt theorists, the material conditions in an inegalitarian class society create suffering. Yet the unequal social conditions that lead to unequal pain are not natural or inevitable. They can be changed. Moral outrage over suffering leads not only to sympathy and compassion, but to efforts to transform the particular social conditions that give rise to pain. Freedom from pain and suffering can be achieved in a just society, and it is critical theory's goal to envision that society and its attainment through the fulfillment of human needs and potentials.

CRITICAL THEORY AND THE DOMINATION OF NATURE

Working in New York City and in Los Angeles, California, in the 1940s, Horkheimer and Adorno turned their attention to the problem of the domination of nature and human beings. In the ancient world the emergence of a sense of self as distinct from the external natural world entailed a denial of internal nature in the human being. "In class history, the enmity of the self to sacrifice implied a sacrifice of the self, inasmuch as it was paid for by a denial of nature in man for the sake of domination over non-human nature and over other men. . . . With the denial of nature in man not merely the telos of the outward control of nature, but the telos of man's own life is distorted and befogged."[3] Odysseus in the ancient world and Francis Bacon at the onset of modernity epitomized the break with the enchanted past of myth and mimesis (imitation). Mimesis is participation in nature through identification with it. Odysseus represents the struggle to overcome the imitation of nature and immersion in the pleasures of animal life and tribal society. Through the emergence of his own identity as an individual self, he is able to break the hold of the mythic past and to control his animal instincts, his men, his wife, and other women. He becomes alienated from his own emotions, bodily pleasures, other human beings, and nature itself.

Tribal societies pursued their needs through the imitation of nature. Human beings became as much like the animals they hunted as possible. Power over nature, hence self-preservation, was achieved through imitative magic. Enlightenment thinking disenchants nature by removing that magic and turning the subject into an object. The process of objectification distances subject and object. In the early modern era, the domination of nature reached a new level in the thought of Francis Bacon. In the early seventeenth century, Bacon advocated extending the dominion of "man" over the entire universe through an experimental science that extracted secrets from nature. Power over nature was subsequently extended to mathematics and physics by René Descartes and Isaac Newton.

Horkheimer and Adorno used critical philosophy to expose the underlying instrumental reasoning behind both scientific thought and capitalist society. The disenchantment of external nature was achieved by despiritualizing a human being's internal nature. "The subjective spirit which cancels the animation of nature can master a despiritualized nature only by imitating its rigidity and despiritualizing itself in turn."[4] Bourgeois society lived according to quantification, calculation, profits, exchange, and the logic of identity. But so deeply embedded does this way of thinking become that it is presumed to be reality by mainstream society. So powerful is the mystique of reason as instrument in the control of nature and human bodies that it banishes other modes of participating in the world to the periphery of society. Logic and mathematics along with the calculus of capital become privileged modes of thought, defining the very meaning of truth. The identity of $a = a$ in logic and between the two sides of an equation in mathematics allows instrumental reason to describe the natural world. Describing the world through logic and mathematics in turn leads to prediction and hence to the possibility of controlling nature. Instrumental reason and enlightenment are thus synonymous with domination.

The domination of external nature by internal nature exacts a cost. The unrestrained use of nature destroys its own conditions for continuation, as the inexorable expansion of capital undercuts its own natural resource base. Similarly, the repression of human emotions and animal pleasures leads not to human happiness, but to anguish. But the tighter the rein, the greater the potential for rebellion. The revolt of nature is thus contained within the enlightenment project. Internal nature rebels psychically, spiritually, and bodily. External nature revolts ecologically. Here critical theory and the ecology movement intersect.

Horkheimer and Adorno entitled their book *Dialectic of Enlightenment* to emphasize the flux between mimesis and enlightenment. Mimesis, like enlightenment, depended on identity. Identifying with another subject through imitation allowed for self-preservation in tribal societies. Naming of plants and animals entailed the identification of characteristics that remain constant through change. Names in turn bring power in ritual magic. Yet at the same time, mimesis retains its original sense of participation in nature. On the other hand, enlightenment is also myth. Its historical specificity, its contextuality within modernity, and its limited truth domain destroy its claims to absolute knowledge. The myth of enlightenment is the guiding story of modernity.

The domination of internal nature makes possible the domination of external nature, which in turn leads to the domination of human beings. The relationship between the domination of nature and the domination of human beings was explicitly stated by Adorno in a 1967 essay on Oswald Spengler. "The confrontation of man with nature, which first produces the tendency to dominate nature, which in turn results in the domination of men by other men, is nowhere to be seen in the *Decline of the West*."[5] The domination of external nature therefore precedes and is a condition for the domination of human beings in society. Horkheimer and Adorno's analysis of the dialectics of various forms of domination—internal, external, human, and nonhuman—thus sets up a problematic for future work on the roots of the ecological crisis. It opens room for discussion of the many aspects of the relationships between domination and the life world—women, minorities, technology, economy, justice, spirit, and science—the topics considered in this book.

Another member of the Institute for Social Research who took up residence in California (at the University of California, San Diego), following the World War II exodus, was Herbert Marcuse. His works *Eros and Civilization* (1955), *One Dimensional Man* (1964), *Counterrevolution and Revolt* (1972) applied the approach of critical theory to social reproduction by examining the role of mass media, the control of information, and the decline of the family in maintaining the culture of needs necessary to the success of capitalism. Although Marcuse was not an ecological philosopher, he did address the issue of ecology in a special symposium on "Ecology and Revolution," published in *Liberation* magazine in 1972.[6] He saw ecology as a revolutionary force of life against which the counterrevolutionary forces of ecocide were destroying "the sources and resources of life itself" in the service of monopoly capital. Writing in the context of the Vietnam War, Marcuse connected the genocide and the ecocide in North Vietnam, which at the time were being carried out through modern science and technology's showcase weapons—the "electronic battlefield" and chemical defoliants—with the wider war against nature. Similarly, the process of transforming people into objects in market society also transforms nature into commodities, leaving little natural beauty, tranquility, or untouched space. The ecology movement exposes this war against nature and attacks the space capital carves out for its own. The movement attempts to defend what is left of untouched nature and argues that the entire production-consumption model of "war, waste, and gadgets" must be halted in the name of the very survival of life itself.

The same year that Marcuse published his brief article, William Leiss, who had obtained his doctoral degree in philosophy from the University of California, San Diego, in 1969, published a major treatise on *The Domination of Nature* (1972). Leiss explored the legacy of Horkheimer and Adorno, who had returned to Frankfurt after the war. Elaborating on the role of Francis Bacon in setting the modern agenda of power over nature through science and technology, Leiss analyzed technology's role in mastering both the external world of nature and the human being. In the modern world, nature has already been subjected to widespread exploitation, and for many in the Western world, material well-being has been achieved. Yet the production of an endless parade of technological improvements maintains the subjection of people's internal natures by chaining them to the manufacturing process. As aggression against external nature accelerates, the technological connection renders people's internal natures increasingly passive. At the same time, the potential for social conflict increases because of growing wealth differentials and the specialized geographical locations of essential resources such as oil. "The cunning of unreason takes its revenge," writes Leiss.

"In the process of globalized competition men become the servants of the very instruments fashioned for their own mastery over nature." The result is the revolt of nature, both internal and external, in a "blind irrational outbreak of human nature" and the concomitant collapse of ecological systems.[7]

Yet critical theory's astute analysis of the human domination of nature, according to Australian political theorist Robyn Eckersley, fails to offer any real alternative to an anthropocentric approach. Critical theory and the green movement have not been as mutually supportive as their common German origins and interests in new social movements might suggest. Horkheimer and Adorno did not develop their promising insights into the liberatory aspects of critical reason, the ecological dimensions of the revenge of nature, or the voice of nonhuman nature as speech for all that is mute. Marcuse's analysis of the ecology movement as a revolutionary force and his concept of nature as an opposing partner did not move toward an ecocentric ethic that gave value or rights to nonhuman organisms and inanimate entities. Jürgen Habermas, who in 1964 took over Horkheimer's chair in philosophy at Frankfurt, retained his alignment with the Social Democrats rather than lending support to the German Greens in the 1980s and viewed the radical ecology movement as neo-romantic rather than emancipatory. According to Eckersley, political theory needs to move toward an ecocentric approach to the human-nature relationship. "Ecocentric theorists," she writes, "are concerned to develop an ecologically informed approach that is able to value (for their own sake) not just individual living organisms, but also ecological entities at different levels of aggregation, such as populations, species, ecosystems, and the ecosphere (Gaia)."[8]

GLOBALIZATION

The ecology movement of the 1960s and '70s extended the critique of the domination of nature and human beings by industrial capitalism begun by Marx, Engels, and the Frankfurt theorists. Political economists looked at the relationships between first world capitalism and third world colonialism through Immanual Wallerstein's model of core and peripheral economies.[9] While the industrial revolution in eighteenth-century Europe and nineteenth-century America had stimulated economic production, raised living standards, reduced death rates, and led to smaller family sizes (the demographic transition), it did so at the expense of third world peoples and resources. To rebuild Europe after the Second World War, the Bretton Woods institutions of the World Bank and International Monetary Fund (IMF) were established. The core economies of Europe, North America, and Japan spearheaded resource extraction loops in the third world in which soils, forests, and mines were used to produce export crops such as coffee, sugar, hemp, beef, copper, and aluminum. Globalization is the expansion of first world capitalism into third world countries, where resources and labor are cheap, environmental regulations are weak, and free trade is promoted through tariff reductions.

Rosemary Radford Ruether characterizes corporate globalization as the latest stage of colonial imperialism. Loans to third world countries (primarily in the Southern hemisphere) for projects such as dams, roads, electricity, telephones, and industry center on exports of commodities to the first world countries of the Northern hemisphere. Wealthy landowners in the South import luxury items from the North, raising their own standard of living, in turn pushing the vast majority of people into deeper poverty on marginal lands and in urban slums. Through free trade agreements that protect property rights and patents, transnational corporations become increasingly powerful and wealthy at the expense of nature and the poor.

Joel Kovel sees global capitalism as spelling the end of nature and indeed the end of the world as we know it today. Capitalism as the enemy of nature continually expels the by-products of production—for example, carbon dioxide, toxic chemicals, waste products, and junk—into ecosystems that cannot absorb or recycle the molecules to maintain healthy systems. The growth in wastes is inextricably spurred by the growth of capital. For Koval, capital is a cancer on nature. We cannot overcome the ecological crisis without overcoming capital.

Global capitalism comprises a set of economic forces that systematically interact with ecological forces inexorably degrading nature's buffering capacity to resist the deleterious effects of human production. Unchecked economic growth depletes water resources, oil reserves, food sources, and air quality, threatening bodily health and human survival. While population growth is slowing down and leveling off, economic growth is expanding. The hope for a steady-state or low-growth economy or even the acceptance of "limits to growth" finds little support among capitalist producers. For Kovel, humanity is not only the instigator, but ultimately the victim of economic growth.

Green theorist Brian Tokar, however, sees hope in global ecological movements. One avenue for bringing about a sustainable society and a socially just world is through green politics advanced on a worldwide front. International organizations such as Greenpeace, the Rainforest Action Network, and Native Forest Network are instrumental in promoting ecological actions to reverse the effects of capitalist waste, uneven development, and biopiracy. Most third world movements combine ecological, social, political, and cultural factors. Creating a green future starts by looking outside the capitalist, industrialist, consumerist model of the North. Traditional knowledge at the global grassroots level, combined with local empowerment, coalition building, and living within the means of the local watershed, are ways to discover what is sustainable in the diverse cultures of the world. A new model that flows from South to North may reverse the ecologically unsustainable trajectory created by colonialist, imperialist development.

Deeply interconnected with colonialism and the political economy of the third world are issues of expanding population. As Barry Commoner argues, population growth rates decline when countries undergo the demographic transition: declining death rates, resulting from rising standards of living, are followed by declining birthrates, as large numbers of children are no longer needed for rural labor and old age security. But unlike industrial development in the first world, third world dependency relations have prevented a rapid reduction in population growth rates. Much depends, therefore, on how development occurs and whether that development will be environmentally sustainable and socially just.[10]

Awareness of the delicacy of the biosphere must go hand in hand with democratically planned production geared both for nature's needs and for human needs. Fulfillment of basic needs entails individual human health and autonomy. For individual needs to be fulfilled, societies have to be able to reproduce themselves. They have to produce the basic needs of material life—food, clothing, shelter, and energy—and to provide the conditions for physical and emotional health. Social justice involves a guarantee of basic liberties, such as free speech and the right to assemble, the removal of social inequalities that prevent equitable distribution of social goods, and equality of opportunity. How these basic needs are to be satisfied in a sustainable world entails a viable relationship between the state and human liberty and resolving the problem of the human domination of nature.

Paul Hawken, Amory Lovins, and L. Hunter Lovins advocate a new economy rooted in natural capitalism. They argue that ecosystem services are not only essential to maintaining a healthy human economy, but also from an economic standpoint alone they can be estimated to provide between $36 and $58 trillion per year (as estimated in 1998 dollars) compared to a gross world product of $39 trillion. Services provided by ecosystems include regulating atmospheric gasses, processing wastes, maintaining nutrient flows, and storing and purifying water. Natural capital provides these life-support services invisibly and reliably until disrupted. We do not give a second thought to them until they break down, in some cases irrevocably (such as soil erosion, desertification, melting of glaciers, or extinction of pollinators). Investing in natural capital means maintaining and increasing the present stock to offset population doubling, increased pollution, and resource demand. Only by recognizing the significance of both natural capital and human-made capital to future economic projections can we anticipate a green and sustainable future.

DEEP, SOCIAL, AND SOCIALIST ECOLOGY

While green economics and politics provide a practical means for moving toward global sustainability, overcoming the deeply engrained anthropocentrism at the root of the domination of nature by humans requires a new philosophy and ethics. Deep ecologists argue that mainstream environmentalism is limited and incremental in scope and that what is needed before real change can occur is a transformation in consciousness. In 1973 Norwegian philosopher Arne Naess published a now-famous article on "The Shallow, and the Deep, Long-Range Ecology Movement."[11] The idea of Deep Ecology was taken up by California sociologist Bill Devall and philosopher George Sessions who promoted and developed it into a series of newsletters and articles. They argued that the anthropocentric core of mainstream Western philosophy must be overturned and replaced with a new metaphysics, psychology, ethics, and science. They looked for alternatives within the Western traditions, Eastern philosophy, and the insights of indigenous peoples. A major compilation of sources was published by Devall and Sessions under the title *Deep Ecology: Living as If Nature Mattered* in 1985 and elaborated further by Naess in *Ecology, Community, and Lifestyle* in 1989.[12] In 1990 Australian philosopher Warwick Fox linked Deep Ecology with transpersonal psychology through the ideas of identification with the nonhuman world and the notion of an expanded self that was capable of moving beyond the atomized, isolated ego.[13] Fundamental to the deep ecological approach is the need to overcome the narrow-minded, suicidal anthropocentrism on which the future of civilization as we know it seems to be currently foundering.

An alternative to Deep Ecology's placement of the blame for the ecological crisis at the doorstep of anthropocentrism is social ecology.[14] Defined and defended by social philosopher Murray Bookchin, social ecology, like critical theory, grounds its analysis in domination.[15] But for Bookchin, as opposed to deep ecologists, critical theorists, and Marxists, the domination of human beings is historically and causally prior to the domination of nature. Bookchin writes, "By the early sixties, my views could be summarized in a fairly crisp formulation: the very notion of the domination of nature by man stems from the very real domination of human by human."[16] Whereas early tribal societies were basically egalitarian and lived within nature, the increasing prestige of male elders created social hierarchies and inequalities that led to power over other human beings, especially women, and ultimately over nature. The growth of ancient city-states, medieval walled towns, and state capitalism depended on increasingly entrenched hierarchies of elders over other tribal members, men over women, and elites over laborers and slaves. This social domination led to the domination of people over nature. The goal of social ecology is to remove hierarchy and domination from society and as a result the domination of people over nature. Bookchin's ecological anarchism envisions an ecological society to be achieved through reliance on the resources and energy of the local bioregion, face-to-face grassroots democracy within libertarian municipalities linked together in a confederation, and the dissolution of the state as a source of authority and control.

Socialist ecology, as conceptualized by Marxist economist James O'Connor, contrasts with Bookchin's social ecology in that it is grounded not in the concept of domination but in political economy.[17] Only the Marxist categories of labor exploitation, production, the profit rate, capital circulation and accumulation, and so on, can adequately account for the degradation of nature under capitalism.

Socialist ecology is distinguished from state socialism as represented by the failed Soviet Union and Eastern bloc countries whose industrial growth models resulted in environmental disaster. Instead it looks toward new forms of eco-socialism brought about by green social movements, with commitments to democracy, internationalism, and ways to overcome the dualism of local versus state control and administration. But like deep and social ecology, it recognizes the autonomy of nonhuman nature, ecological diversity, and the science of ecology as the basic science of survival for the twenty-first century.

ECOFEMINISM

Deep ecology's efforts to place the blame for ecological deterioration on the domination of nature by human beings (anthropocentrism) meets resistance from ecofeminists, who see the domination of both nature and women by men (andro-centrism) as the root cause of the modern crisis. French feminist Françoise d'Eaubonne set up *Ecologie-Féminisme* in 1972 as part of the project of "launching a new action: ecofeminism," and in 1974 published a chapter entitled "The Time for Ecofeminism" in her book *Feminism or Death*.[18] In that chapter, translated by French feminist scholar Ruth Hottell, d'Eaubonne states that women in the "Feminist Front" separated from the movement and founded the information center, called the "Ecology-Feminism Center." Their new action was christened ecofeminism, and it attempted a synthesis "between two struggles pre-viously thought to be separated, feminism and ecology." The goal was to "remake the planet around a totally new model," for it was "in danger of dying, and we along with it." They called for a mutation of the world that would allow the human species to escape from death and to continue to have a future. Writing as a militant radical feminist, d'Eaubonne placed the problem of the death of the planet squarely on the shoulders of men. The slogan of the "Ecology-Feminism Center" was "to tear the planet away from the male today in order to restore it for humanity of tomorrow. . . . If the male society persists there will be no tomorrow for humanity."[19]

D'Eaubonne presented a litany of planetary ills ranging from the global pop-ulation explosion to worldwide pollution and American consumption, urban crowding, and violence. Both capitalism and socialism were scenes of ecological disasters. The most immediate death threats to the planet were overpopulation (a glut of births) and the destruction of natural resources (a glut of products). Although many men attempted to label overpopulation "a third world problem," the real cause of the sickness was patriarchal power. D'Eaubonne followed the analysis of nineteenth- and early twentieth-century proponents of ancient matri-archal societies, such as Johann Bachofen, Friedrich Engels, Robert Briffault, and August Bebel, who saw "the worldwide defeat of the female sex" some five thousand years ago that initiated an age of patriarchal power.[20] It was the male system created five thousand years ago, not capitalism or socialism, that gave men the power to sow both the earth (fertility) and women (fecundity). The iron age of the second sex began, women were caged, and the earth appropriated by males. The male society "built by males and *for* males" that took over running the planet did so in terms of competition, aggression, and sexual hierarchy, "allocated in such a way to be exercised by men over women." Patriarchal power produced agricultural overexploitation and industrial overexpansion. "The Earth, symbol and former preserve of the Great Mothers, has had a harder life and has resisted longer; today, her conqueror has reduced her to agony. This is the price of phallocracy."[21]

If women had not lost the war of the sexes when phallocracy was born, d'Eaubonne maintained, "Perhaps we would have never known either the jukebox or a spaceship landing on the moon, but the environment would have never known the current massacre." Pollution, environmental destruction, and runaway demography are men's words spawned by a male culture. They would have no place in a female culture linked to the "ancient ancestry of the Great Mothers. . . . A culture of women would have never been this, this extermination of nature, this systematic destruction—with maximum profit in mind—of all the nourishing resources." If women were returned to the power they lost, their first act would be to limit and space out births as they had done in the agricultural past. Demographic problems are created by men, especially in the Catholic countries. Husbands who control women's bodies and implant them with their seed, doctors who examine them, and male priests who call for large families are bearers of male power over women's wombs.[22]

D'Eaubonne saw ecofeminism as a new humanism that put forth the goals of the "feminine masses" in an egalitarian administration of a reborn world. A society in the feminine would not mean power in the hands of women, but no power at all. The human being would be treated as a human being, not as a male or female. Women's personal interests join those of the entire human community, while individual male interests are separate from the general interests of the community. The preservation of the earth was a question not just of change or improvement, but of life or death. The problem, she said, paraphrasing Marx, is "to change the world . . . *so that there can still be a world.*" But only the feminine, which is concerned with all levels of society and nature, can accomplish "the ecological revolution." She concluded her foundational essay with these telling words: "And the planet placed in the feminine will flourish for all."[23]

In the United States, the term ecofeminism was used at Murray Bookchin's Institute for Social Ecology in Vermont around 1976 to identify courses as ecological, namely, ecotechnology, ecoagriculture, and ecofeminism. The course on ecofeminism was taught by Ynestra King, who used the concept in 1980 as a major theme for the conference "Women and Life on Earth: Ecofeminism in the '80s," held in Amherst, Massachusetts. King published "Feminism and the Revolt of Nature" in 1981 in a special issue of *Heresies* on "Feminism and

Ecology." Her approach reflects the Frankfurt school's conceptual framework of the disenchantment of the world and the revolt of nature. The promise of ecological feminism lies in its liberatory potential to create a rational reenchantment that integrates the spiritual with the material and being with knowing.[24]

King conceptualized ecological feminism as a transformative feminism drawing on the insights of both radical cultural feminism and socialist feminism. Radical cultural feminists such as Mary Daly in *Gyn-ecology* (1978) and Susan Griffin in *Woman and Nature* (1978) linked together the domination of women and nature under patriarchy. Men use both to defy death and attain immortality. Woman's oppression is rooted in her biological difference from men who use women to secure their own immortality through childbearing. Nature's oppression is rooted in its biological otherness from men who secure immortality as rational creators of human culture. For radical feminists, women and nature can be liberated only through a feminist separatist movement that fights their exploitation through the overthrow of patriarchy. Socialist feminists, however, ground their analysis, not in biological difference, but in the historically constructed material conditions of production and reproduction as a base for the changing superstructure of culture and consciousness. Underlying both positions, King argues, is a false separation of nature from culture. Instead, a transformative feminism offers an understanding of the dialectic between nature and culture that is the key to overcoming the domination of both women and nature. Such a position is needed if an ecological culture that reconnects nature and culture is to emerge.

In 1984 Australian sociologist Ariel Kay Salleh published a fundamental critique of Arne Naess's foundational 1973 paper on Deep Ecology and Bill Devall's "The Deep Ecology Movement" from an ecofeminist perspective. Taking issue with each of Naess's seven foundational points, she exposed the male bias in the framing of Deep Ecology. Naess's use of the term "man" obscures important differences between the sexes in which woman's bodily and biological experiences already ground her in a coterminous relationship with nature. Second, the role of domination expressed in the master-slave relationship of "man" to nature is replicated in "man's" historically based patriarchal relationship to woman, a relationship that negates the very principle of biological egalitarianism on which Naess seeks to ground Deep Ecology. Naess's third principle, that of diversity and symbiosis which seeks coexistence among all living forms, further ignores the patriarchal annihilation of women's manifest creativity and cultural inventiveness. Nor is the sexual and social oppression experienced by women recognized in Naess's fifth point—anticlass posture—a position that seeks to realize all human potential regardless of class or social status. Naess's sixth point, "complexity, not complication" and its supporting arguments reflect the masculine norm of a systems-theoretical, instrumental, rational, scientist approach, as opposed to a nurturant environmentalism grounded in alternative gender roles. In place of Naess's final assumption—local autonomy and decentralization—Salleh would substitute small collectivities where spontaneous communication results in a less competitive, more participatory process of decision making. Deep ecology, she concludes, is just another reformist movement that suppresses the feminine and ignores the woman inside all human beings.

Australian philosopher Val Plumwood extends the analysis of domination initiated by d'Eaubonne, King, and Salleh by comparing the debates between deep ecologists, social ecologists, and ecofeminists. Each group of ecophilosophers makes valid points, she argues, but in so doing seeks to reinforce its own standpoint by rejecting those of its rivals. Thus Deep Ecology is correct to challenge the human centeredness of social ecology, but social ecology is also right in its analysis that hierarchical differences within human society affect the character of environmental problems. Thus an alternative, cooperative approach is needed. Ecofeminism with its emphasis on relations has the potential to see connections among various forms of oppression such as those affecting women, minorities, the colonized, animals, and nature. Recognition of the weblike character of various forms of domination suggests a cooperative strategy of web repair. The ecofeminist approach focuses on relations and interconnections among the various ecology movements and leads to the possibility of a more comprehensive and cooperative theory and practice.

A major problem for ecofeminist theory is essentialism. Do women (and men) have innate unchanging characteristics (or essences), or are all male and female qualities historically contingent? Do women's reproductive biological functions (ovulation, menstruation, and the potential for pregnancy, childbearing, and lactation) that make them different from men constitute their essence? The essentialist perception of women as closer to nature, as a result of their biological functions of reproduction, has historically been used in the service of domination to limit their social roles to childbearers, childrearers, caretakers, and housekeepers.[25] Furthermore, do women have a special relationship to nature that men cannot share? If women declare that they are different from men and as ecofeminists set themselves up as caretakers of nature, they would seem to cement their own oppression and thwart their hopes for liberation and equality. The contradiction between essentialism and ecofeminist empowerment is addressed by Noël Sturgeon.

Sturgeon argues that ecofeminism is a complex political movement that reflects a tension between the essentialist implications of the woman-nature connection and a deep desire for a positive empowerment of women to bring about environmental change. Women's status in patriarchal society as inferior to that of men often means that women are the first to experience and protest against life-threatening assaults on both their own bodies and on nature's body. Some ecofeminists find the inspiration to protest these assaults and to promote social change in images of female deities that reflect female spiritual power. The contradiction between essentialism and feminist empowerment is further complicated by ecofeminism's antiracist tenor. Although primarily a white feminist movement, ecofeminism has from its beginnings made a conscious attempt to be inclusive and to work with and recognize women of color as role models. Many women of color who are environmental activists, however, do not identify themselves as ecofeminists. Nevertheless, the antiracism of ecofeminism provides an important resource for the environmental justice movement discussed in part 5.

While Noël Sturgeon's ultimate goal is to examine the tensions within the ecofeminist movement, Mary Mellor, in "Towards a Feminist Green Socialism," advocates a new society entirely. The WE world based on "women's-experience" would supersede the ME world that has developed historically out of "men's-experience." That new society would be feminist in that women's life-producing and life-sustaining work would replace life-destructive institutions. It would be green in that it would balance human needs for life against the planet's need for life. It would be socialist because it would acknowledge that all people should have the right to an equitable society rooted in social justice. Such a society could be brought into existence by restructuring social institutions from below through the struggles of workers, environmental activists, feminists, and peace activists. In these ways, ecofeminists seek to overcome both the domination of women and the domination of nature.

ENVIRONMENTAL JUSTICE

Reversing the domination of nature and human beings requires environmental justice. Environmental justice entails the fulfillment of basic needs through the equitable distribution and use of natural and social resources and freedom from the effects of environmental misuse, scarcity, and pollution. According to philosopher Peter Wenz, environmental justice is a problem of distributive justice. In some cases, such as the Garrett Hardin's tragedy of the commons, in which herdsmen overgraze a pasture, lumber companies deplete a timber supply, or chemical companies pollute a river, the problem is one of coordinated or mandated restraint. In other cases, where a basic need, such as water or food, is in short supply, the scarce resource must be equitably distributed through tribal or governmental law and policy. Democratic societies that are relatively nonrepressive and open need to be perceived as being socially just, or civil rebellion may occur. People who share the benefits, responsibilities, and burdens of living in an open society may have to make sacrifices to maintain those advantages. Environmental laws and policies, therefore, must embody principles of environmental justice that a majority of people agree are reasonable.

Problems of environmental justice arise, however, when past decisions and practices disproportionately affect certain groups of people. Robert Bullard argues that in the United States, minorities, women, and the poor are often the victims of environmental racism. Internal colonies of minorities have experienced racism in the form of polluted air and water, toxins from landfills and incinerators, and hazardous waste facilities. Pollution goes hand in hand with poverty, deteriorating buildings, poor schools, and inadequate healthcare, and is reinforced by government neglect. The mainstream environmental movement has ignored urban ecology, choosing to focus on wilderness preservation, pollution abatement, and population control. For environmental justice to occur, the effects of decade- or century-old decisions must be ameliorated and new socially just policies put in place. People of color have come together in a powerful movement to reverse environmental racism and promote environmental justice through grassroots organizations, conferences, protests, and demonstrations. The 1987 United Church of Christ "Report on Toxic Wastes and Race in the United States," the Mothers of East Los Angeles coalition, the Race, Poverty, and Environment Newsletter, and the 1991 First National People of Color Environmental Leadership Summit, with its "Principles of Environmental Justice," are but a few manifestations of a powerful new movement for environmental justice.

Luke Cole and Sheila Foster see the movement for environmental justice as being framed by several predecessor movements. Perhaps most important, the civil rights movement of the 1960s and 1970s contributed political and church-based organizing skills, direct action, and civil disobedience, while the anti-toxics movement of the 1980s added scientific, technical, and legal analysis along with grassroots activism. Academics, Native Americans, the labor movement, and farm workers all fed into the 1991 People of Color Environmental Leadership Summit that codified the various strands into a new movement rooted in social justice and transcending traditional environmentalism. The new activism goes beyond toxic dumping and environmental pollution to demand structural reforms that address poverty, segregation, health disparities, unemployment, and education.

Environmental justice for third world peoples involves a wholesale questioning and restructuring of Western-style development projects. Since World War II, argues Indian physicist and philosopher Vandana Shiva, development has actually been maldevelopment. Development, which was supposed to have been a postcolonial project, is rooted in the domination of women, tribal peoples, and nature by patriarchy and capitalism. Cash-cropping undermines traditional subsistence and land rights, destroys soil, water, and forests, and renders people and nature passive. Maldevelopment subverted the traditional feminine principle in nature and resulted in the feminization of poverty. It named nature's own reproductive systems nonproductive unless they produced surplus capital and profits. Maldevelopment, according to Shiva, violates the integrity of interconnected, interdependent communities and initiates a process of exploitation and violence. Recovering the feminine principle in nature, which produces diversity and women's traditional connections to the ecosystems that produce life, would mean rethinking development in the name of environmental justice.

The political consequences, not only of corporate development but also of the actions of large international conservation organizations, are of vital importance to indigenous peoples affected by worldwide conservation projects, argues

Mark Dowie. The preservation of wilderness by cordoning off huge tracts of lands from use by human beings subverts the very real needs for survival by indigenous peoples. People whose ancestors have lived on lands for millennia are removed to the edges of the new parks, where they often exist in poverty without access to land, resources, or even water. Setting aside parks by organizations such as the Nature Conservancy (TNC), the World Wildlife Fund (WWF), the Wildlife Conservation Society (WCS), and the International Union for the Conservation of Nature (IUCN) has created conservation refuges for the purpose of preserving biodiversity and promoting ecotourism. In many cases the result has been decreased, not increased biodiversity. Urged by indigenous spokespersons for the displaced peoples, conservation organizations are beginning to confront the contradictions inherent in wilderness preservation and to work with indigenous organizations. Environmental justice is therefore an international issue.

SPIRITUAL ECOLOGY

In a foundational article published in 1967, historian Lynn White Jr. argued that the domination of nature stemmed from the Judeo-Christian mandate expressed in Genesis 1:28, "to increase, multiply, replenish the earth and subdue it." "Christianity," he declared, "is the most anthropocentric religion the world has ever seen. . . . By destroying pagan animism, Christianity made it possible to exploit nature in a mood of indifference to the feelings of natural objects."[26] Since then, hundreds of articles debating the validity of White's thesis have been written and numerous alternative interpretations and practices offered. Mainstream religions of all denominations have reread and reinterpreted their traditional relationships to nature in an effort to offer alternatives to the Genesis mandate. Ecologically based forms of spirituality that draw inspiration from Eastern religions, Native American philosophies, and pagan traditions offer connections to the natural world that engender appreciation, care, and environmental action on behalf of the planet.

Carolyn Merchant sees the Christian story as propelling a worldwide effort to recover the Garden of Eden lost in the Fall from the original mythical garden. Through exploration, colonization, and industrialization, forests have been cut and deserts irrigated in an effort to turn the whole earth into farms, gated communities, and shopping malls. In its secular form, this progressive Recovery Narrative has shaped human science, technology, and capitalist development since the scientific revolution and continues until today. But an environmentalist and feminist counternarrative reveals the desecration of the planet brought about by the conquest of the earth. These narratives likewise call for a recovery but one oriented toward restoration of the planet's ecosystems and social justice for its displaced peoples. For environmentalists, the devastating loss of forests, species, soils, and glaciers can be reversed by worldwide efforts to combat global warming and to promote conservation and restoration.

For feminists, the domination of women by men in society is legitimated by religions that worship a male God. Goddess symbols affirm female power as well as women's cultural and social heritage of spirituality and bonding, as well as the celebration of women, the life cycle of nature, and the female body. The goddess for some women is the female aspect of God, while for others she is a powerful presence within nature, and for still others the vibrant sexuality of nature's reproductive capacities. The counternarrative thus includes female spirituality.

Joanna Macy draws on Deep Ecology, systems theory, and Bodhisattva Buddhism to develop a sense of a self interconnected with the natural world. Deep ecology's expanded self (John Seed's "I am the rainforest protecting myself"), Gregory Bateson's systems theory (the self as individual plus its environment—the entire pattern that connects), and Buddhism's sense of interconnectedness (the dependent co-arising of phenomena) help us break out of the imprisonment of the self-contained ego. The new ecological self is a self connected to the world, bringing new resources of courage, ingenuity, and endurance to combat despair and engage in healing our body—the world.[27]

Charlene Spretnak relates the efforts of mainstream religions to find ecological alternatives to the Judeo-Christian mandate to subdue the earth by looking within their own traditions for alternative interpretations and passages. The stewardship ethic of humans as caretakers of the rest of creation, ecologically based worship services, respect for women and the female principle in creation, care and concern for homeless and poor people, and development of worker-owned and community-based economic alternatives in depressed areas are but a few examples. Creation spirituality that emphasizes the interconnectedness of all creation, revivals of the "old religion" (paganism and Wicca), and learning respect for the land from indigenous peoples are other examples from outside the mainstream. These spiritual approaches are consistent with principles of green politics, such as ecology, nonviolence, post-patriarchal principles, and grassroots democracy.[28]

John Cobb Jr. offers another ecological approach to spirituality via process theology, based on the philosophy of Alfred North Whitehead's *Process and Reality* (1928). This philosophy views the world as an organism comprising individual organisms existing in relationship to the environment. Every organism is constituted by its set of relations with the rest of the world. The ecological model is a relational science that is consistent with relativity and quantum field theory. An organism is a series of events that Whitehead called occasions; the world is a vast field of occasions in enduring patterns. God for process theologians is a name for the ecological model in which all the relations are complete and perfect. This God is not a domineering God imposing his will on a separate creation, but an open, receptive, and responsive God. Process theologians and their followers are deeply involved in issues of ecology, feminism, peace, social justice, liberation, and freedom.

Among the North American groups longest and most virulently affected by Western religious dominance are Native Americans, who, according to Anishinaabeg activist Winona LaDuke, often feared to practice their traditional forms of spirituality for fear of reprisals. Native people relied on the animate world for sustenance, and their rituals reaffirmed their place in the larger creation and the gifts given them by the creator. Origin stories connect people to the land and renewal ceremonies reaffirm connections and responsibilities. While the right of native peoples to practice their own religions has been officially recognized since the 1970s, in many cases sacred sites are threatened, inaccessible, or lost to resource extraction. But to turn the world around, says LaDuke, we need the leadership of indigenous and marginalized peoples. These are the people who know how to survive and who can bring a vision to the future.

The reciprocity between Indians and nature contrasts with the capitalist-industrial model of accumulation. The impact of industrialization on Native American communities has been to create toxic and nuclear waste dumps on reservations, especially when white communities resist them. Uranium and coal mining employed Indians on their own lands, but left them sick with lung and skin cancers. The shift from nuclear to hydropower (such as the James Bay hydro project) affected many northern US and Canadian tribes. Instead of participating in hydropower, and oil, coal, and uranium extraction, the Intertribal Council on Utility Power (COUP) proposes to develop wind power as a gift from Taté, the wind spirit. The wind is the power given by the creator and the coalition is an opportunity to bring tribes together in supplying renewable energy.

The forms of spirituality offered by Macy, Spretnak, Cobb, and LaDuke help to overcome the Genesis 1:28 version of nature domination characterized by Merchant as the mainstream Recovery Narrative. In so doing, human domination of the earth can be transformed into a partnership with nature through the participatory, mimetic modes of relating to nature identified by the Frankfurt theorists.

POSTMODERN SCIENCE

From the point of view of science, as well as religion, our ways of relating to the planet are undergoing a significant transformation. The Enlightenment ethic of the domination of nature fostered by mechanistic science's reduction of the world to dead atoms moved by external forces is being replaced by a postmodern, ecological worldview based on interconnectedness, process, and open systems.

Fritjof Capra contends that physics is in the midst of a paradigm shift to a new set of assumptions about reality and new ways of representing the world that will replace those of the modern era. This transformation in physics mirrors a much larger cultural and social transformation resulting from dislocations such as the environmental, nuclear, and poverty crises. The problems facing science and society today reflect the inadequacy of the structures of modernism—mechanistic physics, industrialization, and the inequalities of class, race, and gender—that are fundamental to the domination of nature and human beings. Solving them requires a new social paradigm—"a [new] constellation of concepts, values, perceptions, and practices shared by a community, which form a particular vision of reality that is the basis of the way the community organizes itself."[29]

Capra proposes that the ecological systems view of life is the new paradigm emerging to replace the mechanistic worldview. This includes an emphasis on the whole over the parts, on process over structure, on the relative knowability of the external world, on the idea of networks of knowledge and information, and on the recognition of the necessity of approximation. The assumptions of the systems approach entail a new ethic that is life affirming rather than life destroying, and it recognizes the interconnectedness of all things and the human place in the network.

What that ecological systems view entails, however, is not a straightforward proposition. The ecological paradigm is itself undergoing an evolution. Donald Worster argues that the concept of the ecosystem that inspired the ecology movement of the 1960s and 1970s is moving away from an ideal of stability and order and toward a science that more readily admits of chaos and instability. The ecosystem approach held that humans were outsiders who disturbed the system through resource extractions and environmental pollution. By the same token however, humanity could control its destructive tendencies and restore the system to equilibrium. But by challenging mechanistic science's assumption of predictability, chaos theory renders such management far more problematical. Rather than a system of biotic and abiotic components operating predictably in accordance with the laws of thermodynamics, today's ecologists are far more prone to seeing chaos and unpredictability where order once reigned. By accepting unpredictability as the usual situation rather than the unusual, humanity must change its relationship to nature. Rather than lordly dominators over nature entailed by mechanistic science, humans must forgo the hubris of control and accept nature on more nearly equal terms. Such a change in the human relationship to nature is also entailed by developments in postmodern, postmechanistic physics.

Postmodern science challenges the limitations of modern science, first by early twentieth-century developments in relativity and quantum mechanics and then through postmechanistic physics. Physicist David Bohm has developed a theory of unbroken wholeness as the ground of matter, energy, and life that he argues resolves many of the contradictory assumptions underlying relativity theory and quantum mechanics. Bohm's theory is based on the underlying holomovement—a flow of energy in multidimensional space-time out of which unfolds the three-dimensional world described by Newtonian physics. The mechanistic world is implicit within or enfolded into a higher order of reality. This visible world is the explicate order on which most successful work in physics has been done. "In my technical writings," says Bohm, "I have sought to show that the mathematical laws of quantum theory can be understood as describing the holomovement, in which the whole is enfolded in each region, and the region is unfolded into the whole."[30] The holomovement is life implicit and consciousness implicit. This view militates against the fragmentation of the mechanistic world-view and suggests an integration between matter and consciousness, value and fact, ethics and science.

The postmodern, ecological worldview, unlike the modern mechanistic worldview, is based on the impossibility of completely predicting the behavior of the natural world. Chaos theory suggests that most environmental and biological systems, such as weather, noise, population, and ecological patterns, cannot be described accurately by the linear equations of mechanistic science and may be governed by nonlinear chaotic relationships.[31] To atmospheric physicist Edward Lorenz is attributed the famous metaphor of the butterfly effect, which was influential in the early work in chaos theory for its description of sensitive dependence on initial conditions. This approach questions the ability of science to make predictions in all but the most unusual situations, namely, the limited number of closed, isolated systems successfully described by mechanistic science.

In his 1972 paper presented to the annual meeting of the American Association for the Advancement of Science, Lorenz asked whether the flap of a butterfly's wings in Brazil could set off a tornado in Texas. His point was that we cannot predict the results of small effects such as a butterfly on the weather, because the atmosphere is unstable with respect to perturbations of small amplitude. We don't know how many small effects there are (such as butterflies) or even where they are located. We can't even set up a controlled experiment to find out if the atmosphere is unstable because we can never know what might have happened if we hadn't disturbed it. Most important is the problem of the "inevitable approximations which must be introduced in formulating . . . [the governing physical] principles as procedures which the human brain or the computer can carry out." The rapid doubling of errors precludes great accuracy in real world forecasting. The best that we can hope for is to make the "best forecasts which the atmosphere is willing to have us make."[32]

The problem of predictability is pushed further by Ilya Prigogine in his work with Isabelle Stengers on *Order Out of Chaos* and the concept of self-organization.[33] Classical thermodynamic processes discovered in the nineteenth century, such as equilibrium and near-equilibrium cases that describe the steam engine and the refrigerator, suggest that the universe is running down and becoming more disorderly and chaotic. Yet in Prigogine's far-from-equilibrium thermodynamics, in situations found in hydrodynamics, many chemical processes, and evolution, a new reorganization can occur in which order can emerge out of chaos. In such situations irreversibility and nonlinearity can lead to self-organization. Irreversibility and nonlinearity increase the role of fluctuations and lead to bifurcations (divisions) in which the system can go in several directions, that is, the nonlinear equations have several different solutions. The outcome cannot be predicted with certainty. Unstable dynamic systems, such as the weather systems described by Lorenz, behave differently than stable systems, such as the planetary systems described by Newton. Prigogine suggests that the unpredictability of systems holds implications for the domination of nature and leads to alternative strategies for human interactions with natural systems:

> The important element is that unstable systems are not controllable. . . . The classical view on the laws of nature, on our relation with nature, was domination. That we can control everything. If we change our initial conditions, the trajectories slightly change. . . . But that is not the general situation. . . .
>
> We see in nature the appearance of spontaneous processes which we cannot control in [the] strict sense in which it was imagined to be possible in classical mechanics. . . . The world in which we are living is highly unstable. However, what I want to emphasize is that this knowledge of the instability may lead to other types of strategies, may lead to other ways of interacting systems.[34]

What is remarkable about a number of the advances in modern science at the end of the twentieth century, Prigogine goes on, is their appearance "at the very moment where our humanity is going through an age of transition, where instability, irreversibility, fluctuation, amplification, is found in every human activity. . . . What is so interesting, is that there is a kind of overall atmosphere, [an] overall cultural atmosphere, be it in science, be it in human science, which [is still developing today]."[35]

CONCLUSION

The work of postmodern scientists on unpredictability implies that human beings must give up the possibility of totally dominating and controlling nature. Because ecological and social systems are open, interacting, and unpredictable, we must allow for the possibility of surprise. Global weather patterns that include hurricanes and tornados; geological changes, such as earthquakes and volcanoes; and ecological and evolutionary processes cannot be predicted with sufficient certainty to give human beings complete control over nonhuman nature. We must leave room within our planning of industry, agriculture, forestry, and water usage and in our construction of dams, factories, and housing developments for nature's unpredictable events. We cannot dam every wild river, cut every old-growth forest, irrigate every desert, or build homes in every flood plain.

The reenchantment of nature called forth by the Frankfurt school's analysis of domination implies a partnership with nonhuman nature. Nature is an equal subject, not an object to be controlled. A partnership ethic means that a human community is in a sustainable ecological relationship with its surrounding natural community. Human beings are neither inferior to nature and dominated by it as in premodern societies, nor superior to it through their science and technology as in modern societies. Rather, human beings and nonhuman nature are equal partners in survival. The continuance of life as we know it requires an ethic of restraint, a holding back in implementing and producing some of the things potentially possible through science and technology (nuclear bombs and power, for example). It requires an active effort to restore the earth through replanting prairies, forests, and meadows. It requires reparations to the earth and social justice for human beings oppressed by the colonization of their lands, bodies, and hearts. As the First National People of Color Environmental Summit put it in their "Principles of Social Justice" in 1991:

> Environmental justice affirms the sacredness of Mother Earth, ecological unity and the interdependence of all species, and the right to be free from ecological destruction. . . . Environmental justice requires that we, as individuals, make personal and consumer choices to consume as little of Mother Earth's resources and to produce as little waste as possible; and make the conscious decision to challenge and reprioritize our lifestyles to insure the health of the natural world for present and future generations.[36]

NOTES

1. The following discussion of critical theory and the Frankfurt School draws on Douglas Kellner, *Critical Theory, Marxism, and Modernity* (Baltimore: Johns Hopkins University Press, 1989), esp. chs. 1–5; Martin Jy, *The Dialectical Imagination* (Boston: Little, Brown, 1973); Martin Jay, *Marxism and Totality: The Adventures of a Concept from Lukács to Habermas* (Berkeley: University of California Press, 1984), esp. chs. 6–8; Martin Jay, *Force Fields: Between Intellectual History and Cultural Critique* (New York: Routledge, 1993), esp. chs. 1, 2, 8, 9, 10; Mark Poster, *Critical Theory and Poststructuralism: In Search of a Context* (Ithaca, NY: Cornell University Press, 1989).

2. The extent to which Marx was "ecological" is debatable. See Albert Schmidt, *The Concept of Nature in Marx*, trans. Ben Fowkes (London: New Left Review, 1971); Howard Parsons, *Marx and Engels on Ecology* (Westport, CT: Greenwood Press, 1977); Donald C. Lee, "On the Marxian View of the Relationship between Man and Nature," *Environmental Ethics* 2 (Spring 1980): 3–16, and "Toward a Marxian Ecological Ethic: A Response to Two Critics," *Environmental Ethics* 4 (Winter 1982): 339–43; Val Routley (Plumwood), "On Karl Marx as an Environmental Hero," *Environmental Ethics* 3 (Fall 1981): 237–44; Charles Tolman, "Karl Marx, Alienation, and the Mastery of Nature," *Environmental Ethics* 3 (Spring 1981): 63–74; Hwa Yol Jung, "Marxism, Ecology, and Technology," *Environmental Ethics* (Summer 1983): 169–71; Reiner Grundmann, *Marxism and Ecology* (New York: Oxford University Press, 1991).

3. Max Horkheimer and Theodor Adorno, *Dialectic of Enlightenment*, trans. John Cummings (New York: Herder and Herder, 1972), p. 54.

4. Horkheimer and Adorno, *Dialectic of Enlightenment*, p. 57.

5. Theodor Adorno, "Spengler after the Decline" in *Prisms*, trans. Samuel and Shierry Weber (1967; Cambridge, MA: MIT Press, 1981), p. 67. I thank Murray Bookchin for this reference.

6. Marcuse's relationship to ecology has been debated. See Andrew Light, "Rereading Bookchin and Marcuse as Environmentalist Materialists," *Capitalism,*

Nature, Socialism 4, no. 1 (March 1993); Murray Bookchin, "Response to Andrew Light's 'Rereading Bookchin and Marcuse as Environmentalist Materialists,'" *Capitalism, Nature, Socialism* 4, no. 2 (June 1993): 101–20; Tim Luke, "Marcuse and Ecology" in *Marcuse Revised*, ed. John Bokina and Timothy Luke (Lawrence: University of Kansas Press, 1993).

    7. William Leiss, *The Domination of Nature* (New York: George Braziller, 1972), pp. 158, 164. On William Leiss, see Koula Mellos, "Leiss's Critical Theory of Human Needs," in *Perspectives on Ecology:A Critical Essay* (New York: St. Martin's Press, 1988), pp. 129–43.

    8. Robyn Eckersley, *Environmentalism and Political Theory* (Albany: State University of New York Press, 1992), p. 47.

    9. Immanual Wallerstein, *The Modern World-System* (New York: Academic Press, 1974;1980).

    10. On the debate over population, see Paul Ehrlich, *The Population Explosion* (New York: Simon and Schuster, 1990); "The Population Bomb: An Explosive Issue for the Environmental Movement?" *Utne Reader* (May/June 1988): 78–88; "Is AIDs Good for the Earth?" *Utne Reader* (November/December, 1987): 14, and "Letters to the Editor," *Utne Reader* (January/February): 4–7; David Harvey, "Population, Resources, and the Ideology of Science," *Economic Geography* 50, no. 3 (July 1974): 256–77; Murray Bookchin, "The Population Myth—Part I" *Green Perspectives* 8, (July 1988), and "The Population Myth—Part II" *Green Perspectives* 15 (April 1989).

    11. Arne Naess, "The Shallow and the Deep, Long-Range Ecology Movement," *Inquiry* 16 (1973): 95–100.

    12. Bill Devall and George Sessions, *Deep Ecology: Living as if Nature Mattered* (Salt Lake City, UT: Peregrine Smith Books, 1985); Arne Naess, *Ecology, Community, and Lifestyle*, trans. David Rothenburg (Cambridge: Cambridge University Press, 1989). See also Bill Devall, *Simple in Means, Rich in Ends: Practicing Deep Ecology* (Salt Lake City, UT: Pregrine Smith Books, 1988).

    13. Warwick Fox, *Toward a Transpersonal Ecology: Developing New Foundations for Environmentalism* (Boston: Shambala, 1990).

    14. On the debate between social ecology and Deep Ecology, see Murray Bookchin, "Social Ecology versus Deep Ecology: A Challenge for the Ecology Movement," (1987) republished in *Socialist Review* 18, no. 3 (July/September 1988); Kirkpatrick Sale, "Deep Ecology and Its Critics," *Nation*, May 14, 1988: 670–75; Murray Bookchin, "As If People Mattered," a response to Kirkpatrick Sale's "Deep Ecology and Its Critics," *Nation*, October 10, 1988; Ynestra King, "Letter to the Editor," *Nation*, December 12, 1987; George Bradford, "How Deep Is Deep Ecology?" *Fifth Estate* 22, no. 3 (1978): 3–30; Stephan Elkins, "The Politics of Mystical Ecology," *Telos*, no. 82 (Winter 1989–1990): 52–70; Tim Luke, "The Dreams of Deep Ecology," *Telos* 76 (Summer 1988): 65–92; Robyn Eckersley, "Divining Evolution: The Ecological Ethics of Murray Bookchin," *Environmental Ethics* 11 (Summer 1989): 99–116; Murray Bookchin, "Recovering Evolution: A Reply to Eckersley and Fox," *Environmental Ethics* 12 (Fall 1990): 253–73.

    15. Murray Bookchin, *Our Synthethic Environment* (New York: Harper Colophon, 1974; originally published under pseudonym Lewis Herber, New York: Knopf, 1962); *Post-Scarcity Anarchism* (San Francisco: Ramparts Books, 1971); *Toward an Ecological Society* (Montreal: Black Rose Books, 1981); *The Ecology of Freedom* (Palo Alto, CA: Cheshire

Books, 1982); *The Modern Crisis* (Philadelphia: New Society Publishers, 1986); *Remaking Society: Pathways to a Green Future* (Montreal: Black Rose Books, 1989); *The Philosophy of Social Ecology: Essays on Dialectical Naturalism* (Montreal: Black Rose Books, 1990).

16. Murray Bookchin, *The Ecology of Freedom: The Emergence and Dissolution of Hierarchy* (Palo Alto, CA: Cheshire Books, 1982), p. 1. Bookchin reverses the causal sequence between the domination of nature and the domination of man as stated by Adorno in *Prisms* (quoted above) and contrasts his position with that of Marx and the Frankfurt school: "However much they opposed domination, neither Adorno or Horkheimer signaled out hierarchy as an underlying problematic in their writings. Indeed, their residual Marxian premises led to a historical fatalism that saw any liberatory enterprise (beyond art, perhaps) as hopelessly tainted by the *need* to dominate nature and *consequently* 'man.' This position stands completely at odds with my own view that the *notion*—and no more than an *unrealizable* notion—of dominating nature stems from the domination of human by human. . . . The Frankfurt School, no less than Marxism, in effect, placed the onus for domination on a 'blind,' 'mute,' 'cruel,' and 'stingy,' nature, not (let me emphasize) only society. My own writings . . . argue that the domination of nature first arose within *society* as part of its institutionalization into gerontocracies that placed the young in varying degrees of servitude to the old and in patriarchies that placed women in varying degrees of servitude to men—not in any endeavour to 'control' nature or natural forces." (Murray Bookchin, "Thinking Ecologically: A Dialectical Approach," *The Philosophy of Social Ecology: Essays on Dialectical Naturalism* [Montreal: Black Rose Books, 1990], pp. 188–89.) Bookchin also contrasts the classical Marxist and liberal formulation of the desirability of the domination of nature by humans with that of certain ecological theorists who desire to continue the domination of humans by nonhuman nature. "Classically, the counterpart of the 'domination of nature by man' has been the 'domination of man by nature.' (Ibid., p. 163.)

17. For an introduction to socialist ecology, see the journal *Capitalism, Nature, Socialism*, especially James O'Connor, *Capitalism, Nature, Socialism* 1 (Fall 1988): 1–38; Alexander Cockburn, "Socialist Ecology: What It Means, Why No Other Will Do?" *Zeta* (February 1989): 15–21; Alexander Cockburn, "Whose Better Nature? Socialism, Capitalism, and the Environment," *Zeta* (June 1989): 27–32. On the debate between social and socialist ecology see Murray Bookchin, "Letters to Z," *Zeta* (April 1989): 3.

18. On Françoise d'Eaubonne's founding of the "Ecologie-Féminisme" Center in 1972, see the chronology in Françoise d'Eaubonne, "Feminism or Death," in *New French Feminisms: An Anthology*, ed. Elaine Marks and Isabelle de Courtivron (Amherst: University of Massachusetts Press, 1980), p. 25. On the center's "launching of a new action: *ecofeminism*," see ch. 16, "The Time for Ecofeminism," trans. Ruth Hottell, p. 201. The original French version is Françoise D'Eaubonne, *Le Féminisme ou la Mort* (Paris: Pierre Horay, 1974), pp. 215–52.

19. D'Eaubonne, "The Time for Ecofeminism," quotations on p. 212.

20. Johann Jakob Bachofen, *Myth, Religion, and Mother Right*, trans. Rudolf Marx (1863; Princeton, NJ: Princeton University Press, 1973); Friedrich Engels, "Origin of the Family, Private Property, and the State," in *Selected Works* (1884; New York: International Publishers, 1968); Robert Briffault, *The Mothers*, abridged ed. (3 vols., 1927; New York: Atheneum, 1977); August Bebel, *Woman in the Past, Present, and Future* (San Francisco: G. B. Benham, 1987).

21. For quotations from d'Eaubonne's, "The Time for Ecofeminism" on the labeling of population as a third world problem, see p. 208, on the male system and its power over women, see p. 208, on the price of phallocracy being paid by the earth, see p. 210.

22. Ibid., quotations on p. 209.

23. Ibid., quotations on pp. 211, 212.

24. Ynestra King, "Feminism and the Revolt of Nature," *Heresies* 13 (1981): 12–16.

25. Sherry Ortner, "Is Female to Male as Nature is to Culture?" in *Women, Culture, and Society*, ed. Michelle Rosaldo and Louise Lamphere (Stanford, CA: Stanford University Press, 1974), pp. 67–87.

26. Lynn White Jr., "The Historical Roots of Our Ecologic Crisis," *Science* 155 (March 10, 1967): 1203–207, reprinted in *Western Man and Environmental Ethics*, ed. Ian G. Barbour (Reading, MA: Addison Wesley, 1973), pp. 18–30, quotations on p. 25.

27. Joanna Macy, "Toward a Healing of Self and World," *Human Potential Magazine* 17, no. 1 (Spring 1992): 10–13, 29–33; Joanna Macy, *Despair and Personal Power in the Nuclear Age* (Philadelphia: New Society Publishers, 1983); Joanna Macy, *World as Lover; World as Self* (Berkeley, CA: Parallax Press, 1991); Joanna Macy, *Mutual Causality in Buddhism and General Systems Theory: The Dharma of Natural Systems* (Albany: State University of New York Press, 1991); John Seed, Joanna Macy, Pat Flemming, and Arne Naess, *Thinking Like a Mountain: Towards a Council of All Beings* (Philadelphia: New Society Publishers, 1988).

28. Charlene Spretnak, *The Spiritual Dimension of Green Politics* (Santa Fe, NM: Bear & Co., 1986); Spretnak, *States of Grace* (San Francisco: Harper and Row, 1991). The infusion of spirituality into green politics is controversial. For critiques, see Janet Biehl, "Goddess Mythology in Ecological Politics," *New Politics* 2 (Winter 1989): 84–105; Murray Bookchin, "Will Ecology Become 'the Dismal Science'?" *Progressive* (December 20, 1991): 18–21.

29. Fritjof Capra, "Systems Theory and the New Paradigm," p. 365.

30. David Bohm, "Postmodern Science and a Postmodern World," quotation on p. 394. David Bohm, *Wholeness and the Implicate Order* (Boston: Routledge and Kegan Paul, 1980). See also John P. Briggs and F. David Peat, *The Looking Glass Universe: The Emergence of Wholeness* (New York: Simon and Schuster, 1984).

31. James Gleick, *Chaos: The Making of a New Science* (New York: Viking, 1987); M. Mitchell Waldrop, *Complexity: The Emerging Science at the Edge of Order and Chaos* (New York: Simon and Schuster, 1992).

32. Edward N. Lorenz, "Predictability: Does the Flap of a Butterfly's Wings in Brazil Set Off a Tornado in Texas?" quotations on p. 399; Edward N. Lorenz, "Crafoord Prize Lecture," *Tellus* (1984): 36A, 98–110.

33. Ilya Prigogine and Isabelle Stengers, *Order Out of Chaos: Man's New Dialogue with Nature* (New York: Bantam, 1984). On Prigogine's work, see Erich Jantsch, *The Self-Organizing Universe* (New York: Pergamon Press, 1980).

34. Ilya Prigogine, "Science in a World of Limited Predictability," pp. 404–405.

35. Ibid.

36. First National People of Color Environmental Leadership Summit, "Principles of Environmental Justice" adopted October 27, 1991, principles 1 and 17.

PART I

CRITICAL THEORY AND THE DOMINATION OF NATURE

1

MARX AND ENGELS ON ECOLOGY

KARL MARX AND FRIEDRICH ENGELS
EDITED BY HOWARD L. PARSONS

I. DIALECTICS

Man's collective and immediate material dealings with nature bring him into a dialectical relation with it, into a dynamic and potentially developing interaction with it. Such a relation, when critically analyzed, reveals nature as continuous motion, interconnection, and transformation. Nature is a ceaseless series of unities of opposites, which are mutually creative, mutually destructive, and mutually transforming.—H. L. P.

Dialectics versus Metaphysics

When we reflect on nature, or the history of mankind, or our own intellectual activity, the first picture presented to us is of an endless maze of relations and interactions, in which nothing remains what, where and as it was, but everything moves, changes, comes into being and passes out of existence. This primitive, naive, yet intrinsically correct conception of the world was that of ancient Greek philosophy, and was first clearly formulated by Heraclitus: everything is and also is not, for everything is in flux, is constantly changing, constantly coming into being and passing away. But this conception, correctly as it covers the general character of the picture of phenomena as a whole, is yet inadequate to explain the details of which this total picture is composed; and so long as we do not understand these, we also have no clear idea of the picture as a whole. In order to understand these details, we must detach them from their natural or his-

Howard L. Parsons, ed., *Marx and Engels on Ecology* (Westport, CT: Greenwood Press, 1977), pp. 129–85, selections.

43

torical connections, and examine each one separately, as to its nature, its special causes and effects, etc. This is primarily the task of natural science and historical research; branches of science which the Greeks of the classical period, on very good grounds, relegated to a merely subordinate position, because they had first of all to collect materials for these to work upon. The beginnings of the exact investigation of nature were first developed by the Greeks of the Alexandrian period, and later on, in the Middle Ages, were further developed by the Arabs. Real natural science, however, dates only from the second half of the fifteenth century, and from then on it has advanced with constantly increasing rapidity.

The analysis of Nature into its individual parts, the grouping of the different natural processes and natural objects in definite classes, the study of the internal anatomy of organic bodies in their manifold forms—these were the fundamental conditions of the gigantic strides in our knowledge of Nature which have been made during the last four hundred years. But this method of investigation has also left us as a legacy the habit of observing natural objects and natural processes in their isolation, detached from the whole vast interconnection of things; and therefore not in their motion, but in their repose; not as essentially changing, but as fixed constants; not in their life, but in their death. And when, as was the case with Bacon and Locke, this way of looking at things was transferred from natural science to philosophy, it produced the specific narrow-mindedness of the last centuries, the metaphysical mode of thought. . . .

Friedrich Engels, *Herr Eugen Dühring's Revolution in Science*, pp. 26–29[1]

II. THE INTERDEPENDENCE OF MAN
AS A LIVING BEING WITH NATURE

Man depends on nature for vital substances (such as food, water, and oxygen) and processes (such as photosynthesis), and nature in turn is affected by man's activities, such as the use of fire and the domestication of plants and animals. Marx described man's interdependence with nature at both the biochemical and the psychological levels.—H. L. P.

Nature Is Man's Body, on Which He Lives

The life of the species, both in man and in animals, consists physically in the fact that man (like the animal) lives on inorganic nature; and the more universal man is compared with an animal, the more universal is the sphere of inorganic nature on which he lives. Just as plants, animals, stones, air, light, etc., constitute theoretically a part of human consciousness, partly as objects of natural science, partly as objects of art—his spiritual inorganic nature, spiritual nourishment which he must first prepare to make palatable and digestible—so also in the realm of practice they constitute a part of human life and human activity. Physically man lives only on these products of nature, whether they appear in the form of food, heating, clothes, a dwelling, etc. The universality of man appears in practice precisely in the universality which makes all nature his *inorganic body*—both inasmuch as nature is (1) his direct means of life, and (2) the material, the object, and the instrument of his life activity. Nature is man's *inorganic body*—nature, that is, in so far as it is not itself the human body. Man *lives* on nature means that nature is his *body*, with which he must remain in continuous interchange if he is not to die. That man's physical and spiritual life is linked to nature means simply that nature is linked to itself, for man is a part of nature.

Karl Marx, *The Economic and Philosophic Manuscripts of 1844*, p. 112

Man's Essential Interdependence with Nature

. . . Man is directly a *natural being*. As a natural being and as a living natural being he is on the one hand endowed with *natural powers of life*—he is an *active* natural being. These forces exist in him as tendencies and abilities—as *instincts*. On the other hand, as a natural corporeal, sensuous, objective being he is a *suffering*, conditioned and limited creature, like animals and plants. That is to say, the *objects* of his instincts exist outside him, as *objects* independent of him; yet these objects are *objects* that he *needs*—essential *objects*, indispensable to the manifestation and confirmation of his essential powers. To say that man is a *corporeal*, living, real, sensuous, objective being full of natural vigor is to say that he has *real, sensuous, objects* as the objects of his life, or that he can only express his life in real, sensuous objects. To *be* objective, natural and sensuous, and at the same time to have object, nature and sense outside oneself, or oneself to be object, nature and sense for a third party, is one and the same thing. *Hunger* is a natural *need*; it therefore needs a *nature* outside itself, an *object* outside itself, in order to satisfy itself, to be stilled. Hunger is an acknowledged need of my body for an *object* existing outside it, indispensable to its integration and to the expression of its essential being. The sun is the *object* of the plant—an indispensable object to it, confirming its life—just as the plant is an object of the sun, being an expression of the life-awakening power of the sun, of the sun's *objective* essential power. . . .

Karl Marx, *The Economic and Philosophic Manuscripts of 1844*,
pp. 180–82

Matter as Motion, Creativity, and Sensuous Quality

The real founder of *English materialism* and all *modern experimental* science was *Bacon*. For him natural science was true science and *physics* based on perception was the most excellent part of natural science. *Anaxagoras* with his *homoeomeria* and *Democritus* with his atoms are often the authorities he refers to. According to his teaching the *senses* are infallible and are the *source* of all knowledge. Science is *experimental* and consists in applying a *rational method* to the data provided by the senses. Induction, analysis, comparison, observation and experiment are the principal requisites of rational method. The first and most important of the inherent qualities of *matter* is *motion*, not only *mechanical* and *mathematical* movement, but still more *impulse, vital life-spirit, tension,* or, to use Jacob Bohme's expression, the throes [*Qual*] of matter. The primary forms of matter are the living, individualizing *forces of being* inherent in it and producing the distinction between the species.

In *Bacon*, its first creator, materialism contained latent and still in a naive way the germs of all-round development. Matter smiled at man with poetical sensuous brightness. The aphoristic doctrine itself, on the other hand, was full of the inconsistencies of theology.

Karl Marx and Friedrich Engels, *The Holy Family*, p. 172

III. Man's Interdependence with Nature As a Being That Makes a Living

> As distinct from other living beings, man not only "lives on inorganic nature," but he also makes a living of nature, interacting with it by means of his brain, hand, and tools in order to subsist. Thus, far more than any other species, the human species puts its "stamp on nature." Marx and Engels addressed themselves to the question of how the human species is differentiated from other animals, because on the one hand they wished to refute the idealists who repudiated man's animal nature and because on the other hand they wanted to demonstrate the great possibilities for fulfillment in man's evolutionary relation to nature.—H. L. P.

By Productively Interacting with Nature, Man Subsists

The first premise of all human history is, of course, the existence of living human individuals. Thus the first fact to be established is the physical organization of these individuals and their consequent relation to the rest of nature. Of course, we cannot here go either into the actual physical nature of man, or into the natural conditions in which man finds himself—geological, oro-hydrographical, climatic and so on. The writing of history must always set out from these natural bases and their modification in the course of history through the action of man.

Men can be distinguished from animals by consciousness, by religion or anything else you like. They themselves begin to distinguish themselves from animals as soon as they begin to *produce* their means of subsistence, a step which is conditioned by their physical organization. By producing their means of subsistence men are indirectly producing their actual material life.

The way in which men produce their means of subsistence depends first of all on the nature of the actual means they find in existence and have to reproduce.

This mode of production must not be considered simply as being the reproduction of the physical existence of the individuals. Rather it is a definite form of activity of these individuals, a definite form of expressing their life, a definite *mode of life* on their part. As individuals express their life, so they are. What they are, therefore, coincides with their production, both with *what* they produce and with *how* they produce. The nature of individuals thus depends on the material conditions determining their production.

Karl Marx and Friedrich Engels, *The German Ideology*, p. 7

Man Has Impressed His Stamp on Nature through Hand and Brain

When after thousands of years of struggle the differentiation of hand from foot, and erect gait, were finally established, man became distinct from the ape and the basis was laid for the development of articulate speech and the mighty development of the brain that has since made the gulf between man and the ape an unbridgeable one. The specialization of the hand—this implies the *tool*, and the tool implies specific human activity, the transforming reaction of man on nature, production. Animals in the narrower sense also have tools, but only as limbs of their bodies: the ant, the bee, the beaver; animals also produce, but their productive effect on surrounding nature in relation to the latter amounts to nothing at all. Man alone has succeeded in impressing his stamp on nature, not only by so altering the aspect and climate of his dwelling-place, and even the plants and animals themselves, that the consequences of his activity can disappear only with the general extinction of the terrestrial globe. And he has accomplished this primarily and essentially by means of *the hand*. Even the steam-engine, so far his most powerful tool for the transformation of nature, depends, because it is a tool, in the last resort on the hand. But step by step with the development of the hand went that of the brain; first of all came consciousness of the conditions for separate practically useful actions, and later, among the more favoured peoples and arising from that consciousness, insight into the natural laws governing them. And with the rapidly growing knowledge of the laws of nature the means for reacting on nature also grew; the hand alone would never have achieved the steam-engine if, along with and parallel to the hand, and partly owing to it, the brain of man had not correspondingly developed.

With man we enter *history*. Animals also have a history, that of their descent and gradual evolution to their present position. This history, however, is made for them, and in so far as they themselves take part in it, this occurs without their knowledge and desire. On the other hand, the more that human beings become removed from animals in the narrower sense of the word, the more they make their history themselves, consciously, the less becomes the influence of unforeseen effects and uncontrolled forces on this history, and the more accurately does the historical result correspond to the aim laid down in advance.

Friedrich Engels, *Dialectics of Nature*, pp. 46–48

Man's Reaction on Nature

Natural science, like philosophy, has hitherto entirely neglected the influence of men's activity on their thought; both know only nature on the one hand and thought on the other. But it is precisely *the alteration of nature by men,* not solely nature as such, which is the most essential and immediate basis of human thought, and it is in the measure that man has learned to change nature that his intelligence has increased. The naturalistic conception of history, as found, for instance, to a greater or lesser extent in Draper and other scientists, as if nature exclusively reacts on man, and natural conditions everywhere exclusively determined his historical development, is therefore one-sided and forgets that man also reacts on nature, changing it and creating new conditions of existence for himself. There is devilishly little left of "nature" as it was in Germany at the time when the Germanic peoples immigrated into it. The earth's surface, climate, vegetation, fauna, and the human beings themselves have infinitely changed, and all this owing to human activity, while the changes of nature in Germany which have occurred in this period of time without human interference are incalculably small.

Friedrich Engels, *Dialectics of Nature*, p. 306

The Laws of Human Life Are Different from the Laws of Animal Life

The struggle for life. Until Darwin, what was stressed by his present adherents was precisely the harmonious co-operative working of organic nature, how the plant kingdom supplies animals with nourishment and oxygen, and animals supply plants with manure, ammonia, and carbonic acid. Hardly was Darwin recognized before these same people saw everywhere nothing but *struggle.* Both views are justified within narrow limits, but both are equally one-sided and prejudiced. The interaction of bodies in non-living nature includes both harmony and collisions, that of living bodies conscious and unconscious co-operation as well as conscious and unconscious struggle. Hence, even in regard to nature, it is not permissible one-sidedly to inscribe only "struggle" on one's banners. But it is absolutely childish to desire to sum up the whole manifold wealth of historical evolution and complexity in the meagre and one-sided phrase "struggle for existence." That says less than nothing.

The whole Darwinian theory of the struggle for existence is simply the transference from society to organic nature of Hobbes' theory of *bellum omnium contra omnes* and of the bourgeois economic theory of competition, as well as the Malthusian theory of population. When once this feat has been accomplished (the unconditional justification for which, especially as regards the Malthusian theory, is still very questionable), it is very easy to transfer these theories back again from natural history to the history of society, and altogether too naive to maintain that thereby these assertions have been proved as eternal natural laws of society.

Let us accept for a moment the phrase "struggle for existence," for argument's sake. The most that the animal can achieve is to *collect*; man *produces*, he prepares the means of life, in the widest sense of the words, which without him nature would not have produced. This makes impossible any unqualified transference of the laws of life in animal societies to human society. Production soon brings it about that the so-called struggle for existence no longer turns on pure means of existence, but on means of enjoyment and development. Here—where the means of development are socially produced—the categories taken from the animal kingdom are already totally inapplicable. Finally, under the capitalist mode of production, production reaches such a high level that society can no longer consume the means of life, enjoyment and development that have been produced, because for the great mass of producers access to these means is artificially and forcibly barred; and therefore every ten years a crisis restores the equilibrium by destroying not only the means of life, enjoyment and development that have been produced, but also a great part of the productive forces themselves. Hence the so-called struggle for existence assumes the form: to protect the products and productive forces produced by bourgeois capitalist society against the destructive, ravaging effect of this capitalist social order, by taking control of social production and distribution out of the hands of the ruling capitalist class, which has become incapable of this function, and transferring it to the producing masses—and that is the socialist revolution.

The conception of history as a series of class struggles is already much richer in content and deeper than merely reducing it to weakly distinguished phases of the struggle for existence.

Friedrich Engels, *Dialectics of Nature*, pp. 404–405

The Passage from Animal Necessity to Human Freedom

The seizure of the means of production by society puts an end to commodity production, and therewith to the domination of the product over the producer. Anarchy in social production is replaced by conscious organization on a planned basis. The struggle for individual existence comes to an end. And at this point, in a certain sense, man finally cuts himself off from the animal world, leaves the conditions of animal existence behind him and enters conditions which are really human. The conditions of existence forming man's environment, which up to now have dominated man, at this point pass under the dominion and control of man, who now for the first time becomes the real conscious master of Nature, because and in so far as he has become master of his own social organization. The laws of his own social activity, which have hitherto confronted him as external, dominating laws of Nature, will then be applied by man with complete understanding, and hence will be dominated by man. Men's own social organization which has hitherto stood in opposition to them as if arbitrarily decreed by Nature and history, will then become the voluntary act of men themselves. The objective, external forces which have hitherto dominated history, will then pass under the control of men themselves. It is only from this point that men, with full consciousness, will fashion their own history; it is only from this point that the social causes set in motion by men will have, predominantly and in constantly increasing measure, the effects willed by men. It is humanity's leap from the realm of necessity into the realm of freedom.

To carry through his world-emancipating act is the historical mission of the modern proletariat. And it is the task of scientific socialism, the theoretical expression of the proletarian movement, to establish the historical conditions and, with these, the nature of this act, and thus to bring to the consciousness of the now oppressed class the conditions and nature of the act which it is its destiny to accomplish.

Friedrich Engels, *Herr Eugen Dühring's Revolution in Science*, pp. 309–10

Man's Brain and Thoughts Are the Products of Nature and Are in Correspondence with It

But if the further question is raised: what then are thought and consciousness, and whence they come, it becomes apparent that they are products of the human brain and that man himself is a product of Nature, which has been developed in and along with its environment: whence it is self-evident that the products of the human brain, being in the last analysis also products of Nature, do not contradict the rest of Nature but are in correspondence with it.

Friedrich Engels, *Herr Eugen Dühring's Revolution in Science*, pp. 42–43

IV. PRECAPITALIST RELATIONS OF MAN TO NATURE

Marx's investigation of the history of man's relation to nature through communal labor led to his concentration on the modes of production and distribution in precapitalist societies, as contrasted with what occurs in capitalist societies.—H. L. P.

Communal Labor Appropriates Nature as Common Property

In the first form of this landed property, an initial, naturally arisen spontaneous [*naturwüchsiges*] community appears as first presupposition. Family, and the family extended as a clan [*Stamm*], or through intermarriage between families, or combination of clans. Since we may assume that *pastoralism*, or more generally a *migratory* form of life, was the first form of the mode of existence, not that the clan settles in a specific site, but that it grazes off what it finds—humankind is not settlement-prone by nature (except possibly in a natural environment so especially fertile that they sit like monkeys on a tree; else roaming like the animals)—then the *clan community*, the natural community, appears not as a *result* of, but as a *presupposition for the communal appropriation* (temporary) *and utilization of the land.* When they finally do settle down, the extent to which this original community is modified will depend on various external, climatic, geographic, physical, etc., conditions as well as on their particular natural predisposition—their clan character. This naturally arisen clan community, or, if one will, pastoral society, is the first presupposition—the communality [*Gemeinschaftlichkeit*] of blood, language customs—for the *appropriation of the objective conditions* of their life, and of their life's reproducing and objectifying activity (activity as herdsmen, hunters, tillers, etc.). The earth is the great workshop, the arsenal which furnishes both means and material of labour, as well as the seat, the *base* of the community. They relate naively to it as the *property of the community,* of the community producing and reproducing itself in living labour. Each individual conducts himself only as a link, as a member of this community as *proprietor* or *possessor.*

Karl Marx, Grundrisse, p. 472

Appropriation of the Earth by the Commune

The main point here is this: In all these forms—in which landed property and agriculture form the basis of the economic order, and where the economic aim is hence the production of use values, i.e., the *reproduction of the individual* within the specific relation to the commune in which he is its basis—there is to be found: (1) Appropriation not through labour, but presupposed to labour; appropriation of the natural conditions of labour, of the *earth* as the original instrument of labour as well as its workshop and repository of raw materials. The individual relates simply to the objective conditions of labour as being his; [relates] to them as the inorganic nature of his subjectivity in which the latter realizes itself; the chief objective condition of labour does not itself appear as a *product* of labour, but is already there as *nature*; on one side the living individual, on the other the earth, as the objective condition of his reproduction; (2) but this *relation* to land and soil, to the earth, as the property of the labouring individual—who thus appears from the outset not merely as labouring individual, in this abstraction, but who has an *objective mode of existence* in his ownership of the land, an existence *presupposed* to his activity, and not merely as a result of it, a presupposition of his activity just like his skin, his sense organs, which of course he also reproduces and develops, etc., in the life process, but which are nevertheless presuppositions of this process of his reproduction—is instantly mediated by the naturally arisen, spontaneous, more or less historically developed and modified presence of the individual as *member of a commune*—his naturally arisen presence as member of a tribe, etc. An isolated individual could no more have property in land and soil than he could speak. He could, of course, live off it as substance, as do the animals. The relation to the earth as property is always mediated through the occupation of the land and soil, peacefully or violently, by the tribe, the commune, in some more or less naturally arisen or already historically developed form.

Karl Marx, *Grundrisse*, p. 485

V. CAPITALIST POLLUTION AND THE RUINATION OF NATURE

> The development of primitive societies into class societies and eventually into capitalist society has produced a transformation in man's relation to nature. Under capitalism this relation is determined primarily by the ruling class of capitalists. Marx and Engels undertook a thorough examination of this new relation. Like the relation that they bear to workers, the relation of capitalists to nature is marked by exploitation, pollution, and ruination. Marx and Engels observed that the capitalists appropriate the resources of the earth without any cost to themselves, that they transform the earth into an "object of huckstering," that under capitalism the original unity of man and nature is breached as well as the unity of manufacture and agriculture, that man's precipitous alteration of nature issues in unforeseen and harmful consequences, and that the capitalists' wastage and exhaustion of the soil, deforestation, disruption of nature's cycle of matter, greedy policy toward nature, and neglect of man's welfare are ruinous to both nature and man.—H. L. P.

The Productive Forces of Man, Science, and Nature Cost Capital Nothing

We saw that the productive forces resulting from co-operation and division of labour cost capital nothing. They are natural forces of social labour. So also physical forces, like steam, water, &c., when appropriated to productive processes cost nothing. But just as a man requires lungs to breathe with, so he requires something that is work of man's hand, in order to consume physical forces productively. A water-wheel is necessary to exploit the force of water, and a steam-engine to exploit the elasticity of steam. Once discovered, the law of the deviation of the magnetic needle in the field of an electric current, or the law of the magnetization of iron, around which an electric current circulates, cost never a penny.[2] But the exploitation of these laws for the purposes of telegraphy, &c., necessitates a costly and extensive apparatus. The tool, as we have seen, is not exterminated by the machine. From being a dwarf implement of the human organism, it expands and multiplies into the implement of a mechanism created by man. Capital now sets the labourer to work, not with a manual tool, but with a machine which itself handles the tools. Although, therefore, it is clear at the first glance that, by incorporating both stupendous physical forces, and the natural sciences, with the process of production, Modern Industry raises the productiveness of labour to an extraordinary degree, it is by no means equally clear, that this increased productive force is not, on the other hand, purchased by an increased expenditure of labour. Machinery, like every other component of constant capital, creates no new value, but yields up its own value to the product that it serves to beget. In so far as the machine has value, and, in consequence, parts with value to the product, it forms an element in the value of that product. Instead of being cheapened, the product is made dearer in proportion to the value of the machine. And it is clear as noon-day, that machines and systems of machinery, the characteristic instruments of labour of Modern Industry, are incomparably more loaded with value than the implements used in handicrafts and manufactures.

In the first place, it must be observed that the machinery, while always entering as a whole into the labour-process, enters into the value-begetting process only by bits. It never adds more value than it loses, on an average, by wear and tear. Hence there is a great difference between the value of a machine, and the value transferred in a given time by that machine to the product. The longer the life of the machine in the labour-process, the greater is that difference. It is true, no doubt, as we have already seen, that every instrument of labour enters as a whole into the labour-process, and only piece-meal, proportionally to its average daily loss by wear and tear, into the value-begetting process. But this difference between the instrument as a whole and its daily wear and tear, is much greater in a machine than in a tool, because the machine, being made from more durable material, has a longer life; because its employment, being regulated by strictly scientific laws, allows of greater economy in the wear and tear of its parts, and in the materials it consumes; and lastly, because its field of production is incomparably larger than that of a tool.

Karl Marx, *Capital* 1:386–88

In the extractive industries, mines, &c., the raw materials form no part of the capital advanced. The subject of labour is in this case not a product of previous labour, but is furnished by Nature gratis, as in the case of metals, minerals, coal, stone, &c.

Karl Marx, *Capital* 1:603

Natural elements entering as agents into production, and which cost nothing, no matter what role they play in production, do not enter as components of capital, but as a free gift of Nature to capital, that is, as a free gift of Nature's productive power to labour, which, however, appears as the productiveness of capital, as all other productivity under the capitalist mode of production.

Karl Marx, *Capital* 3:745

Capitalist Huckstering of People and Nature

. . . To make earth an object of huckstering—the earth which is our one and all, the first condition of our existence—was the last step toward making oneself an object of huckstering. It was and is to this very day an immorality surpassed only by the immorality of self-alienation. And the original appropriation—the monopolization of the earth by a few, the exclusion of the rest from that which is the condition of their life—yields nothing in immorality to the subsequent huckstering of the earth.

Friedrich Engels, *Outlines of a Critique of Political Economy*, in *The Economic and Philosophic Manuscripts of 1844*, by Karl Marx, p. 210

Capitalism Rends the Unity of Agriculture and Industry, Disturbs Man's Relation to the Soil, Wastes Workers, and Robs Laborers and Soil

In the sphere of agriculture, modern industry has a more revolutionary effect than elsewhere, for this reason, that it annihilates the peasant, that bulwark of the old society, and replaces him by the wage-labourer. Thus the desire for social changes, and the class antagonisms are brought to the same level in the country as in the towns. The irrational, old-fashioned methods of agriculture are replaced by scientific ones. Capitalist production completely tears asunder the old bond of union which held together agriculture and manufacture in their infancy. But at the same time it creates the material conditions for a higher synthesis in the future, viz., the union of agriculture and industry on the basis of the more perfected forms they have each acquired during their temporary separation. Capitalist production, by collecting the population in great centres, and causing an ever increasing preponderance of town population, on the one hand concentrates the historical motive power of society; on the other hand, it disturbs the circulation of matter between man and the soil, i.e., prevents the return to the soil of its elements consumed by man in the form of food and clothing; it therefore violates the conditions necessary to lasting fertility of the soil. By this action it destroys at the same time the health of the town labourer and the intellectual life of the rural labourer. But while upsetting the naturally grown conditions for the maintenance of that circulation of matter, it imperiously calls for its restoration as a system, as a regulating law of social production, and under a form appropriate to the full development of the human race. In agriculture as in manufacture, the transformation of production under the sway of capital, means, at the same time, the martyrdom of the producer; the instrument of labour becomes the means of enslaving, exploiting, and impoverishing the labourer; the social combination and organization of labour-processes is turned into an organized mode of crushing out the workman's individual vitality, freedom, and independence. The dispersion of the rural labourers over larger areas breaks their power of resistance while concentration increases that of the town operatives. In modern agriculture, as in the urban industries, the increased productiveness and quantity of the labour set in motion are bought at the cost of laying waste and consuming by disease labour-power itself. Moreover, all progress in capitalistic agriculture is a progress in the art, not only of robbing the labourer, but of robbing the soil; all progress in increasing the fertility of the soil for a given time, is a progress towards ruining the lasting sources of that fertility. The more a country starts its development on the foundation of modern industry, like the United States, for example, the more rapid is this process of destruction. Capitalist production, therefore, develops technology, and the combination together of various processes into a social whole, only by sapping the original sources of all wealth—the soil and the labourer.

Karl Marx, Capital, 1:505–507

Capitalist Failure to Utilize the Waste Products of Industry, Agriculture, and Human Consumption, and Ways of Utilizing Waste

The capitalist mode of production extends the utilization of the excretions of production and consumption. By the former we mean the waste of industry and agriculture, and by the latter partly the excretions produced by the natural exchange of matter in the human body and partly the form of objects that remains after their consumption. In the chemical industry, for instance, excretions of production are such by-products as are wasted in production on a smaller scale; iron filings accumulating in the manufacture of machinery and returning into the production of iron as raw material, etc. Excretions of consumption are the natural waste matter discharged by the human body, remains of clothing in the form of rags, etc. Excretions of consumption are of the greatest importance for agriculture. So far as their utilization is concerned, there is an enormous waste of them in the capitalist economy. In London, for instance, they find no better use for the excretion of four and a half million human beings than to contaminate the Thames with it at heavy expense.

Rising prices of raw materials naturally stimulate the utilization of waste products.

Karl Marx, Capital 3:101–103

In Altering Nature Man Produces Unforeseen and Harmful Consequences

Animals, as already indicated, change external nature by their activities just as man does, even if not to the same extent, and these changes made by them in their environment, as we have seen, in turn react upon and change their originators. For in nature nothing takes place in isolation. Everything affects every other thing and vice-versa, and it is mostly because this all-sided motion and interaction is forgotten that our natural scientists are prevented from clearly seeing the simplest things. We have seen how goats have prevented the regeneration of forests in Greece; on St. Helena, goats and pigs brought by the first navigators to arrive there succeeded in exterminating almost completely the old vegetation of the island, and so prepared the ground for the spreading of plants brought by later sailors and colonists. But if animals exert a lasting effect on their environment, it happens unintentionally, and as far as the animals themselves are concerned, it is an accident. The further men become removed from animals, however, the more their effect on nature assumes the character of premeditated, planned action directed toward definite ends known in advance. The animal destroys the vegetation of a locality without realizing what it is doing. Man destroys it in order to sow field crops on the soil thus released, or to plant trees or vines which he knows will yield many times the amount sown. He transfers useful plants and domestic animals from one country to another and thus changes the flora and fauna of whole continents. More than this. Through artificial breeding, both plants and animals are so changed by the hand of man that they become unrecognizable. The wild plants from which our grain varieties originated are still being sought in vain. The question of the wild animal from which our dogs are descended, the dogs themselves being so different from one another, or our equally numerous breeds of horses, is still under dispute. . . .

In short, the animal merely *uses* external nature, and brings about changes in it simply by his presence; man by his changes makes it serve his ends, *masters* it. This is the final, essential distinction between man and other animals, and once again it is labour that brings about this distinction.

Let us not, however, flatter ourselves overmuch on account of our human conquests over nature. For each such conquest takes its revenge on us. Each of them, it is true, has in the first place the consequences on which we counted, but in the second and third places it has quite different, unforeseen effects which only too often cancel out the first. The people who, in Mesopotamia, Greece, Asia Minor, and elsewhere, destroyed the forests to obtain cultivable land, never dreamed that they were laying the basis for the present devastated condition of those countries, by removing along with the forests the collecting centres and reservoirs of moisture. When, on the southern slopes of the mountains, the Italians of the Alps used up the fir forests so carefully cherished on the northern slopes, they had no inkling that by doing so they were cutting at the roots of the dairy industry in their region; they had still less inkling that they were thereby depriving their mountain springs of water for the greater part of the year, making it possible for these to pour still more furious flood torrents on the plains during the rainy season. Those who spread the potato in Europe were not aware that with these farinaceous tubers they were at the same time spreading the disease of scrofula. Thus at every step we are reminded that we by no means rule over nature like a conqueror over a foreign people, like someone standing outside nature— but that we, with flesh, blood, and brain, belong to nature, and exist in its midst, and that all our mastery of it consists in the fact that we have the advantage over all other creatures of being able to know and correctly apply its laws.

And, in fact, with every day that passes we are learning to understand these laws more correctly, and getting to know both the more immediate and the more remote consequences of our interference with the traditional course of nature. . . .

Friedrich Engels, *Dialectics of Nature*, pp. 239–46

Capitalism Wastes and Exhausts the Soil

Here, in small-scale agriculture, the price of land, a form and result of private land ownership, appears as a barrier to production itself. In large-scale agriculture, and large estates operating on a capitalist basis, ownership likewise acts as a barrier, because it limits the tenant farmer in his productive investment of capital which in the final analysis benefits not him, but the landlord. In both forms, exploitation and squandering of the vitality of the soil (apart from making exploitation dependent upon the accidental and unequal circumstances of individual produce rather than the attained level of social development) takes the place of conscious rational cultivation of the soil as eternal communal property, an inalienable condition for the existence and reproduction of a chain of successive generations of the human race.

Karl Marx, *Capital* 3:812

Deforestation under Capitalism

The long production time (which comprises a relatively small period of working time) and the great length of the periods of turnover entailed make forestry an industry of little attraction to private and therefore capitalist enterprise, the latter being essentially private even if the associated capitalist takes the place of the individual capitalist. The development of culture and of industry in general has ever evinced itself in such energetic destruction of forests that everything done by it conversely for their preservation and restoration appears infinitesimal.

Karl Marx, *Capital* 2:244

Capitalism Ruins the Worker's Health and the Soil's Fertility

But in its blind unrestrainable passion, its were-wolf hunger for surplus-labour, capital oversteps not only the moral, but even the merely physical maximum bounds of the working-day. It usurps the time for growth, development, and healthy maintenance of the body. It steals the time required for the consumption of fresh air and sunlight. It higgles over a meal-time, incorporating it where possible with the process of production itself, so that food is given to the labourer as to a mere means of production, as coal is supplied to the boiler, grease and oil to the machinery. It reduces the sound sleep needed for the restoration, reparation, refreshment of the bodily powers to just so many hours of torpor as the revival of an organism, absolutely exhausted, renders essential. It is not the normal maintenance of the labour-power which is to determine the limits of the working-day; it is the greatest possible daily expenditure of labour-power, no matter how diseased, compulsory, and painful it may be, which is to determine the limits of the labourers' period of repose. Capital cares nothing for the length of life of labour-power. All that concerns it is simply and solely the maximum of labour-power, that can be rendered fluent in a working-day. It attains this end by shortening the extent of the labourer's life, as a greedy farmer snatches increased produce from the soil by robbing it of its fertility.

Karl Marx, *Capital* 1:264–65

NOTES

1 The quotations from the works of Marx and Engels are from the following:

Friedrich Engels, *Dialectics of Nature* (New York: International Publishers, 1954).
———. *Herr Eugen Dühring's Revolution in Science (Anti-Dühring)*, ed. C. P. Dutt and trans. Emile Burns (New York: International Publishers, 1939, 1966).
Karl Marx, *Capital: A Critique of Political Economy*, vol. 1, trans. Samuel Moore and Edward Aveling and ed. Friedrich Engels; vol. 2, trans. Ernest Untermann and ed. Friedrich Engels; vol. 3, trans. Ernest Untermann and ed. Friedrich Engels (New York: International Publishers, 1967).
———. *The Economic and Philosophic Manuscripts of 1844*, trans. Martin Milligan and ed. Dirk J. Struik (New York: International Publishers, 1964).
———. *Grundrisse: Foundations of the Critique of Political Economy* (rough draft), trans. Martin Nicolaus (New York: Random House, 1973).
Karl Marx and Friedrich Engels, *The German Ideology,* pts. 1 and 3, trans. and ed. R. Pascal (New York: International Publishers, 1947).
———. *The Holy Family* (Los Angeles: Progress Publishers, 1956).

2. Science, generally speaking, costs the capitalist nothing, a fact that by no means hinders him from exploiting it. The science of others is as much annexed by capital as the labour of others. Capitalistic appropriation and personal appropriation, whether of science or of material wealth, are, however, totally different things. Dr. Ure himself deplores the gross ignorance of mechanical science existing among his dear machinery-exploiting manufacturers, and Liebig can a tale unfold about the astounding ignorance of chemistry displayed by English chemical manufacturers.

2

THE CONCEPT OF ENLIGHTENMENT

MAX HORKHEIMER
AND THEODOR ADORNO
Translated by John Cumming

In the most general sense of progressive thought, the Enlightenment has always aimed at liberating men from fear and establishing their sovereignty. Yet the fully enlightened earth radiates disaster triumphant. The program of the Enlightenment was the disenchantment of the world, the dissolution of myths, and the substitution of knowledge for fancy. Bacon, the "father of experimental philosophy,"[1] had defined its motives. He looked down on the masters of tradition, the "great reputed authors" who first

> believe that others know that which they know not; and after themselves know that which they know not. But indeed facility to believe, impatience to doubt, temerity to answer, glory to know, doubt to contradict, end to gain, sloth to search, seeking things in words, resting in part of nature; these and the like have been the things which have forbidden the happy match between the mind of man and the nature of things; and in place thereof have married it to vain notions and blind experiments: and what the posterity and issue of so honorable a match may be, it is not hard to consider. Printing, a gross invention; artillery, a thing that lay not far out of the way; the needle, a thing partly known before: what a change have these three things made in the world in these times; the one in state of learning, the other in the state of war, the third in the state of treasure, commodities, and navigation! And those, I say, were but stumbled upon and lighted upon by chance. Therefore, no doubt, the sovereignty of man lieth hid in knowledge; wherein many things are reserved, which kings with their treasure cannot buy, nor with their force command; their spials and intelligencers can give no news of them, their seamen and discoverers cannot sail where they

From: *Dialectic of Enlightenment* (1944; repr., New York: Herder and Herder, 1972), pp. 3–28, 42, excerpts.

grow: now we govern nature in opinions, but we are thrall unto her in necessity: but if we would be led by her in invention, we should command her by action.[2]

Despite his lack of mathematics, Bacon's view was appropriate to the scientific attitude that prevailed after him. The concordance between the mind of man and the nature of things that he had in mind is patriarchal: the human mind, which overcomes superstition, is to hold sway over a disenchanted nature. Knowledge, which is power, knows no obstacles: neither in the enslavement of men nor in compliance with the world's rulers. As with all the ends of bourgeois economy in the factory and on the battlefield, origin is no bar to the dictates of the entrepreneurs: kings, no less directly than businessmen, control technology; it is as democratic as the economic system with which it is bound up. Technology is the essence of this knowledge. It does not work by concepts and images, by the fortunate insight, but refers to method, the exploitation of others' work, and capital. The "many things" which, according to Bacon, "are reserved," are themselves no more than instrumental: the radio as a sublimated printing press, the dive bomber as a more effective form of artillery, radio control as a more reliable compass. What men want to learn from nature is how to use it in order wholly to dominate it and other men. That is the only aim. Ruthlessly, in despite of itself, the Enlightenment has extinguished any trace of its own self-consciousness. The only kind of thinking that is sufficiently hard to shatter myths is ultimately self-destructive. In face of the present triumph of the factual mentality, even Bacon's nominalist credo would be suspected of a metaphysical bias and come under the same verdict of vanity that he pronounced on scholastic philosophy. Power and knowledge are synonymous.[3] . . .

The disenchantment of the world is the extirpation of animism. Xenophanes derides the multitude of deities because they are but replicas of the men who produced them, together with all that is contingent and evil in mankind; and the most recent school of logic denounces—for the impressions they bear—the words of language, holding them to be false coins better replaced by neutral counters. The world becomes chaos, and synthesis salvation. There is said to be no difference between the totemic animal, the dreams of the ghost-seer, and the absolute Idea. On the road to modern science, men renounce any claim to meaning. They substitute formula for concept, rule and probability for cause and motive. Cause was only the last philosophic concept which served as a yardstick for scientific criticism: so to speak because it alone among the old ideas still seemed to offer itself to scientific criticism, the latest secularization of the creative principle. . . .

The pre-Socratic cosmologies preserve the moment of transition. The moist, the indivisible, air, and fire, which they hold to be the primal matter of nature, are already rationalizations of the mythic mode of apprehension. Just as the images of generation from water and earth, which came from the Nile to the Greeks, became here hylozoistic principles, or elements, so all the equivocal multitude of mythical demons were intellectualized in the pure form of ontological essences. Finally, by means of the Platonic ideas, even the patriarchal gods of Olympus were absorbed in the philosophical logos. The Enlightenment, however, recognized the old powers in the Platonic and Aristotelian aspects of metaphysics, and opposed as superstition the claim that truth is predicable of universals. It asserted that in the authority of universal concepts, there was still discernible fear of the demonic spirits which men sought to portray in magic rituals, hoping thus to influence nature. From now on, matter would at last be mastered without any illusion of ruling or inherent powers, of hidden qualities. For the Enlightenment, whatever does not conform to the rule of computation and utility is suspect. So long as it can develop undisturbed by any outward repression, there is no holding it. In the process, it treats its own ideas of human rights exactly as it does the older universals. Every spiritual resistance it encounters serves merely to increase its strength.[4] Which means that enlightenment still recognizes itself even in myths. Whatever myths the resistance may appeal to, by virtue of the very fact that they become arguments in the process of opposition, they acknowledge the principle of dissolvent rationality for which they reproach the Enlightenment. Enlightenment is totalitarian.

Enlightenment has always taken the basic principle of myth to be anthropomorphism, the projection onto nature of the subjective.[5] In this view, the supernatural, spirits and demons, are mirror images of men who allow themselves to be frightened by natural phenomena. Consequently, the many mythic figures can all be brought to a common denominator, and reduced to the human subject. . . .

Formal logic was the major school of unified science. It provided the Enlightenment thinkers with the schema of the calculability of the world. The mythologizing equation of Ideas with numbers in Plato's last writings expresses the longing of all demythologization: number became the canon of the Enlightenment. The same equations dominate bourgeois justice and commodity exchange. . . . To the Enlightenment, that which does not reduce to numbers, and ultimately to the one, becomes illusion; modern positivism writes it off as literature. Unity is the slogan from Parmenides to [Bertrand] Russell. The destruction of gods and qualities alike is insisted upon. . . . In place of the local spirits and demons there appeared heaven and its hierarchy; in place of the invocations of the magician and the tribe, the distinct gradation of sacrifice and the labor of the unfree mediated through the word of command. The Olympic deities are no longer directly identical with elements, but signify them. In Homer, Zeus represents the sky and the weather, Apollo controls the sun, and Helios and Eos are already shifting to an allegorical function. The gods are distinguished from material elements as their quintessential concepts. From now on, being divides into the logos (which with the progress of philosophy contracts to the monad, to a mere point of reference), and into the mass of all things and creatures without. This single distinction between existence proper and reality engulfs all others. Without regard to distinctions, the world becomes subject to man. In this the

Jewish creation narrative and the religion of Olympia are at one: "... and let them have dominion over the fish of the sea, and over the fowl of the air, and over the cattle, and over all the earth, and over every creeping thing that creepeth upon the earth."[6] "O Zeus, Father Zeus, yours is the dominion of the heavens, and you oversee the works of man, both wicked and just, and even the wantonness of the beasts; and righteousness is your concern."[7]. . .

Myth turns into enlightenment, and nature into mere objectivity. Men pay for the increase of their power with alienation from that over which they exercise their power. Enlightenment behaves toward things as a dictator toward men. He knows them in so far as he can manipulate them. The man of science knows things in so far as he can make them. In this way their potentiality is turned to his own ends. In the metamorphosis the nature of things, as a substratum of domination, is revealed as always the same. This identity constitutes the unity of nature. It is a presupposition of the magical invocation as little as the unity of the subject. The shaman's rites were directed to the wind, the rain, the serpent without, or the demon in the sick man, but not to materials or specimens. Magic was not ordered by one, identical spirit: it changed like the cultic masks which were supposed to accord with the various spirits. Magic is utterly untrue, yet in it domination is not yet negated by transforming itself into the pure truth and acting as the very ground of the world that has become subject to it. The magician imitates demons; in order to frighten them or to appease them, he behaves frighteningly or makes gestures of appeasement. Even though his task is impersonation, he never conceives of himself as does the civilized man for whom the unpretentious preserves of the happy hunting-grounds become the unified cosmos, the inclusive concept for all possibilities of plunder. The magician never interprets himself as the image of the invisible power; yet this is the very image in which man attains to the identity of self that cannot disappear through identification with another, but takes possession of itself once and for all as an impenetrable mask. It is the identity of the spirit and its correlate, the unity of nature, to which the multiplicity of qualities falls victim. Disqualified nature becomes the chaotic matter of mere classification, and the all-powerful set becomes mere possession—abstract identity. . . .

Abstraction, the tool of enlightenment, treats its objects as did fate, the notion of which it rejects: it liquidates them. Under the leveling domination of abstraction (which makes everything in nature repeatable), and of industry (for which abstraction ordains repetition), the freedom themselves finally came to form that "herd," which Hegel has declared to be the result of the Enlightenment.[8]

The distance between subject and object, a presupposition of abstraction, is grounded in the distance from the thing itself, which the master achieved through the mastered. The lyrics of Homer and the hymns of the Rig-Veda date from the time of territorial dominion and the secure locations in which a dominant warlike race established themselves over the mass of vanquished natives.[9] . . . A proprietor like Odysseus "manages from a distance a numerous, carefully gradated staff of cowherds, shepherds, swineherds and servants. In the evening, when he has seen from his castle that the countryside is illumined by a thousand fires, he can compose himself for sleep with a quiet mind: he knows that his upright servants are keeping watch lest wild animals approach, and to chase thieves from the preserves which they are there to protect."[10] The universality of ideas as developed by discursive logic, domination in the conceptual sphere, is raised up on the basis of actual domination. The dissolution of the magical heritage, of the old diffuse ideas, by conceptual unity, expresses the hierarchical constitution of life determined by those who are free. The individuality that learned order and subordination in the subjection of the world, soon wholly equated truth with the regulative thought without whose fixed distinctions universal truth cannot exist. Together with mimetic magic, it tabooed the knowledge which really concerned the object. Its hatred was extended to the image of the vanquished former age and its imaginary happiness. The chthonic gods of the original inhabitants are banished to the hell to which, according to the sun and light religion of Indra and Zeus, the earth is transformed. . . .

For science the word is a sign: as sound, image, and word proper it is distributed among the different arts, and is not permitted to reconstitute itself by their addition, by synesthesia, or in the composition of the *Gesamtkunstwerk*. As a system of signs, language is required to resign itself to calculation in order to know nature, and must discard the claim to be like her. As image, it is required to resign itself to mirror imagery in order to be nature entire, and must discard the claim to know her. With the progress of enlightenment, only authentic works of art were able to avoid the mere imitation of that which already is. The practicable antithesis of art and science, which tears them apart as separate areas of culture in order to make them both manageable as areas of culture ultimately allows them, by dint of their own tendencies, to blend with one another even as exact contraries. In its neopositivist version, science becomes aestheticism, a system of detached signs devoid of any intention that would transcend the system: it becomes the game which mathematicians have for long proudly asserted is their concern. . . .

For enlightenment is as totalitarian as any system. Its untruth does not consist in what its romantic enemies have always reproached it for: analytical method, return to elements, dissolution through reflective thought; but instead in the fact that for enlightenment the process is always decided from the start. When in mathematical procedure the unknown becomes the unknown quantity of an equation, this marks it as the well-known even before any value is inserted. Nature, before and after the quantum theory, is that which is to be comprehended mathematically; even what cannot be made to agree, indissolubility and irrationality, is converted by means of mathematical theorems. In the anticipatory identification of the wholly conceived and mathematized world with truth, enlightenment intends to secure itself against the return of the mythic. It confounds thought and mathematics. In this way the latter is, so to speak, released and made into an absolute instance. "An infinite world, in this case a world of idealities, is conceived as one whose objects do not accede singly, imperfectly, and as if by chance to our cognition, but are attained by a rational, systematically unified method—in a process of infinite progression—so that each object is ultimately apparent according to its full inherent being. . . . In the Galilean mathematization of the world, however, *this selfness* is idealized under the guidance of the new mathematics: in modern terms, it becomes itself a mathematical multiplicity."[11] Thinking objectifies itself to become an automatic, self-activating process; an impersonation of the machine that it produces itself so that ultimately the machine can replace it. . . .

Mathematical procedure became, so to speak, the ritual of thinking. In spite of the axiomatic self-restriction, it establishes itself as necessary and objective: it turns thought into a thing, an instrument—which is its own term for it. . . .

In the enlightened world, mythology has entered into the profane. In its blank purity, the reality which has been cleansed of demons and their conceptual descendants assumes the numinous character which the ancient world attributed to demons. Under the title of brute facts, the social injustice from which they proceed is now as assuredly sacred a preserve as the medicine man was sacrosanct by reason of the protection of his gods. It is not merely that domination is paid for by the alienation of men from the objects dominated: with the objectification of spirit, the very relations of men—even those of the individual to himself—were bewitched. The individual is reduced to the nodal point of the conventional responses and modes of operation expected of him. Animism spiritualized the object; whereas industrialism objectifies the spirits of men. Automatically, the economic apparatus, even before total planning, equips commodities with the values which decide human behavior. Since, with the end of free exchange, commodities lost all their economic qualities except for fetishism, the latter has extended its arthritic influence over all aspects of social life. Through the countless agencies of mass production and its culture the conventionalized modes of behavior are impressed on the individual as the only natural, respectable, and rational ones. . . .

While bourgeois economy multiplied power through the mediation of the market, it also multiplied its objects and powers to such an extent that for their administration not just the kings, not even the middle classes are no longer necessary, but all men. They learn from the power of the things to dispense at last with power. Enlightenment is realized and reaches its term when the nearest practical ends reveal themselves as the most distant goal now attained, and the lands of which "their spials and intelligencers can give no news," that is, those of the nature despised by dominant science, are recognized as the lands of origin. Today, when Bacon's utopian vision that we should "command nature by action"—that is, in practice—has been realized on a tellurian scale, the nature of the thralldom that he ascribed to unsubjected nature is clear. It was domination itself. And knowledge, in which Bacon was certain the "sovereignty of man lieth hid," can now become the dissolution of domination. But in the face of such a possibility, and in the service of the present age, enlightenment becomes wholesale deception of the masses.

NOTES

1. François de Voltaire, "Lettre XII," *Oevres complètes, Lettres Philosophiques* (Paris: Garnier, 1879), vol. 22: 118; François de Voltaire, "Letter Twelve, on Chancellor Bacon," *Philosophical Letters* (New York: Bobbs-Merrill, 1961), pp. 46–51; see p. 48.

2. Francis Bacon, "In Praise of Human Knowledge" (*Miscellaneous Tracts upon Human Knowledge*), in *The Works of Francis Bacon*, ed. Basil Montagu (London, 1825), vol. 1:254ff.

3. Cf. Francis Bacon, *Novum Organum, Works* 14:31.

4. Cf. G. W. F. Hegel, *Phänomenologie des Geistes* (The Phenomenology of Spirit) in *Werke* (Frankfurt am Main: Suhrkampf, 1969–1971), 2:410ff.

5. Xenophanes, Montaigne, Hume, Feuerbach, and Salomon Reinach are at one here. See, for Reinach, *Orpheus*, trans. F. Simmons (London, 1909), p. 9ff.

6. Genesis 1:26 (Authorized Version).

7. Archilochos, *Carmina: The Fragments of Archilochus*, trans. Guy Davenport (Berkeley: University of California Press, 1964), fragment 87; quoted by Deussen, *Allgemeine Geschichte der Philosophie* (Leipzig: 1911), vol. 2, pt. 1, p. 18.

8. Hegel, *Phänomenologie des Geistes*, p. 424.

9. Cf. W. Kirfel, *Geschichte Indiens*, in *Propyläenweltgeschichte*, 3:261ff; and G. Glotz, *Histoire Grècque*, vol. 1 of *Histoire ancienne* (Paris: 1938), p. 137ff.

10. Glotz, *Histoire Grècque*, 1:140.

11. Edmund Husserl, "Die Krisis europäischen Wissenschaften und die transzendentale Phänomenologie," *Philosophia* 1 (1936): 95ff.

3

ECOLOGY AND REVOLUTION

HERBERT MARCUSE

Coming from the United States, I am a little uneasy discussing the ecological movement, which has already by and large been co-opted [there]. Among militant groups in the United States, and particularly among young people, the primary commitment is to fight, with all the means (severely limited means) at their disposal, against the war crimes being committed against the Vietnamese people. The student movement—which had been proclaimed to be dead or dying, cynical and apathetic—is being reborn all over the country. This is not an organized opposition at all, but rather a spontaneous movement which organizes itself as best it can, provisionally, on the local level. But the revolt against the war in Indochina is the only oppositional movement the establishment is unable to co-opt because neocolonial war is an integral part of that global counterrevolution, which is the most advanced form of monopoly capitalism.

So, why be concerned about ecology? Because the violation of the earth is a vital aspect of the counterrevolution. The genocidal war against people is also "ecocide" insofar as it attacks the sources and resources of life itself. It is no longer enough to do away with people living now; life must also be denied to those who aren't even born yet by burning and poisoning the earth, defoliating the forests, blowing up the dikes. This bloody insanity will not alter the ultimate course of the war, but it is a very clear expression of where contemporary capitalism is at: the cruel waste of productive resources in the imperialist homeland goes hand in hand with the cruel waste of destructive forces and consumption of commodities of death manufactured by the war industry.

In a very specific sense, the genocide and ecocide in Indochina are the cap-

From: "Ecology and Revolution," *Liberation* 16 (September 1972): 10–12.

italist response to the attempt at revolutionary ecological liberation: the bombs are meant to prevent the people of North Vietnam from undertaking the economic and social rehabilitation of the land. But in a broader sense, monopoly capitalism is waging a war against nature—human nature as well as external nature. For the demands of ever more intense exploitation come into conflict with nature itself, since nature is the source and locus of the life-instincts which struggle against the instincts of aggression and destruction. And the demands of exploitation progressively reduce and exhaust resources: the more capitalist productivity increases, the more destructive it becomes. This is one sign of the internal contradictions of capitalism.

One of the essential functions of civilization has been to change the nature of man and his natural surroundings in order to "civilize" him—that is, to make him the subject-object of the market society, subjugating the pleasure principle to the reality principle and transforming man into a tool of ever more alienated labor. This brutal and painful transformation has crept up on external nature very gradually. Certainly, nature has always been an aspect (for a long time the only one) of labor. But it was also a dimension beyond labor, a symbol of beauty, of tranquility, of a nonrepressive order. Thanks to these values, nature was the very negation of the market society, with its values of profit and utility.

However, the natural world is a historical, a social world. Nature may be a negation of aggressive and violent society, but its pacification is the work of man (and woman), the fruit of his/her productivity. But the structure of capitalist productivity is inherently expansionist: more and more, it reduces the last remaining natural space outside the world of labor and of organized and manipulated leisure.

The process by which nature is subjected to the violence of exploitation and pollution is first of all an economic one (an aspect of the mode of production), but it is a political process as well. The power of capital is extended over the space for release and escape represented by nature. This is the totalitarian tendency of monopoly capitalism: in nature, the individual must find only a repetition of his own society; a dangerous dimension of escape and contestation must be closed off.

At the present stage of development, the absolute contradiction between social wealth and its destructive use is beginning to penetrate people's consciousnesses, even in the manipulated and indoctrinated conscious and unconscious levels of their minds. There is a feeling, a recognition that it is no longer necessary to exist as an instrument of alienated work and leisure. There is a feeling and a recognition that well-being no longer depends on a perpetual increase in production. The revolt of youth (students, workers, women), undertaken in the name of the values of freedom and happiness, is an attack on all the values which govern the capitalist system. And this revolt is oriented toward the pursuit of a radically different natural and technical environment; this perspective has become the basis for subversive experiments such as the attempts by

American "communes" to establish nonalienated relations between the sexes, between generations, between man and nature—attempts to sustain the consciousness of refusal and of renovation.

In this highly political sense, the ecological movement is attacking the "living space" of capitalism, the expansion of the realm of profit, of waste production. However, the fight against pollution is easily co-opted. Today, there is hardly an ad which doesn't exhort you to "save the environment," to put an end to pollution and poisoning. Numerous commissions are created to control the guilty parties. To be sure, the ecological movement may serve very well to spruce up the environment, to make it pleasanter, less ugly, healthier and hence, more tolerable. Obviously, this is a sort of co-optation, but it is also a progressive element because, in the course of this co-optation, a certain number of needs and aspirations are beginning to be expressed within the very heart of capitalism and a change is taking place in people's behavior, experience, and attitudes towards their work. Economic and technical demands are transcended in a movement of revolt which challenges the very mode of production and model of consumption.

Increasingly, the ecological struggle comes into conflict with the laws which govern the capitalist system: the law of increased accumulation of capital, of the creation of sufficient surplus value, of profit, of the necessity of perpetuating alienated labor and exploitation. Michel Bosquet put it very well: the ecological logic is purely and simply the negation of capitalist logic; the earth can't be saved within the framework of capitalism, the third world can't be developed according to the model of capitalism.

In the last analysis, the struggle for an expansion of the world of beauty, nonviolence and serenity is a political struggle. The emphasis on these values, on the restoration of the earth as a human environment, is not just a romantic, aesthetic, poetic idea which is a matter of concern only to the privileged; today, it is a question of survival. People must learn for themselves that it is essential to change the model of production and consumption, to abandon the industry of war, waste, and gadgets, replacing it with the production of those goods and services which are necessary to a life of reduced labor, of creative labor, of enjoyment.

As always, the goal is well-being, but a well-being defined not by ever-increasing consumption at the price of ever-intensified labor, but by the achievement of a life liberated from the fear, wage slavery, violence, stench, and infernal noise of our capitalist industrial world. The issue is not to beautify the ugliness, to conceal the poverty, to deodorize the stench, to deck the prisons, banks, and factories with flowers; the issue is not the purification of the existing society but its replacement.

Pollution and poisoning are mental as well as physical phenomena, subjective as well as objective phenomena. The struggle for an environment ensuring a happier life could reinforce, in individuals themselves, the instinctual roots of their own liberation. When people are no longer capable of distinguishing between beauty and ugliness, between serenity and cacophony, they no longer understand the essential quality of freedom, of happiness. Insofar as it has become the territory of capital rather than of man, nature serves to strengthen human servitude. These conditions are rooted in the basic institutions of the established system, for which nature is primarily an object of exploitation for profit.

This is the insurmountable internal limitation of any capitalist ecology. Authentic ecology flows into a militant struggle for a socialist politics which must attack the system at its roots, both in the process of production and in the mutilated consciousnesses of individuals.

4

THE DOMINATION OF NATURE

WILLIAM LEISS

Technology reveals the active relation of man to nature, the immediate process of production of his life, and thereby also his social life relationships and the cultural representations that arise out of them.

—Marx, *Capital*

1. INTRODUCTION

Technology has been described as the concrete link between the mastery of nature through scientific knowledge and the enlarged disposition over the resources of the natural environment, which supposedly constitutes mastery of nature in the everyday world. Normally the rubric "conquest of nature" is applied to modern science and technology together, simply on account of their manifest interdependence in the research laboratory and industry. When they are considered in isolation, as two related aspects of human activity among many others, the fact of their necessary connection must indeed be recognized if their progress in modern times is to be understood. But it does not follow automatically that they function as a unity with respect to the mastery of nature, since they are not identical with the latter: mastery of nature develops also in response to other aspects in the social dynamic, for example, the process whereby new human needs are formed, and therefore its meaning with respect to technology may be quite different than it is in the case of science.

Nature per se is not the object of mastery; . . . instead various senses of mas-

---

From: "Technology and Domination," in *The Domination of Nature* (New York: George Braziller, 1972), pp. 145–55, 161–65.

tery are appropriate to various perspectives on nature. If this proposition is correct, then the converse is likewise true, namely, that mastery of nature is not a project of science per se but, rather, a broader social task.[1] In this larger context, technology plays a far different role than does science, for it has a much more direct relationship to the realm of human wants and thus to the social conflicts which arise out of them. This is what Marx meant in referring to the "immediate" process of production in which technology figures so prominently—the direct connection between men's technical capacities and their ability to satisfy their desires, which is a constant feature of human history and is not bound to any specific form of scientific knowledge. On the other hand, science, like similar advanced cultural formations (religion, art, philosophy, and so forth), is indirectly related to the struggle for existence: in technical language, these are all *mediated* by reflective thought to a far greater extent. Of course this by no means implies that they lack a social content altogether, but only that it is present in highly abstract form and that by virtue of their rational impulse they transcend to some extent the specific historical circumstances which gave them birth. Therefore scientific rationality and technological rationality are not the same and cannot be regarded as the complementary bases of something called the domination of nature.

The character of technological rationality . . . must . . . be explored. Two considerations are especially relevant to the discussion. In the first place, the immediate connection of technology with practical life activity determines a priori the kind of mastery over nature that is achieved through technological development: caught in the web of social conflict, technology constitutes one of the means by which mastery of nature is linked to mastery over man. Secondly, the employment of technological rationality in the extreme forms of social conflict in the twentieth century—in weapons of mass destruction, techniques for the control of human behavior, and so forth—precipitates a crisis of rationality itself; the existence of this crisis necessitates a critique of reason that attempts to discover (and thus to aid in overcoming) the tendencies uniting reason with irrationalism and terror. These two themes have been brilliantly presented in the work of the contemporary philosopher Max Horkheimer.

The attempt to understand the significance of the domination of nature is a problem with which Horkheimer has been concerned during his entire career: one can find scattered references to it in books and essays of his spanning a period of forty years. In my estimation his analysis, although quite unsystematic, contains greater insight into the full range of the problem under discussion here than does any other single contribution, although one can of course find many affinities between his work and that of others. . . . The fundamental question which he poses is similar to the one raised by Husserl (although from a very different philosophical point of view) in *The Crisis of European Sciences*: What is the concept of rationality that underlies modern social progress? He also shares with Husserl the conviction that the social dilemmas related to scientific and technological progress have reached a critical point in the twentieth century. Horkheimer describes this situation as one in which "the antagonism of reason and nature is in an acute and catastrophic phase."[2]

2. THE CRITIQUE OF REASON

Horkheimer follows Nietzsche's pathbreaking thought and argues that the domination of nature or the expansion of human power in the world is a universal characteristic of human reason rather than a distinctive mark of the modern period:

> If one were to speak of a disease affecting reason, this disease should be understood not as having stricken reason at some historical moment, but as being inseparable from the nature of reason in civilization as we have known it so far. The disease of reason is that reason was born from man's urge to dominate nature. . . . One might say that the collective madness that ranges today, from the concentration camps to the seemingly most harmless mass-culture reactions, was already present in germ in primitive objectivization, in the first man's calculating contemplation of the world as a prey.[3]

Again like Nietzsche he finds the first clear expression of this will to power in the rationalist conceptions of ancient Greek philosophy, conceptions which determined the predominant course of subsequent Western philosophy. The concept (*Begriff*) itself, and especially the concept of the "thing," serves as a tool through which the chaotic, disorganized data of sensations and perceptions can be organized into coherent structures and thus into forms of experience that can give rise to exact knowledge. The deductive form of logic which emerges in Greek philosophy reinforces this original tendency and raises it to a much higher level. The deductive form of thought "mirrors hierarchy and compulsion" and first clearly reveals the social character of the structure of knowledge: "The generality of thought, as discursive logic develops it—domination in the realm of concepts—arises on the basis of domination in reality." The categories of logic suggest the power of the universal over the particular and in this respect they testify to the "thoroughgoing unity of society and domination," that is, to the ubiquity of the subjection of the individual to the whole in human society.[4]

Horkheimer differs from Nietzsche in attempting to distinguish two basic types of reason. Although in themselves all structures of logic and knowledge reflect a common origin in the will to domination, there is one type of reason in which this condition is transcended and another in which it is not: the former he calls objective reason, the latter, subjective reason.[5] The first conceives of human reason as a part of the rationality of the world and regards the highest expression of that reason (truth) as an ontological category, that is, it views truth as the grasping of the essence of things. Objective reason is represented in the philosophies of Plato and Aristotle, the Scholastics, and German idealism. It includes the specific rationality of man (subjective reason) by which man defines himself and his goals, but not exclusively, for it is oriented toward the whole of the realm of beings; it strives to be, as Horkheimer remarks, the voice of all that is mute in nature. On the other hand, subjective reason exclusively seeks mastery over things and does not attempt to consider what extrahuman things may be in and for themselves. It does not ask whether ends are intrinsically rational, but only how means may be fashioned to achieve whatever ends may be selected; in effect it defines the rational as that which is serviceable for human interests. Subjective reason attains its most fully developed form in positivism.

The two concepts do not exist for Horkheimer as static historical constants. Objective reason both undergoes a process of self-dissolution and also succumbs to the attack of subjective reason, and this fate in a sense represents necessary historical progress. The conceptual framework and hierarchy established by objective reason is too static, condemning men (as the subjects of historical change) to virtual imprisonment in an order which valiantly tries to maintain its traditional foundations intact. By the seventeenth century, for example, the combination of Aristotelianism and Christian dogma in late medieval philosophy had become intellectually sterile, a system in which the repetition of established formulas was substituted for original thought. The declining period of great philosophical systems is characterized by the increasingly effective onslaught of skepticism, such as that which struck Greek metaphysics in Hellenistic times and scholastic philosophy in the sixteenth century. The skeptics represent "movements of enlightenment"—a recurring pattern of which the eighteenth-century French Enlightenment is the most famous example—wherein thinkers undermine concepts and dogmas that were once vital but have subsequently ossified, often in the process turning into ideological masks for the material interests of certain social groups.

But in its later stages the movement of enlightenment reveals its own eternal contradictions, represented by Horkheimer and Adorno in their famous notion of the "dialectic of enlightenment."[6] What marks the general program of enlightenment as a unitary phenomenon despite its various historical guises is the "demythologization of the world." In its earliest period it combated the traditional religious mythologies (for example, in Greek civilization); in the modern West it takes the form of a struggle against demystification in religion and philosophy, and in its most advanced stages—in positivism—it carries this campaign to the heart of conceptual thought itself, finally upholding the position that only propositions conforming to one particular notion of "verifiable knowledge" have any meaning at all. The sustained effort of demythologizing in modern times ends by stripping the world of all inherent purpose. Nature, for example, appears to scientific thought only as a collection of bodies in eternally lawful motion, and the social reflection of this scientific vision is the idea of a set of natural laws of economic behavior which blindly follow their established course and which inhibit no inherent rationality. The consequence of this view is to set the relationship of man and the world inescapably in the context of domination: man must either meekly submit to these natural laws (physical and economic) or attempt to master them; for since they possess no purpose, or at least none that he can understand, there is no possibility of reconciling his objectives with those of the natural order.

The dialectical nature of enlightenment becomes clear only at this advanced level. All other purposes having been driven out of the world, only one value remains as evident and fundamental: self-preservation. This is sought through mastery of the world to assure the self-preservation of the species and, within the species, through mastery of the economic process to assure the self-preservation of the individual. Yet the puzzling fact remains that adequate security (as the goal of self-preservation) is never attained, either for the species or the individual, and sometimes seems to be actually diminishing for both. Thus the struggle for mastery tends to perpetuate itself endlessly and to become an end in itself.

In the course of enlightenment the predominant function of reason is to serve as an instrument in the struggle for mastery. Reason becomes above all the tool by which man seeks to find in nature adequate resources for self-preservation. It separates itself from the nature given in sense perception and finds a secure point in the thinking self (the *ego cogito*), on the basis of which it tries to discover the means for subjecting nature to its requirements. In the new natural philosophy of the seventeenth century, this procedure is adopted as the theme of science, and as the principal mode of behavior through which the mastery of nature is pursued, science assumes an increasingly influential role in society. For Horkheimer the attributes of the modern scientific conception of nature which predispose it for the purposes of mastery are, in part: the principle of the uniformity of nature, the inherent technological applicability of its findings, the reduction of nature to pure "stuff" or abstract matter through the elimination of qualities as essential features of natural phenomena, and especially the primacy of mathematics in the representation of natural processes.[7]

Horkheimer does not present this picture of science in isolation, but rather tries to understand what complementary conditions are necessary in order for that science to become, as it has, a historical reality of great dimensions. He contends that the "mastery of inner nature" is a logical correlate of the mastery of external nature; in other words, the domination of the world that is to be carried out by subjective reason presupposes a condition under which man's reason is already master in its own house, that is, in the domain of human nature. The prototype of this connection can be found in Cartesian philosophy, where the ego appears as dominating internal nature (the passions) in order to prevent the emotions from interfering with the judgments that form the basis of scientific knowledge. The culmination of the development of the transcendental subjectivity inaugurated by Descartes is to be found in Fichte, in whose early works "the relationship between the ego and nature is one of tyranny," and for whom the "entire universe becomes a tool of the ego, although the ego has no substance or meaning except in its own boundless activity."[8]

In the social context of competition and cooperation the abstract possibilities for an increase in the domination of nature are transformed into actual technological progress. But in the ongoing struggle for existence the desired goal (security) continues to elude the individual's grasp, and the technical mastery of nature expands as if by virtue of its own independent necessity, with the result that what was once clearly seen as a means gradually becomes an end in itself:

> As the end result of the process, we have on the one hand the self, the abstract ego emptied of all substance except its attempt to transform everything in heaven and on earth into means for its preservation, and on the other hand an empty nature degraded to mere material, mere stuff to be dominated, without any other purpose than that of this very domination.[9]

On the empirical level the mastery of inner nature appears as the modern form of individual self-denial and instinctual renunciation required by the social process of production. For the minority this is the voluntary, calculating self-denial of the entrepreneur; for the majority, it is the involuntary renunciation enforced by the struggle for the necessities of life.

The crucial question is: What is the historical dynamic that spurs on the mastery of internal and external nature in the modern period? Two factors shape the answer. One is that the domination of nature is conceived in terms of an intensive exploitation of nature's resources, and the other is that a level of control over the natural environment which would be sufficient (given a peaceful social order) to assure the material well-being of men has already been attained. But external nature continues to be viewed primarily as an object of potentially increased mastery, despite the fact that the level of mastery has risen dramatically. The instinctual renunciation—the persistent mastery and denial of internal nature—which is required to support the project for the mastery of external nature (through the continuation of the traditional work process for the sake of the seemingly endless productive applications of technological innovations) appears as more and more irrational in view of the already attained possibilities for the satisfaction of needs.

Horkheimer answers the question posed in the preceding paragraph, as follows: "The warfare among men in war and in peace is the key to the insatiability of the species and to its ensuing practical attitudes, as well as to the categories and methods of scientific intelligence in which nature appears increasingly under the aspect of its most effective exploitation."[10] The persistent struggle for existence, which manifests itself as social conflict both within particular societies and also among societies on a global scale, is the motor which drives the mastery of nature (internal and external) to ever greater heights and which precludes the setting of any a priori limit on this objective in its present form. Under these pressures the power of the whole society over the individual steadily mounts and is exercised through techniques uncovered in the course of the increasing mastery of nature. Externally, this means the ability to control, alter, and destroy larger and larger segments of the natural environment. Internally, terroristic and nonterroristic measures for manipulating consciousness and for internalizing heteronomous needs (where the individual exercises little or no independent reflective judgment) extend the sway of society over the inner life of the person. In both respects the possibilities and the actuality of domination over men have been magnified enormously.

As a result of its internal contradictions, the objectives which are embodied in the attempted domination of nature are thwarted by the enterprise itself. For the mastery of nature has been and remains a social task, not the appurtenance of an abstract scientific methodology or the happy (or unhappy) coincidence of scientific discovery and technological application. As such its dynamic is located in the specific societal process in which those objectives have been pursued, and the overriding feature of that context is bitter social conflict. Even in those nations, the fruits of technical progress are most evident.

> Despite all improvements and despite fantastic riches there rules at the same time the brutal struggle for existence, oppression, and fear. That is the hidden basis of the decay of civilization, namely that men cannot utilize their power over nature for the rational organization of the earth but rather must yield themselves to blind individual and national egoism under the compulsion of circumstances and of inescapable manipulation.[11]

The more actively is the pursuit of the domination of nature undertaken, the more passive is the individual rendered; the greater the attained power over nature, the weaker the individual vis-à-vis the overwhelming presence of society. . . .

3. THE REVOLT OF NATURE

The growing domination of men through the development of new techniques for mastering the natural environment and for controlling human nature does not go unresisted. Horkheimer analyzes the reaction to it under the heading of the "revolt of nature," a brilliant and original conception that has never received the attention it deserves.[12] The revolt of nature means the rebellion of human nature which takes place in the form of violent outbreaks of persistently repressed instinctual demands. As such it is of course not at all unique to modern history, but is rather a recurrent feature of human civilization. What is new in the twentieth century, on the other hand, is the fact that the potential scope of destructiveness which it entails is so much greater.

There are many reasons for this. The most fundamental is the fact that—at least in the industrially advanced countries—the traditional grounds for the repression of instinctual demands have been vitiated but yet continue to operate: the denial of gratification, the requirements of the work process, and the struggle for existence persist almost unchanged despite the feasibility of gradually mitigating those conditions by means of a rational organization of the available productive forces. The reality principle which has prevailed throughout history has lost most of its rational basis but not its force, and this introduces an element of irrationality into the very core of human activity: "Since the subjugation of nature, in and outside of man, goes on without a meaningful motive, nature is not really transcended or reconciled but merely repressed."[13] The denial of instinctual gratification—the subjugation of internal nature—is enforced in the interests of civilization; release from this harsh regime of the reality principle was to be found in the subjugation of external nature, which would permit the fuller satisfaction of instinctual demands while preserving the order of civilization. The persistence of social conflict thwarts that objective, however, and prompts a search for new means of repressing the sources of conflict in human nature. Security is sought in the power over external nature and over other men, a power that seems possible on the basis of the remarkable accomplishments of scientific and technological rationality, but the need for security, arising always afresh out of the irrational structure of social relations, is never appeased. The dialectic of rationality and irrationality feeds the periodic outbursts of destructive passions with ever more potent fuels.

Secondly, the revolt of human nature, directed against the structure of domination and its rationality, is proportional in intensity to that of the prevailing domination itself. Greater pressures produce correspondingly more violent explosions; the magnified level, of domination in modern society, achieved in respect to both external and internal nature (as we have seen, both make their effects felt in everyday social life despite their differing immediate objects), is also a measure of the heightened potential of the revolt of nature. Thirdly, in recent times this revolt itself has been manipulated and encouraged by ruling social forces as an element in the struggle for sociopolitical mastery. Horkheimer refers here to fascism, which cultivates the latent irrationalities in modern society as material to be managed by rational techniques (propaganda, rallies, and so forth) in the service of political objectives.

The idea of the revolt of nature suggests that there may be an internal limit within the process of enlarging domination that was outlined above. Certainly, at every level of technological development the irrationalities present in the structure of social relations have prevented the realization of the full benefits that might have been derived from the instruments (including human labor) available for the exploitation of nature's resources. The misuse, waste, and destruction of these resources at every stage is at least partially responsible for the continued search for new technological capabilities, as if the possession of more refined techniques could somehow compensate for the misapplication of the existing ones. And because of the lasting institutional frameworks through which particular groups control the behavior of others, the new techniques are utilized sooner or later in the service of domination.

Yet it does not seem possible that this process can continue indefinitely, for at the higher levels the gap between the rational organization of labor and instrumentalities on the one hand, and the irrational uses to which that organization is put on the other, widens to the point where the objectives themselves are called into question. The problem is not only that the degree of waste and misuse of resources has increased enormously, but also that the implements of destruction now threaten the biological future of the species as a whole. This is the point beyond which the nexus of rational techniques and irrational applications ceases to have any justification at all; it represents the internal limit in the exercise of domination over internal and external nature, to exceed which entails that the intentions are inevitably frustrated by the chosen means.

The purpose of mastery over nature is the security of life—and its enhancement—alike for individuals and the species. But the means presently available for pursuing these objectives encompass such potential destructiveness that their full employment in the struggle for existence would leave in ruins all the advantages so far gained at the price of so much suffering. In the intensified social conflicts of the contemporary period, and especially in the phenomenon of fascism, Horkheimer sees this dialectic at work, and this is what he has tried to describe in the notion of the revolt of nature. The use of the most advanced rational techniques of domination over external and internal nature to prevent the emergence of the free social institutions envisaged in the utopian tradition represents for him the blind, irrational outbreak of human nature against a process of domination that has become self-destructive.

In the interval since Horkheimer first presented this notion a related aspect of the problem has been recognized: in a different sense the concept of the revolt of nature may be applied in relation to ecological damage in the natural environment. There is also an inherent limit in the irrational exploitation of external nature itself, for under present conditions the natural functioning of various biological ecosystems is threatened. It is possible that permanent and irreversible damage to some parts of the major planetary ecosystems may have already occurred; the consequences of this are not yet clear.[14] If it is the case that the natural environment cannot tolerate the present level of irrational technological applications without suffering breakdowns in the mechanisms that govern its cycles of self-renewal, then we would be justified in speaking of a revolt of external nature which accompanies the rebellion of human nature.

The dialectic of reason and unreason in our time is epitomized in the social dynamic which sustains the scientific and technological progress through which the resources of nature are ever more artfully exploited. This triumph of human rationality derives its impetus from the uncontrolled interaction of processes that are rooted in irrational social behavior: the wasteful consumption of the advanced capitalist societies, the fearful military contest between capitalist and socialist blocs, the struggles within and among socialist societies concerning the correct road to the future, and the increasing pressure on third world nations and their populations to yield fully to economic development and ideological commitment. In the passions that prompt such behavior are forged the ineluctable chains which bind together technology and political domination at present.

NOTES

1. Max Horkheimer, "Soziologie und Philosophie," in *Sociologica II: Reden und Vorträge* by Max Horkheimer and Theodor Adorno (Frankfurt: Europäische Verlagsanstalt, 1962).

2. Max Horkheimer, *Eclipse of Reason* (New York: Columbia University Press, 1947), p. 177.

3. Ibid., p. 176.

4. Max Horkheimer and Theodor Adorno, *Dialektik der Aufklärung* (Frankfurt: S. Fischer, 1969), pp. 20, 27–29.

5. Horkheimer, *Eclipse of Reason,* ch. 1; and Max Horkheimer, "Zum Begriff der Vernunft," in Horkheimer and Adorno, *Sociologica II.*

6. Horkheimer and Adorno, *Dialektik der Auflklärung,* 9ff; a brief outline is given in Horkheimer, *Eclipse of Reason.*

7. Horkheime and Adorno, *Dialektii der Aufklärung,* pp. 31ff, 189–91.

8. Horkheimer, *Eclipse of Reason,* p. 108. In one of his essays, Marcuse cited the following passage from Fichte's *Staatslehre* (1813): "The real station, the honor and worth of the human being, and quite particularly of man in his morally natural existence, consists without doubt in his capacity as original progenitor to produce out of himself new men, new commanders of nature: beyond his earthly existence and for all eternity to establish new masters of nature." Quoted in "On Hedonism," in *Negations: Essays in Critical Theory* (Boston: Beacon Press, 1968), p. 186.

9. Horkheimer, *Eclipse of Reason,* p. 97.

10. Ibid., p. 109.

11. Max Horkheimer, "Zum Begriff des Menschen," in *Zur Kritik der instrumentellen Vernunft,* ed. A. Schmidt, 198 (Frankfurt: S. Fischer, 1967).

12. See generally Horkheimer, *Eclipse of Reason,* ch. 3, esp. p. 109ff.

13. Ibid., p. 94.

14. See UNESCO, Intergovernmental Conference of Experts on the Scientific Basis for Rational Use and Conservation of the Resources of the Biosphere, "Final Report," UNESCO Document SC/MD/9 (1969).

5

THE FAILED PROMISE
OF CRITICAL THEORY

ROBYN ECKERSLEY

The critical theory developed by the members of the Frankfurt Institute of Social Research ("the Frankfurt school") has revised the humanist Marxist heritage in ways that directly address the wider emancipatory concerns of eco-centric theorists [i.e., the valuing (for their own sake) not just of individual living organisms, but also of ecological entities at different levels of aggregation].[1] In particular, Critical Theorists have laid down a direct challenge to the Marxist idea that "true freedom" lies beyond socially necessary labor. They have argued that the more we try to "master necessity" through the increasing application of instrumental reason to all spheres of life, the *less* free we will become.

Critical Theory represents an important break with orthodox Marxism—a break that was undertaken in order to understand, among other things, why Marx's original emancipatory promise had not been fulfilled. Like many other strands of Western Marxism, Critical Theory turned away from the scientism and historical materialism of orthodox Marxism. In the case of the Frankfurt school, however, it was not through a critique of political economy but rather through a critique of culture, scientism, and instrumental reason that Marxist debates were entered. One of the enduring contributions of the first generation of Frankfurt school theorists (notably, Max Horkheimer, Theodor Adorno, and Herbert Marcuse) was to show that there are different levels and dimensions of domination and exploitation *beyond* the economic sphere and that the former are no less important than the latter. The most radical theoretical innovation concerning this broader understanding of domination came from the early Frankfurt school the-

---

From: *Environmentalism and Political Theory* (Albany: State University of New York Press, 1992), pp. 97–106.

orists' critical examination of the relationship between humanity and nature. This resulted in a fundamental challenge to the orthodox Marxist view concerning the progressive march of history, which had emphasized the liberatory potential of the increasing mastery of nature through the development of the productive forces. Far from welcoming these developments as marking the "ascent of man from the kingdom of necessity to the kingdom of freedom" (to borrow Engels's phrase), Horkheimer, Adorno, and Marcuse saw them in essentially negative terms, as giving rise to the domination of both "outer" and "inner" nature.[2]

These early Frankfurt school theorists regarded the rationalization process set in train by the Enlightenment as a "negative dialectics." This was reflected, on the one hand, in the apprehension and conversion of nonhuman nature into resources for production or objects of scientific inquiry (including animal experimentation) and, on the other hand, in the repression of humanity's joyful and spontaneous instincts brought about through a repressive social division of labor and a repressive division of the human psyche. Hence their quest for a human "reconciliation" with nature. Instrumental or "purposive" rationality—that branch of human reason that is concerned with determining the most efficient means of realizing pregiven goals and which accordingly apprehends only the instrumental (i.e., use) value of phenomena—should not, they argued, become the exemplar of rationality for society. Human happiness would not come about simply by improving our techniques of social administration, by treating society and nature as subject to blind, immutable laws that could be manipulated by a technocratic elite.

The early Frankfurt school's critique of instrumental rationality has been carried forward and extensively revised by Jürgen Habermas, who has sought to show, among other things, how political decision making has been increasingly reduced to pragmatic instrumentality, which serves the capitalist and bureaucratic system while "colonizing the life-world."[3] According to Habermas, this "scientization of politics" has resulted in the lay public ceding ever greater areas of system-steering decision making to technocratic elites.

All of these themes have a significant bearing on the green critique of industrialism, modern technology, and bureaucracy, and the green commitment to grassroots democracy. Yet Critical Theory has not had a major direct influence in shaping the theory and practice of the green movement in the 1980s, whether in West Germany or elsewhere.[4] The ideas of Marcuse and Habermas did have a significant impact on the thinking of the New Left in the 1960s and early 1970s and . . . the general "participatory" theme that characterized that era has remained an enduring thread in the emancipatory stream of ecopolitical thought. Yet this legacy is largely an indirect one. Of course, there are some emancipatory theorists who have drawn upon Habermas's social and political theory in articulating and explaining some aspects of the green critique of advanced industrial society.[5] However, this can be contrasted with the much greater general influence of post-Marxist green theorists such as Murray Bookchin, Theodore Roszak, and Rudolf Bahro, and non-Marxist green theories such as bioregionalism, deep/

transpersonal ecology, and ecofeminism—a comparison that further underscores the distance green theory has had to travel away from the basic corpus of Marxism and neo-Marxism in order to find a comfortable "theoretical home."

It is important to understand why Critical Theory has not had a greater direct impact on green political theory and practice given that two of its central problems—the triumph of instrumental reason and the domination of nature—might have served as useful theoretical starting points for the green critique of industrial society. This possibility was indeed a likely one when it is remembered that both the Frankfurt school and green theorists acknowledge the dwindling revolutionary potential of the working class (owing to its integration into the capitalist order); both are critical of totalitarianism, instrumental rationality, mass culture, and consumerism; and both have strong German connections. Why did these two currents of thought not come together?

There are many possible explanations as to why Critical Theory has not been more influential. One might note, for example, the early Frankfurt school's pessimistic outlook (particularly that of Adorno and Horkheimer), its ambivalence toward nature romanticism (acquired in part from its critical inquiry into Nazism), its rarefied language, its distance from the imperfect world of day-to-day political struggles (Marcuse being an important exception here), and its increasing preoccupation with theory rather than praxis (despite its original project of uniting the two). Yet a more fundamental explanation lies in the direction in which Critical Theory has developed since the 1960s in the hands of Jürgen Habermas, who has, by and large, remained preoccupied with and allied to the fortunes of democratic socialism (represented by the Social Democratic party in West Germany) rather than the fledgling green movement and its parliamentary representatives.[6] Of course, the green movement has not escaped Habermas's attention. However, he has tended to approach the movement more as an indicator of the motivational and legitimacy problems in advanced capitalist societies rather than as the historic bearer of emancipatory ideas (this is to be contrasted with Marcuse, who embraced the activities of new social movements).[7] Habermas has analyzed the emergence of new social movements and green concerns as a grassroots "resistance to tendencies to colonize the life-world."[8] With the exception of the women's movement (which Habermas does consider to be emancipatory), these new social movements (e.g., ecology, antinuclear) are seen as essentially *defensive* in character.[9] While acknowledging the ecological and bureaucratic problems identified by these movements, Habermas regards their proposals to develop counterinstitutions and "liberated areas" from *within the life-world* as essentially unrealistic. What is required, he has argued, are "technical and economic solutions that must, in turn, be planned globally and implemented by administrative means."[10] Yet as Anthony Giddens has pointedly observed, if the pathologies of advanced industrialism are the result of the triumph of purposive rationality, how can the life-world be defended against the encroachments of bureaucratic and economic steering mechanisms without transforming those very mechanisms?[11] In defending the revolutionary potential of new social movements, Murray Bookchin has accused Habermas of intellectualizing new social movements, "to a point where they are simply incoherent, indeed, atavistic."[12] According to Bookchin, Habermas has no sense of the potentiality of new social movements.

Yet Habermas's general aloofness from the green movement (most notably, its radical ecocentric stream) goes much deeper than this. It may be traced to Habermas's theoretical break with the "negative dialectics" of the early Frankfurt school theorists and with their utopian goal of a "reconciliation with nature." Habermas has argued that such a utopian goal is neither necessary nor desirable for human emancipation. Instead, he has welcomed the rationalization process set in train by the Enlightenment as a *positive* rather than negative development. This chapter will be primarily concerned to locate this theoretical break and outline the broad contours of the subsequent development of Habermas's social and political theory in order to identify what I take to be the major theoretical stumbling blocks in Habermas's oeuvre. This will help to explain, on the one hand, why Habermas regards the radical ecology movement as defensive and "neo-romantic" and, on the other hand, why ecocentric theorists would regard many of Habermas's theoretical categories as unnecessarily rigid and anthropocentric.

In contrast, a central theme of the early Frankfurt school theorists, namely, the hope for a reconciliation of the negative dialectics of enlightenment that would liberate both human and nonhuman nature, speaks directly to ecocentric concerns. While Adorno and Horkheimer were pessimistic as to the prospect of such a reconciliation ever occurring, Marcuse remained hopeful of the possibility that a "new science" might be developed, based on a more expressive and empathic relationship to the nonhuman world. This stands in stark contrast to Habermas's position—that science and technology can know nature only in instrumental terms since that is the only way in which it can be effective in terms of securing our survival as a species. Unlike Habermas, Marcuse believed that a qualitatively different society might produce a qualitatively different science and technology. Ultimately, however, Marcuse's notion of a "new science" remained vague and undeveloped and, in any event, was finally overshadowed—indeed contradicted—by his overriding concern for the emancipation of the human senses and the freeing up of the instinctual drives of the individual. . . . This required nothing short of the total abolition of necessary labor and the rational mastery of nature, a feat that could be achieved *only* by advanced technology and widespread automation.

THE LEGACY OF HORKHEIMER, ADORNO, AND MARCUSE

The contributions of Horkheimer and Adorno in the 1940s, and Marcuse in the 1950s and 1960s, contain a number of theoretical insights that foreshadowed the ecological critique of industrial society that was to develop from the late 1960s.[13]

Indeed, these insights might have provided a useful starting point for ecocentric theorists by providing a potential theoretical linkage between the domination of the human and nonhuman worlds. By drawing back from the preoccupation with class conflict as the "motor of history" and examining instead the conflict between humans and the rest of nature, Horkheimer and Adorno developed a critique that sought to transcend the socialist preoccupation with questions concerning the control and distribution of the fruits of the industrial order. In short, they replaced the critique of political economy with a critique of technological civilization. As Martin Jay has observed, they found a conflict whose origins predated capitalism and whose continuation (and probable intensification) appeared likely to survive the demise of capitalism.[14] Domination was recognized as increasingly assuming a range of noneconomic guises, including the subjugation of women and cruelty to animals—matters that had been overlooked by most orthodox Marxists.[15] The Frankfurt school also criticized Marxism for reifying nature as little more than raw material for exploitation, thereby foreshadowing aspects of the more recent ecocentric critique of Marxism. Horkheimer and Adorno argued that this stemmed from the uncritical way in which Marxism had inherited and perpetuated the paradoxes of the Enlightenment tradition—their central target. In this respect, Marxism was regarded as no different from liberal capitalism.

Horkheimer and Adorno's contribution was essentially conducted in the form of a philosophical critique of reason. Their goal was to rescue reason in such a way as to bring instrumental reason under the control of what they referred to as "objective" or "critical" reason. By the latter, Adorno and Horkheimer meant that synthetic faculty of mind that engages in critical reflection and goes beyond mere appearances to a deeper reality in order to reconcile the contradictions between reality and appearance. This was to be contrasted with "instrumental reason," that one-sided faculty of mind that structures the phenomenal world in a commonsensical, functional way and is concerned with efficient and effective adaptation, with means, not ends. The Frankfurt school theorists sought to defend reason from attacks on both sides, that is, from those who reacted against the rigidity of abstract rationalism (e.g., the romanticists) and from those who asserted the epistemological supremacy of the methods of the natural sciences (i.e., the positivists). The task of Critical Theory was to foster a mutual critique and reconciliation of these two forms of reason. In particular, reason was hailed by Marcuse as an essential "critical tribunal" that was the core of any progressive social theory; it lay at the root of Critical Theory's utopian impulse.[16]

According to Horkheimer and Adorno, the Age of Enlightenment had ushered in the progressive replacement of tradition, myth, and superstition with reason, but it did so at a price. The high ideals of that period had become grossly distorted as a result of the ascendancy of instrumental reason over critical reason, a process that Max Weber decried as simultaneously leading to the rationalization *and* disenchantment of the world. The result was an inflated sense of human self-importance and a quest to dominate nature. Horkheimer and Adorno argued that this overemphasis on human self-importance and sovereignty led, paradoxically, to a loss of freedom. This arose, they maintained, because the instrumental manipulation of nature that flowed from the anthropocentric view that humans were the measure of all things and the masters of nature inevitably gave rise to the objectification and manipulation of humans:

> Men pay for the increase in their power with alienation from that over which they exercise their power. Enlightenment behaves toward things as a dictator toward men. He knows them in so far as he can manipulate them. The man of science knows things in so far as he can make them. In this way their potentiality is turned to his own ends. In the metamorphosis the nature of things, as the substratum of domination, is revealed as always the same. This identity constitutes the unity of nature.[17]

The first generation of Critical Theorists also argued that the "rational" domination of outer nature necessitated a similar domination of inner nature by means of the repression and renunciation of the instinctual, aesthetic, and expressive aspects of our being. Indeed, this was seen to give rise to the paradox that lay at the heart of the growth of reason. The attempt to create a free society of autonomous individuals via the domination of outer nature was self-vitiating because this very process also distorted the subjective conditions necessary for the realization of that freedom.[18] The more we seek material expansion in our quest for freedom from traditional and natural constraints, the more we become distorted psychologically as we deny those aspects of our own nature that are incompatible with instrumental reason. As Alford has observed, Horkheimer and Adorno condemned "not merely science but the Western intellectual tradition that understands reason as effective adaptation."[19] Whereas Weber had described the process of rationalization as resulting in the disenchantment of the world, Horkheimer and Adorno described it as resulting in the "revenge of nature." This was reflected in the gradual undermining of our biological support system and, more significantly, in a new kind of repression of the human psyche. Such "psychic repression" was offered as an explanation for the modern individual's blind susceptibility, during times of social and economic crisis, to follow those demagogues (Hitler being the prime example) who offer the alienated individual a sense of meaning and belonging. From a Critical Theory perspective, then, just as the totalitarianism of Nazism was premised on the will to *engineer* social problems out of existence, the bureaucratic state and corporate capitalism may be seen as seeking to *engineer* ecological problems out of existence.

Adorno, Horkheimer, and Marcuse longed for "the resurrection of nature"—a new kind of mediation between society and the natural world. Whitebook has described this resurrection as referring to "the transformation of our relation to and knowledge of nature such that nature would once again be taken as purposeful, meaningful or as possessing value."[20] This did not mean a nostalgic regress into primitive animism or pre-Enlightenment mythologies that sacrificed critical reason—the phenomenon of Nazism demonstrated the dangers of such a simplistic solution. Rather, their utopia required the *integrated* recapture of the past. This involved remembering rather than obliterating the experiences and ways of being of earlier human cultures and realizing that the modern rationalization process and the increasing differentiation of knowledge (particularly the factual, the normative, and the expressive) has been both a learning and unlearning process. What was needed, Adorno, Horkheimer, and Marcuse believed, was a new harmonization of our rational faculties and our sensuous nature.[21]

Yet Adorno and Horkheimer recognized that their utopia was very much against the grain of history. Unlike Marx, they stressed the *radical discontinuity* between the march of history and the liberated society they would like to see. As we saw, this sprang from the lack of a revolutionary subject that would be able to usher in the reconciliation of humanity with inner and outer nature. After all, how could there be a revolutionary subject when the individual in mass society had undergone such psychological distortion and was no longer autonomous? Accordingly, they were unable to develop a revolutionary praxis to further their somewhat vague utopian dream. However, they insisted that the utopian impulse that fueled that dream, although never fully realizable, must be maintained as providing an essential source of critical distance that guarded against any passive surrender to the status quo.

Although Marcuse explored the same negative dialectics as Adorno and Horkheimer, he reached a more optimistic conclusion concerning the likelihood of a revolutionary praxis developing. In particular, he saw the counterculture and student movements of the 1960s and early 1970s as developing a more expressive relationship to nature that was cooperative, aesthetic—even erotic. Here, he suggested, were the seeds of a new movement that could expose the ideological functions of instrumental rationality and mount a far-reaching challenge to the "false" needs generated by modern consumer society that had dulled the individual's capacity for critical reflection.[22] Marcuse saw aesthetic needs as subversive force because they enable things to be seen and appreciated *in their own right*.[23] Indeed, he argued that the emancipation of the senses and the release of instinctual needs was a prerequisite to the liberation of nature (both internal and external). In the case of the former, this meant the liberation of our primary impulses and aesthetic senses. In the case of the latter, it meant the overcoming of our incessant struggle with our environment and the recovery of the "life-enhancing forces in nature, the sensuous aesthetic qualities which are foreign to a life wasted in unending competitive performance."[24]

Marcuse also advanced the provocative argument that this kind of "sensuous perception" might form the epistemological basis of a new science that would overcome the one-dimensionality of instrumental reason that he believed underpinned modern science. Under a new science, Marcuse envisaged that knowledge might become a source of pleasure rather than the means of extending human control. The natural world would be perceived and responded to in an open, more passive, and more receptive way and be guided by the object of study (rather than by human purposes). Such a new science might also reveal previously undisclosed aspects of nature that could inspire and guide human conduct.[25] This was to be contrasted with modern "Galilean" science, which Marcuse saw as "the 'methodology' of a pre-given historical reality within whose universe it moves"; it reflects an interest in experiencing, comprehending, and shaping "the world in terms of calculable, predictable relationships among exactly identifiable units. In this project, universal quantifiability is a prerequisite for the *domination* of nature."[26]

Habermas has taken issue with Marcuse, claiming that it is logically impossible to imagine that a new science could be developed that would overcome the manipulative and domineering attitude toward nature characteristic of modern science.[27] There are certainly passages in Marcuse's *One-Dimensional Man* that suggest that it is the scientific method itself that has ultimately led to the domination of humans and that therefore a change in the very method of scientific inquiry is necessary to usher in a liberated society.[28] Against Habermas's interpretation, however, William Leiss has argued that these are isolated, inconsistent passages that run contrary to the main line of Marcuse's argument, which is that the problem is not with science or instrumental rationality per se but "with the repressive social institutions which exploit the achievements of that rationality to preserve unjust relationships."[29]

Yet these inconsistencies in Marcuse's discussion of the relationship between science and liberation do not appear to be resolvable either way. Indeed, it is possible to discern a third position that lies somewhere between Habermas's and Leiss's interpretations (although it is closer to Leiss's): that the fault lies neither with science nor instrumental rationality per se nor repressive social institutions per se but rather with the instrumental and anthropocentric character of the modern worldview. In *One-Dimensional Man*, Marcuse was concerned to highlight the inextricable interrelationship between science and society. He conceded that pure as distinct from applied science "does not project particular practical goals nor particular forms of domination," but it does proceed in a certain universe of discourse and cannot transcend that discourse.[30] According to Marcuse,

> scientific rationality was in itself, in its very abstractness and purity, operational in as much as it developed under an instrumental horizon. This interpretation would tie the scientific project (method and theory), *prior* to all application and utilization, to a specific societal project, and would see the tie precisely in the inner form of scientific rationality, i.e., the functional character of its concepts.[31]

It is clear that Marcuse regarded the scientific method as being dependent on a preestablished universe of ends, in which and *for* which it has developed.[32] It follows, as he points out in *Counterrevolution and Revolt*, that:

A free society may well have a very different a priori and a very different object; the development of the scientific concepts may be grounded in an experience of nature as a totality of life to be protected and "cultivated," and technology would apply this science to the reconstruction of the environment of life.[33]

Marcuse's point is a very general one: that a new or liberatory science can only be inaugurated by a liberatory society. It would be a "new" science because it would serve a new preestablished universe of ends, including a qualitatively new relationship between humans and the rest of nature. This third interpretation is much closer to Leiss's interpretation than Habermas's since it argues that we must reorder social relations before we reorder science if we wish to "resurrect" nature. Only then would we be able to cultivate a liberatory rather than a repressive mastery of nature.

Yet it is important to clarify what Marcuse meant by a "liberatory mastery of nature." As Alford has convincingly shown, Marcuse's new science appears as mere rhetoric when judged against the overall thrust of his writings.[34] As we saw in the previous chapter, Marcuse's principal Marxist reference point was the *Paris Manuscripts*, which Marcuse saw as providing the philosophical grounding for the realization of the emancipation of the senses and the reconciliation of nature. Moreover, his particular Marx/Freud synthesis was concerned to overcome repressive dominance, that is, the repression of the pleasure principle (the gratification of the instincts) by the reality principle (the need to transform and modify nature in order to survive, which is reflected in the work ethic and the growth of instrumental reason). Marcuse saw the reality principle as being culturally specific to an economy of scarcity. In capitalist society, the forces of production had developed to the point where scarcity (which gave rise to the "reality principle") need no longer be a permanent feature of human civilization. That is, the technical and productive apparatus was seen to be capable of meeting basic necessities with minimum toil so that there was no longer any basis for the repression of the instincts via the dominance of the work ethic. The continuance of this ethic must be seen as "surplus repression," which Marcuse maintained was secured, inter alia, by the manipulation of "false" consumer needs.[35] Marcuse ultimately wished to reap the full benefits promised by mainstream science, namely, a world where humans would be spared the drudgery of labor and be free to experience "eros and peace."

However, the necessary quid pro quo for the reassertion of the pleasure principle over the reality principle was that the nonhuman world would continue to be sacrificed in the name of human liberation. Marcuse shared Marx's notion of two mutually exclusive realms of freedom and necessity and, like Marx, he believed that "true freedom" lay beyond the realm of labor. Accordingly, total automation, made possible by scientific and technological progress, was essential on the ground that necessary labor was regarded as inherently unfree and burdensome. It demanded that humans subordinate their desires and expressive instincts to the requirements of the "objective situation" (i.e., economic laws, the market, and the need to make a livelihood.... Socialist stewardship under humanist eco-Marxism would usher in a "reconciliation with nature" of a kind that would see to the total domestication or humanization of the nonhuman world. As Malinovich has observed, "For Marcuse the concept of the 'development of human potentiality for its own sake' became *the* ultimate socialist value."[36] In Marcuse's own words, the emancipation of the human senses under a humanistic socialism would enable

> "the human appropriation of nature," i.e., through the transformation of nature into an environment (medium) for the human being as "species being"; free to develop the specifically human faculties: the creative, aesthetic faculties.[37]

Despite his intriguing discussion of the notion of a new, nondomineering science, then, Marcuse's major preoccupation with human self-expression, gratification, and the free play of the senses ultimately overshadowed his concern for the liberation of nonhuman nature. Any nonanthropocentric gloss that Marcuse may have placed on Marx's *Paris Manuscripts* must be read down in this context. Nonetheless, Marcuse's "ecocentric moments" (i.e., his discussion of a qualitatively different science and society that approach the nonhuman world as a partner rather than as an object of manipulation) serve as a useful foil to Habermas's more limited conceptualization of the scientific project.

NOTES

1. The Frankfurt school was founded in 1923 as an independently endowed institute for the exploration of social phenomena. For a historical overview, see Martin Jay, *The Dialectical Imagination: A History of the Frankfurt School and the Institute of Social Research 1923–1970* (Boston: Little, Brown, 1973). [For a definition and discussion of ecocentrism, see Robyn Eckersley, *Environmentalism and Political Theory* (Albany: State University of New York Press, 1992), p. 47 and ch. 3, "Ecocentrism Explained and Defended."]

2. Friedrich Engels, "Socialism: Utopian and Scientific," in *The Marx-Engels Reader*, ed. Robert C. Tucker, 638 (New York: Norton, 1972).

3. By "life-world" Habermas means "the taken-for-granted universe of daily social activity." Anthony Giddens, "Reason without Revolution? Habermas's *Theories des kommunikativen Handelns*," in *Habermas and Modernity*, ed. Richard J. Bernstein (Cambridge, England: Polity, 1985), p. 101.

4. See, for example, Werner Hülsberg, *The German Greens: A Social and Political Profile* (London: Verso, 1988), pp. 8–9; and John Ely, "Marxism and Green Politics in West Germany," *Thesis Eleven* 1, no. 13 (1986): 27 and n. 11. It should be noted, however, that the themes of the *early* Frankfurt school theorists (Adorno, Horkheimer, and Marcuse) have had an important influence on the writings of Murray Bookchin, who has been an influential figure in the green movement in North America. Bookchin was to

invert the early Frankfurt school's thesis concerning the domination *of* human and non-human nature.

5. For example, William Leiss, *The Domination of Nature* (Boston: Beacon, 1974); Timothy W. Luke and Stephen K. White, "Critical Theory, the Informational Revolution, and an Ecological Path to Modernity," in *Critical Theory and Public Life*, ed. John Forester, 22–53 (Cambridge: MIT Press, 1985); and John Dryzek, *Rational Ecology: Environment and Political Economy* (Oxford: Blackwell, 1987).

6. See, for example, Peter Dews, ed., *Habermas: Autonomy and Solidarity* (London: Verso, 1986), p. 210.

7. Marcuse saw the ecology and feminist movements in particular as the most promising political movements, and he foreshadowed many of the insights of ecofeminism. For example, in *Counterrevolution and Revolt* (London: Allen Lane, 1972), he argued for the elevation of the "female principle," describing the women's movement as a radical force that was undermining the sphere of aggressive needs, the performance principle, and the social institutions by which these are fostered (p. 75).

8. Jürgen Habermas, "New Social Movements," *Telos* 49 (1981): 35. This article is extracted from the final chapter of Jürgen Habermas, *The Theory of Communicative Action*, vol. 2, *Life-World and System: A Critique of Functionalist Reason*, trans. Thomas McCarthy (Boston: Beacon Press, 1987).

9. Habermas, "New Social Movements," p. 34.

10. Ibid., p. 35.

11. See Giddens, "Reason without Revolution?" p. 121.

12. Murray Bookchin, "Finding the Subject: Notes on Whitebook and Habermas Ltd.," *Telos* 52 (1982): 83.

13. Theodor Adorno and Max Horkheimer, *Dialectic of Enlightenment*, trans. John Cumming (London: Verso, 1979). This work was written during the Second World War and first published in 1944. See also Herbert Marcuse, *Eros and Civilization: A Philosophical Inquiry into Freud* (London: Routledge and Kegan Paul, 1956), and *One-Dimensional Man* (London: Routledge and Kegan Paul, 1964; London: Abacus, 1972).

14. Jay, *Dialectical Imagination*, p. 256.

15. Ibid., p. 257 (see Adorno and Horkheimer, *Dialectic of Enlightenment*, pp. 84, 245–55). Friedrich Engels's discussion of the subjugation of women in *The Origin of the Family, Private Property and the State* (London: Lawrence and Wishart, 1940) is, of course, an important exception.

16. See Martin Jay, "The Frankfurt School and the Genesis of Critical Theory," in *The Unknown Dimension: European Marxism Since Lenin*, ed. Dick Howard and Karl E. Klare, 240–41 (New York: Basic, 1972).

17. Adorno and Horkheimer, *Dialectic of Enlightenment*, p. 9.

18. This theme has also been pursued by Eric Fromm in his *Escape from Freedom* (New York: Holt, Rinehart, and Winston, 1969).

19. C. Fred Alford, *Science and the Revenge of Nature: Marcuse and Habermas* (Gainesville: University Presses of Florida, 1985), p. 16.

20. Joel Whitebook, "The Problem of Nature in Habermas," *Telos* 40 (1979): 55.

21. Albrecht Wellmer, "Reason, Utopia, and the Dialectic of Enlightenment," *Praxis International* 3 (1983): 91.

22. Marcuse, *One-Dimensional Man*.

23. Marcuse, *Counterrevolution and Revolt*, p. 74.

24. Ibid., p. 60.

25. Ibid. Marcuse argued that instead of seeing nature as mere utility, "the emancipated senses, in conjunction with a natural science proceeding on their basis, would guide the 'human appropriation' of nature."

26. Marcuse, *One-Dimensional Man*, pp. 133–34.

27. Jürgen Habermas, *Toward a Rational Society: Student Protest, Science, and Politics*, trans. Jeremy J. Shapiro (London: Heinemann Educational Books, 1971), pp. 85–87.

28. For example, Marcuse has stated: "The principles of modern science were *a priori* structured in such a way that they could serve as conceptual instruments for a universe of self-propelling, productive control; theoretical operationalism came to correspond to practical operationalism. The scientific method [which] led to the ever-more-effective domination of nature thus came to provide the pure concepts as well as the instrumentalities for the ever-more-effective domination of man by man *through* the domination of nature" (*One-Dimensional Man*, p. 130). And later: "The point which I am trying to make is that science, *by virtue of its own method* and concepts, has projected and promoted a universe in which the domination of nature has remained linked to the domination of man—a link which tends to be fatal to the universe as a whole" (ibid., p. 136).

29. William Leiss, "Technological Rationality: Marcuse and His Critics," *Philosophy of the Social Sciences 2* (1972): 34–35. This essay also appears as an appendix to Leiss, *Domination of Nature*, pp. 199–212.

30. Marcuse, *One-Dimensional Man*, p. 129.

31. Ibid., pp. 129, 131.

32. Ibid., p. 137.

33. Marcuse, *Counterrevolution and Revolt*, p. 61.

34. Alford, *Science and the Revenge of Nature*, pp. 49–68.

35. Marcuse, *Eros and Civilization*, esp. pp. 35, 37, 87–88.

36. Myriam Miedzian Malinovich, "On Herbert Marcuse and the Concept of Psychological Freedom," *Social Research* 49 (1982): 164.

37. Marcuse, *Counterrevolution and Revolt*, p. 64.

Part II

GLOBALIZATION

6

CORPORATE GLOBALIZATION

ROSEMARY RADFORD RUETHER

COLONIAL ROOTS OF GLOBALIZATION

We need to start this discussion by some definitions of what is meant by the term globalization. For me, what is being discussed today as "globalization" is simply the latest stage of Western colonialist imperialism. We need to see these current patterns of appropriation of wealth and concentration of power in the West, now especially in the hands of the elites of the United States, in this context of more than five hundred years of Western colonialism.[1]

Western colonialism can be divided into three phases. The first phase from the late fifteenth century to the early nineteenth century ended with the independence of most of the colonies in the Americas. The second stage, from the mid-nineteenth century to the 1950s, saw the dividing up of Africa among the European nations, as well as most of Asia and the Middle East. England emerged as the great nineteenth-century imperialist nation, creating the empire on which the sun never set. But the aftermath of the Second World War saw the Dutch, French, and English exhausted by the devastation of their home countries and no longer able to afford the direct occupation of these vast colonial territories.

Thus the 1950s saw a process of political decolonialism in which flag independence was conceded to many of these territories in Africa, Asia, and the Middle East. A few colonial powers refused to let go, such as the Portuguese in Angola and Mozambique. Local white settlers tried to block African majority rule, as in Rhodesia and South Africa, and this sparked long bloody revolutionary

From: *Integrating Ecofeminism, Globalization and World Religions* (Lanham, MD: Roman & Littlefield, 2005), pp. 1–8.

struggles. But the general pattern that emerged from 1950s and '60s decolonialization was neocolonialism, not popular majority rule. England and France sought to negotiate relations with their former colonies that conceded control over foreign policy and economic wealth to the white settlers and former colonial rulers. The masses of people in former colonies remained impoverished and exploited.

The United States emerged from the Second World War as the strongest world military power and quickly assumed a role of reinforcer of the neocolonial system of control by the West. Third world liberation movements, seeking to throw off neocolonial hegemony over their nations' foreign policy and wealth, often adopted a socialist ideology and allied with the socialist world against continuing Western domination. In response, the West, led by the United States, made anticommunism the ruling ideology of its foreign policy.

The critique and rejection of communism, as defined in Marxist-Leninist states in the first half of the twentieth century as a viable alternative to capitalism, are complex phenomena. The adoption of police-state repression of dissent and totalitarian political organization are certainly worthy of criticism by those concerned with democratic liberties. Yet communism, even in its more repressive expressions, also offered some elements of economic egalitarianism that were attractive to those concerned with justice as well.

But the Western ideological use of anticommunism did not allow a balanced discussion of these ambiguities. Rather, anticommunism functioned as a black-and-white rhetoric intended to create a totally negative impression of anything labeled "communism." This ideology was used to justify an attack on any social and political system emerging from the third world that would more justly distribute wealth and political power to the majority. By demonizing communism as atheistic totalitarianism, and pretending to be the champion of "democracy," the West masked the fact that what this crusade was mostly all about was the maintenance of neocolonial Western-controlled capitalism and the prevention of genuine, locally controlled political and economic democracy.

With the collapse of the Soviet Union and the emergence of the United States as the overwhelming leader of global military and economic power, the third phase of colonialism built during the cold war is now coming into greater visibility. This takes the form of a bid for US imperial rule over the rest of the world, not only over the third world, but also seeking to dominate the Middle East and to divide and marginalize the European Economic Union. Britain, ever ambivalent about submerging itself as a small island nation within the European community, seeks to attach itself to the coattails of this American empire, and thus maintain its own global reach. This I think explains the desperate loyalty of Tony Blair to American military adventures around the world.

GLOBALIZATION, THE BRETTON WOODS INSTITUTIONS, AND GROWING POVERTY

To understand this third phase of colonialism, dubbed "globalization," one must look not only at its military expression, concentrated in the hands of the US armed forces, but also at the economic institutions that have been built over the last fifty years to control the wealth of the entire planet. This effort to concentrate economic power in the hands of international and particularly of US elites also demands the marginalization of the United Nations. For US elites, the UN must be prevented from operating in any way as a world body that gives equal voice to the third world or indeed to any nation other than the United States. The world system that has been built as the global extension of US hegemony is what is called the Bretton Woods institutions: the World Bank, International Monetary Fund, and, since 1995, the World Trade Organization.

The World Bank and the International Monetary Fund (IMF) were established in 1944–1947 to rebuild war-torn Europe. They are funded by contributions from member nations, with the United States, with 20 percent of its funds, as the largest donor. The G-8 nations, the United States, plus England, France, Germany, Italy, Canada, Japan, and Russia, monopolize the funding and control the decisions. As Europe quickly rebuilt itself, these financial institutions turned to lending for what came to be called "development" of the third world, actually to consolidate control over the economies of the third world by the West. In the 1970s, continued US military spending, the rise of multinational corporations, and the sudden rise of oil prices by OPEC, the Organization of Petroleum Exporting Countries, caused huge funds to be built up in international banks. Under Robert McNamara's leadership (1968–1981), the former secretary of defense who designed such murderous projects as the electronic battlefield in the Vietnam War, the policy of the World Bank became the pushing of high-volume, low-interest development loans to the third world.[2]

McNamara favored large development projects, such as huge dams. Some of these former colonial states lacked the political and economic capacity to use such large loans for effective national development. Many of these states were in the hands of dictatorships, such as Marcos in the Philippines, who used such funds for showy projects or stashed them in personal bank accounts. Many projects remained unfinished, with the benefits going to multinationals and national elites, not to the local people. Masses of people were displaced by projects such as dams, without ever being appropriately resettled. Little attention was paid to environmental devastation. The mounting debts accrued from such loans began to cause an international debt crisis. This was manifest in 1982 when Mexico announced that it could not pay its debts. International banking institutions feared a general renunciation of debts by poorer nations.

The response to this debt crisis by the international banking system was to shape the program of Structural Adjustment (SA) aimed at forcing third world countries to pay their debts at the expense of internal development. The formula of Structural Adjustment entailed devaluation of local currency; the sharp rise in interest rates on loans; the removal of trade barriers that protected local industries and agriculture; the privatization of public sector enterprises, such as transportation, energy, telephones, and electricity; and the deregulation of goods, services, and labor, that is, the removal of minimum wage laws and state subsidies for basic foods, education, and health services for the poor. Accepting this package of Structural Adjustment was mandatory in order to receive new loans to repay debts. Each country was directed to focus on one or two traditional export commodities, such as coffee, to earn money in international currency (dollars) to repay debts, at the expense of the diversification of agricultural and industrial production for local consumption.[3]

The World Bank and IMF blamed the governments of third world countries for their poor record in development and debt payment. The claim was that local governments were inefficient, wasting money in subsidizing local services. SA programs were billed as "austerity" measures that would cause temporary "pain" (to whom?) but would soon cause the whole economy to adjust and prosper. The reality was largely the opposite of these rosy predictions. By focusing on stepped-up production of a few export products, such as coffee, the international market for such products was glutted, the prices fell, and so even though the countries were producing and exporting more, they were earning less on their exports.

Local wages also fell, while prices rose, especially with devaluation of currency, which overnight made the same money worth a half to a tenth of what it had before. Government subsidies on food, basic commodities, health, education, and transportation were all cut or eliminated, meaning that meeting all these basic needs became much more expensive, often out of the reach of the poorer classes. For example, in post-Sandinista Nicaragua free local health clinics and centers for popular adult education were closed down. Local hospitals no longer had funds to provide medicines and repair equipment. Those going to the hospital often found they had to go out and buy the medicine they needed in pharmacies. Schools were privatized and became very expensive, and even state schools raised tuition beyond the reach of an increasingly impoverished majority. The gains in literacy and health access under the revolutionary regime were rapidly lost. The result was rising poverty, malnutrition, unemployment, homelessness, especially of children, crime, and the turn to drugs for money.[4]

Pushing high-interest loans to repay debts under these conditions of Structural Adjustment created a spiraling upward of the debt trap, even as the poverty of the countries supposed to repay these debts was spiraling downward. Poor countries were able to pay only 30–40 percent of the interest on the loans, with the rest added to the principal owed, so that even though the countries continued to squeeze their resources to repay their loans, their debts mounted year by year. Thus, Structural Adjustment had the effect of creating a net extraction of wealth from poor to rich countries, or rather to international banks. For example, in 1988, $50 billion more was paid by poor countries to banks than were actually loaned to them from banks.

Structural Adjustment also had other major effects. By dismantling trade barriers, local production was devastated. Flooded by cheap products from multinational corporations, local industries and agriculture went out of business. In Nicaragua, peanut farmers and a local peanut butter industry could not compete with Skippy's peanut butter from the United States and went out of business. In Korea, rice farmers were put out of business by cheap rice imports from the United States and lost their land. All this was defended as simply the appropriate workings of market laws. Yet large multinationals enjoy subsidies and tax breaks from their governments, while local industries in third world countries were not similarly allowed to protect their industries and agriculture. American rice is cheap, not because American farmers are more efficient, but because these farmers and multinational rice distributors are subsidized by the US government.[5]

Why did third world governments take such loans in the first place? Even more, why did they accept these conditions for repayment that were devastating their economies? Basically, for three reasons. Although the majority of people were suffering, the wealthy elites, who controlled the governments favored by the United States, were prospering. Development loans were a major way for them to cash in on enormous profits. Second, the economists in these governments were trained in the same schools of economics as those of the World Bank and accepted these theories of market neoliberalism as unquestioned dogma. Finally, any government that resisted the SA package would be made into a pariah, isolated and denied further loans and markets. This was the strategy toward Nicaragua which brought down the Sandinista government and which has been applied for more than forty years against Cuba. These strictures were enough to bring most third world governments into line.[6]

This system of global control by international financial institutions and corporations is being greatly extended since 1995 by the World Trade Organization. The WTO sets market rules that not only prevent any trade barriers that protect local industries, but also enforce new rules that extend the ability of such corporations to exploit local wealth, such as TRIMS and TRIPS, that is, Trade-Related Investment Measures and Trade-Related Intellectual Property laws. These new market rules prevent local governments from protecting their own financial institutions and property ownership against takeovers by foreign corporations. They allow corporations to patent the genetic properties of seeds, plants, and even human DNA, preventing local farmers from producing their own seeds and plants that have been part of local agriculture for thousands of years. Corporations are also buying up watersheds and aquifers, and forcing local people to pay for water that they formerly used free from their own wells and streams.[7]

These market rules of the WTO function on behalf of the unaccountable economic power of transnational corporations. The growth of these huge corporations as a major world power is key to this system of global capitalism. In the 1880s corporations in the United States won the legal status of persons. In the 1950s and 1960s they uprooted themselves from accountability to local communities. The 1980s to the present have seen their concerted effort to dismantle any national or international laws that would regulate their "freedom" of movement and investment.

One major effort to prevent such regulation of transnational corporations was the Multilateral Agreement on Investment (MAI) that was negotiated in secret between 1995 and 1998 by the Organization for Economic Cooperation and Development, an organization representing the transnational corporations. The MAI sought to limit the legal ability of governments at all levels (local, provincial, and national) to regulate foreign investment and the activities of foreign-based corporations. National borders could thus become totally permeable to large corporations who could enter any country and buy up their businesses, banks, and other assets. Governments would not be permitted to pass laws to protect their national assets, businesses, and banks, to regulate labor conditions, or to prevent human rights abuses or environmental damage. Although international outcry prevented this agreement from being accepted, investment liberalization is still the operating agenda of the World Trade Organization, which continues to seek to incorporate its rules into its trade regulations.[8]

Deregulation in trade and investment has also been accompanied by unregulated speculative trade in money. This is made possible by the integration of the world stock market into one system which can be accessed through electronic communication. Such financial speculation does not invest in any actual development; it simply profits from money exchanges. Thus with the flick of a computer signal billions of dollars can be moved around the world, buying stocks and bonds when the market goes up and selling when it goes down, creating vast profits for unaccountable financial traders, but throwing entire countries and regions of the world into financial crisis. The Asian financial crisis that hit Thailand, Japan, and Korea in 1997–1998 was partly caused by such speculative trade in money.[9]

This global system of transnational corporations and the Bretton Woods institutions means third world governments have largely lost their national sovereignty, their right or ability to pass laws to protect their own national industries or shape their own development and foreign policies. Through international banking institutions, global corporations, representing the interests of rich elites in dominant nations, rule the world.

The gap between rich and poor has steadily grown, with some 85 percent of the wealth of the world in the hands of some 20 percent of the world's population, much of that concentrated in the top 1 percent, while the remaining 80 percent share out the remaining 15 percent and the poorest 20 percent, more than a billion people, live in deep misery on the brink of starvation.[10] In 1960 the richest 20 percent had thirty times the wealth of the poorest 20 percent; by 1995 this gap had grown to eighty-two times. The 225 richest people in the world have a combined wealth of over $1 trillion, equal to the annual income of the poorest 50 percent of humanity, or 2.5 billion people, while the richest three people have assets that exceed that of the forty-eight poorest nations. This means, in terms of absolute levels of poverty, that in 1999 almost half of the world's population was living on less than $2 a day, and more than 20 percent of the world, 1.2 billion people, on less than $1 a day, according to World Bank figures.[11]

NOTES

1. See "The Legacy of Inequality: Colonial Roots," in *Rethinking Globalization: Teaching Justice for an Unjust World*, ed. Bill Bigelow and Bob Person, 31–60 (Milwaukee: Rethinking Schools, 2002). See also Edwardo Galeano, *Open Veins of Latin America: Five Centuries of the Pillage of a Continent* (New York: Monthly Review Press, 1998); Adam Hochschild, *King Leopold's Ghost: A Story of Greed, Terror and Heroism in Colonial Africa* (New York: Mariner Books, 1998).

2. See Bruce Rich, *Mortgaging the Earth: The World Bank, Environmental Impoverishment and the Crisis of Development* (Boston: Beacon, 1994), pp. 49–106. See also Susan George and Fabrizio Sabelli, *Faith and Credit: The World Bank's Secular Empire* (San Francisco and Boulder: Westview Press, 1994), pp. 1–57. Similar material is covered in John Cobb, *The Earthist Challenge to Economism: A Theological Critique of the World Bank* (New York: St. Martin's Press, 1999), pp. 61–89.

3. See Rich, *Mortgaging the Earth*, pp. 186–289; George and Sabelli, *Faith and Credit*, pp. 58–72; and Cobb, *Earthist Challenge*, pp. 90–107.

4. See Sharon Hostetler et al., *A High Price to Pay: Structural Adjustment and Women in Nicaragua* (Washington, DC: Witness for Peace, 1995).

5. Critiques of World Bank and WTO policies and structural adjustment abound: see, for example, Ralph Nader et al., *The Case against "Free Trade": GATT, NAFTA and the Globalization of Corporate Power* (San Francisco: Earth Island Press, 1993); Kevin Danahern, *50 Years Is Enough: The Case against the World Bank and the International Monetary Fund* (Cambridge, MA: South End Press, 1994); Debi Barker and Jerry Mander, *Invisible Government: The World Trade Organization—Global Government for a New Millennium* (San Francisco: International Forum on Globalization, 1999); Walden Bello, *The United States, Structural Adjustment and Global Poverty* (London: Pluto Press, 1994).

6. See George and Sabelli, *Faith and Credit*, pp. 112–34, 190–296.

7. See Vandana Shiva, *Biopiracy: The Plunder of Nature and Knowledge* (Boston: South End Press, 1997). See also Vandana Shiva, *Stolen Harvest: The Hijacking of the Global Food Supply* (Boston: South End Press, 1999); Maude Barlow and Tony Clarke, *Blue Gold: The Battle against Corporate Theft of the World's Waters* (New York: New Press, 2002).

8. See Cynthia Moe-Lobeda, *Globalization and God: Healing a Broken World* (Minneapolis: Fortress Press, 2002), pp. 40–41. See also Council of Canadians, *"The MAI Inquiry: Confronting Globalization and Reclaiming Democracy"* (Toronto: Council of Canadians, 1999).

9. See Walden Bello, *The Future in Balance: Essays on Globalization and Resistance* (Oakland, CA: Food First and Focus on the Global South, 2001), pp. 93–94.

10. A useful primer on global inequality in its myriad dimensions is Bob Sutcliffe, *100 Ways of Seeing an Unequal World* (London: Zed Press, 2002).

11. Moe-Lobeda, *Globalization and God*, p. 28. On World Bank figures on global poverty, see www.worldbank.org/poverty.

7

GLOBAL CAPITALISM
AND THE END OF NATURE

JOEL KOVEL

ON HUMAN ECOLOGY AND THE
TRAJECTORY OF THE ECOLOGICAL CRISIS

Ecology takes on a human form, since humans are part of nature, and, like all other creatures, require a pattern of relationships to survive and flourish. Each kind of creature has its ecological signature, which for humans is given in the terms of our peculiar species traits of sociality, language, culture, and the like. Society, which results from the expression of these traits—that is, of our *human nature*—is plainly an ecosystem, since it is internally related and has dynamic boundaries with other, natural ecosystems. . . .

All of the characteristics of ecosystems, including degrees of destabilization and disintegration, apply to societies. But there is one property that human society uniquely possesses as the species-specific expression of human nature, namely, that the boundary between human and natural ecosystems is the site of the peculiarly human activity known as *production*, the conscious transforming of nature for human purposes. All creatures transform others—that is simply another way of expressing the dynamic relations between ecosystems. But only humans do so consciously, with all that entails. . . . From this standpoint, the ecological crisis may be said to be human production gone bad.

Put more formally, the current stage of history can be characterized by *structural forces that sytematically degrade and finally exceed the buffering capacity of nature with respect to human production, thereby setting into motion an unpredictable yet interacting and expanding set of ecosystemic breakdowns.* The eco-

From: *The Enemy of Nature: The End of Capitalism or the End of the World?* (London: Zed Books, 2002), pp. 20–25, 115–16.

103

logical crisis is what is meant by this phase. In it we observe the desynchroniza-
tion of life cycles and the disjointing of species and individuals, resulting in the
fragmentation of ecosystems human as well as nonhuman. For humanity is not
just the perpetrator of the crisis: it is its victim as well. And among the signs of
our victimization is the incapacity to contend with the crisis, or even to become
conscious of it.

Although the essentials of the ecological crisis lie in qualitative relationships, its outcome will turn upon quantity. It does not take the proverbial rocket
scientist to tell us that if the load placed by humans upon the earth's buffering
capacity keeps growing, then collapse will ensue, with consequences that logically include the possibility of extinction. It is not our province to dissect these
buffering mechanisms, or how they are surpassed in the plurality of ecosystemic
insults. Nor do I wish to evoke apocalyptic imagery to make my point, or calculate the number of years we have left until doomsday, if only because the
scenario of apocalypse, with its sudden and total end accompanied by rapture,
retribution, and so on, is not on the cards so much as a kind of steady, deleterious
fraying of ecosystem with incalculable aftereffects. Our job in any case is to
understand the social dynamics of the crisis, and to see whether anything can be
done about them. Here it is useful to consider society from an ecosystemic standpoint, and ponder the meaning of findings such as the fact that even as the situation of global ecology has markedly worsened in the past thirty years, so has the
level of elite responsiveness declined.

Since Plato at least, people have been observing the potential for deleterious
environmental effects, and since the publication of George Perkins Marsh's *Man
and Nature* in 1864, the possibility of systemic ecological damage has been
raised. Marsh, however, was a visionary, and it took another century for the grim
possibility of global ecosystemic decay to enter the general consciousness and
become a concern of elites. In 1970, the notion of the "limits to growth" entered
the collective vocabulary, to be joined as time went on by other buzzwords such
as "sustainability" and "throughput."[1]

For a time it seemed as if humanity had awakened to its own harmfulness. But
then something strange happened. Even as the vocabulary of ecological concern
proliferated, along with a large bureaucratic apparatus, nongovernmental as well as
governmental, for putting it into effect, a shift occurred and the notion of "limits to
growth" became passé. Where once not so very long ago there was substantial concern that some combination of rising population and industrial expansion would
overwhelm the earth with catastrophic consequences for civilization, today
thoughts of the kind are distinctly unfashionable, even if not entirely extinguished.

What is odd is that, as we have already seen, "growth," whether of population or industrial output, certainly did not slacken in this period. The latter is
especially troubling, inasmuch as population, however unacceptably large it may
be, shows signs of leveling across most of the world (even reaching zero or
slightly negative levels in Japan and some Western European countries, and

rather precipitous declines in the former Soviet bloc). Nothing of the sort can be said about the other kind of growth, that pertaining to industrial output or production in general, however this may be measured.[2] According to the Worldwatch Institute, a mainstream organization charged with monitoring the world's ecology, the global economy increased from $2.3 trillion in 1900 to $20 trillion in 1990, and an astounding $39 trillion in 1998. To quote, the "growth in economic output in just three years—from 1995 to 1998—exceeded that during the 10,000 years from the beginning of agriculture until 1990. And growth of the global economy in 1997 alone easily exceeded that during the seventeenth century."[3] This is consistent with the fact that world trade has increased by a factor of twenty-five over the past four decades, all of which lends support to the prediction, made in 1997, that gross world product will *double* within the next twenty years, that is, to some $80 trillion."[4]

The Malthusian principle that population will increase exponentially—a crude reduction of conscious creatures to machines obeying the rules of elementary algebra—has now been empirically as well as theoretically demolished. If there is to be a fatally destabilizing exponential increase of load, it will come in the economic sphere. This is certified by the figures just given, and, more significantly, by the value accorded them in established channels of opinion. We can easily imagine the horror and outrage with which an announcement that population would double in the next twenty years would be greeted. A similar claim made for economic activity, however, not only evades criticism but is greeted as though a sign of the Second Coming. Predictions of growth may or may not turn out to be on schedule. In fact, they got slowed a bit by the Asian financial meltdowns that began even as they were announced, and all the vagaries of the global economy will play a role in their realization. What matters, however, is that the world is run by those who see limits to growth as anathema.

The scenario of ecological collapse holds, in essence, that the cumulative effects of growth eventually overwhelm the integrity of ecosystems on a world scale, leading to a cascading series of shocks. Just how the blows will fall is impossible to tell with any precision, although a number of useful computer models have been assembled.[5] In general terms, we would anticipate interacting calamities that invade and rupture the core material substrata of civilization—food, water, air, habitat, bodily health. Already each of these physical substrata is under stress, and the logic of the crisis dictates that these stresses will increase. Other shocks and perturbations are likely to ensue as resource depletion supervenes—for example, in the supply of petroleum, which is expected to begin leveling off and then decline after the next ten years.[6] Or some unforeseen economic shock will topple the balance: perhaps climatic catastrophes will trigger a collapse of the $2 trillion global insurance industry, with, as Jeremy Leggett has noted, "knock-on economic consequences which are completely ignored in most analyses of climate change."[7] Perhaps famines will incite wars in which rogue nuclear powers will launch their reign of terror. Perhaps a similar fate will come through the eruption of as yet unforeseen global pandemics, such as the return of smallpox, currently considered to be within the range of possibilities open to terrorist groups. Or perhaps a sudden breakup of the Antarctic ice shelf will cause seas to suddenly rise by several meters, displacing hundreds of millions and precipitating yet more violent climatic changes. Or perhaps nothing so dramatic will take place, but only a slow and steady deterioration in ecosystems, associated with a rise in authoritarianism. The apocalyptic scenarios now so commonly making the rounds of films, best-selling novels, comic books, computer games, and television are not so much harbingers of the future as inchoate renderings of the present ecological crisis. With terror in the air, these mass fantasies can become the logos of a new order of fascism—a fascism that, in the name of making the planet habitable, only aggravates the crisis as it further disintegrates human ecologies. . . .

All this brings us to the larger question of just what is growing in the regime of "growth." We can see right away that the answer engages a number of levels. From the standpoint of ecosystems, the concrete agents of breakdown are the material forces thrown into nature by our industrial apparatus, and this is ultimately a question of molecules and energy flows, whether these be organochlorines, carbon dioxide, or the blade of a chainsaw. But although this level grows, it is as a function of another kind of growth. Here we find the true god of society, and the actual subject of growth that its rulers will not compromise. At this level, what grows is the imaginary and purely human entity of money—not money in itself, but money in motion: Capital. The real issue of the ecological crisis resides in this mysterious entity and the social forces established for its nurture and reproduction. We have to ask whether we can overcome the ecological crisis without overcoming Capital. If the answer is no, then the map of the future needs to be redrawn.

NOTES

1. G. P. Marsh, *Man and Nature*, ed. D. Lowenthal (1864; Cambridge, MA: Belknap Press, 1965), about which Andrew Goudie writes that it was "probably the most important landmark in the history of the study of the role of humans in changing the face of the earth" (A. Goudie *The Human Impact on the Natural Environment* [Cambridge, MA: MIT Press, 1991], p. 3). See also D. Meadows, D. Meadows, J. Randers, and W. Behrens, *The Limits to Growth* (London: Earth Island, 1972). Another landmark study of the next decade was the Brundtland Report (G. Brundtland, ed., *Our Common Future: The World Commission on Environment and Development*, 1987).

2. For present purposes, we may regard measures in monetary units, like gross national (or world) product as equivalent to, and moving in tandem with, measures of a directly physical kind, such as resource depletion. There are major problems with this, including the adequacy of GNP as an indicator of economic well-being, and also its equivalency with ecological processes. Thus spending $100 on a psychoanalytic interview and

buying $100 worth of herbicide for one's lawn are not exactly the same ecologically, although both increase GNP by the same amount. In an ecologically sane society, as many have pointed out, indices such as GNP will no longer guide policy. For now, however, it is a useful indicator of the problem.

3. P. Brown, "More Refugees Flee from Environment Than Warfare," *Guardian Weekly*, July 1–7, 1999, 10. See also L. Brown, C. Flavin, and S. Postel, *Saving the Planet* (New York: Norton, 1991), p. 23.

4. From the authoritative voice of Renato Ruggiero, then director of the World Trade Organization, quoted April 23, 1997 in the *Wall Street Journal*.

5. As in D. Meadows, D. Meadows, and J. Randers, *Beyond the Limits* (London: Earthscan, 1992).

6. Less in absolute terms than in the cost of extraction. No effort need be made here to evaluate the precise contours of this looming crisis—or that in other essential resources, such as topsoil. The situation is too complex and unpredictable for that. And it moves in multiple directions. As petroleum is the source of most of the greenhouse gases, it might be hypothesized that its decline as a resource would place less of a load on the ecosphere and perhaps open the way for new and ecological energy replacements. The question, though, is whether the currently installed market system can deal rationally with these and other stresses.

7. Quoted in E. Goldsmith and C. Henderson, "The Economic Costs of Climate Change," *Ecologist* 29, no. 2 (1999): 99.

8

GLOBAL ECOLOGICAL MOVEMENTS

BRIAN TOKAR

International campaigns by organizations such as Greenpeace, the Rainforest Action Network, and the Native Forest Network are unfolding in a climate of growing ecological awareness by people throughout the world. Nowhere is this more apparent than in southern Asia, Latin America, and other so-called less-developed regions. The emergence of articulate and sometimes militant third world voices in the ecology movement during the past decade offers a necessary counterpoint to the persistent myth that such awareness is largely a product of first world affluence.[1] In societies where people still live close to the land, the ecological integrity of that land is far from a luxury. Indeed, for people struggling to sustain traditional ways of life amid sometimes overwhelming development pressures, maintaining their home region's forests, soils, water, and wildlife is a matter of day-to-day survival. The ideas and actions of ecological movements in the third world thus complement the efforts of grassroots ecoactivists in the United States, and also offer an important challenge to the highly publicized international campaigns of mainstream environmental groups.

In third world countries today, growing numbers of people have been forced from the land into crowded, pollution- and disease-ridden cities, where they are utterly dependent on the exigencies and uncertainties of the global marketplace to satisfy their basic needs. Scholars and activists with a critical internationalist perspective have come to see these social and economic dislocations as the underlying cause of poverty, malnutrition, and social decay. These frequent dislocations, in turn, are largely the product of a vicious cycle of highly destructive commercial and industrial development.[2]

From: *Earth for Sale: Reclaiming Ecology in the Age of Corporate Greenwash* (Boston: South End Press, 1997), pp. 159–64, 168–74.

Vandana Shiva, widely acknowledged as the most articulate international voice of the new third world ecological outlook, sees such development—which she defines as "capital accumulation and the commercialization of the economy for the generation of 'surplus' and profits"—as the main cause of poverty and dispossession throughout the world today. For Shiva and other third world ecologists, this is merely the latest stage in a centuries-old process of colonization that degrades the natural world, exploits and excludes women, and causes the erosion of cultures that have thrived for thousands of years in a mutually sustaining relationship with the land upon which they depend. Shiva writes:

> "Development" could not but entail destruction for women, nature and subjugated cultures, which is why, throughout the Third World, women, peasants and tribals are struggling for liberation from "development" just as they earlier struggled for liberation from colonialism.[3]

Since the late 1980s, activists in the North have become increasingly aware of movements of people struggling to protect their traditional lands from the ravages of the global market economy. Rainforest activists have drawn attention to the plight of the Penan on the island of Borneo, the Yanomami of the Brazilian Amazon, and numerous other peoples who have put their lives on the line to resist incursions by multinational timber companies onto their traditional territories. The assassination of prominent labor organizer Francisco (Chico) Mendes in 1988 raised public awareness in the North of the Brazilian rubber tappers, who have intervened against the colonization of the Amazon rainforest by mining and cattle ranching interests, hydroelectric developers, and other destructive enterprises. In Kenya, women organized through the world-renowned Green Belt Movement have planted more than twenty million trees as a measure against deforestation and the destruction of traditional lifeways.[4]

One of the most dramatic of the third world's blossoming ecological movements has been the Chipko, or tree-hugging movement, which was initiated by indigenous women in the Himalayan highlands of northern India in the 1970s and spread rapidly across the country. Merging a traditional Hindu devotion to the integrity of the forests with the more recent tradition of Gandhian nonviolence, the women and men of Chipko have struggled against the exploitation of native forests and the displacement of indigenous ecosystems by plantations of commercially valued trees. Fasting, embracing ancient trees, lying down in front of logging trucks, and removing planted eucalyptus seedlings that strain precious groundwater supplies, the people of Chipko have asserted that the forests' role in replenishing the soil, water, and air must take precedence over their exploitation as a source of exotic timber for export.[5]

Larry Lohmann of the *Ecologist* magazine, who spent many years living and working in Thailand, highlights the importance of understanding the uniqueness and diversity of these movements. Movements such as Chipko are rarely simply "environmental" in the terms used by most Westerners. They emerge from a complex interplay of social, political, cultural, historical, and ecological factors, and more often than not, they defy Western dualisms of public vs. private ownership, morality vs. self-interest, biocentrism vs. anthropocentrism, militancy vs. pragmatism. They emerge from people's determination to protect traditional communal systems of livelihood, production, and allocation, rooted in distinct cultural and social patterns, from the intolerably destabilizing pressures of Western development. Lohmann writes, "Environmental knowledge and action, in Thailand as elsewhere, is locally specific, dependent on a constant, fluid interplay between theory and practice, and embedded in the democratically evolving practices of ordinary people."[6]

"The so-called environmental crisis in Africa," writes development critic Ben Wisner, "far from being a simple matter of population pressure creating vicious cycles of poverty, land degradation, famine, and further spirals of compensatory female fertility, is a crisis caused by the loss of local control over land and labor."[7] This reality, along with the effort to maintain cultural integrity in the face of a homogenizing global culture that denies and devalues traditional systems of knowledge, lies at the root of the struggle of the Ogoni people of Nigeria against the exploitation of their region's oil resources, of the opposition of people in Ghana to the flooding of their land to generate electricity for aluminum smelting, and of the resistance of the Maasai of Tanzania against threats to their pastoral ways by international "ecotourism." South African activist and scholar Yash Tandon writes:

> [P]rivate ownership of land and of nature's resources is, for the African, an unnatural phenomenon. It is profoundly antisocial and antihumanist. Land and its resources should only be held as a trust to the community and to all nature's living creatures. Its entrustment to individuals is an act against humanity and life itself.[8]

GENETIC IMPERIALISM

As multinational corporations have stepped up the extraction of commodities such as timber and minerals—and have made Asia and Africa a dumping ground for the North's toxic wastes—they have also set out to colonize the genetic resources and indigenous knowledge of third world peoples.[9] The pleas of activists in the 1980s that tropical rainforests should be protected for the wealth of useful, largely unknown biological products they contain have been transformed into a new agenda of biological colonialism. Corporations are surveying remote areas of the world for medicinal plants, indigenous relatives of common food crops, exotic sweeteners, sources of naturally occurring pesticides, and even the genetic material of once-isolated indigenous peoples. The biotechnology industry has proved particularly solicitous of plants, animals, and people that display unique genetic traits, which can be transferred—using recombinant DNA technology—to common crop varieties, bacteria, and other life-forms for future study and commercial exploitation.[10]

The traditional peoples who are largely responsible for the centuries-long cultivation and protection of beneficial plants and animals receive little if any benefit from these activities. Samples are collected, often by university-based researchers, with the aid of the local people who are most knowledgeable about local foods and medicines. The samples are sent to laboratories in urban centers where genetic traits are studied, products are developed, and patents are obtained that grant the company that supported the research a proprietary right to its findings. Biotechnology companies often seek broadly sweeping patents that offer a monopoly on all possible products from a given natural source; traditional knowledge is thus transformed into a source of commercial products to be sold worldwide at a substantial profit.

For example, India's neem tree, which has been tapped as a source of insecticidal oil, medicines, and other products since ancient times, is now becoming the proprietary property of Western corporations, such as the US chemical giant W. R. Grace. Anticancer drugs extracted from the rosy periwinkle of Madagascar have produced well over $100 million in profits for the Eli Lilly company, but are unlikely to be made available in their country of origin. The prospecting of living material reached a new height in 1995 when the US National Institutes of Health received a patent on living cells cultured from the tissues of an indigenous Hagahai person from the remote highlands of Papua New Guinea. Activists saw this as an alarming step toward the establishment of a worldwide trade in human genetic material.[11]

For activists around the world, these developments represent a qualitatively new stage in the exploitation of the South's resources by northern economic interests. The protocol on Trade-Related Intellectual Property contained in the 1994 GATT empowers the World Trade Organization to compel countries to enforce patent rules developed by Northern governments, including the widespread patenting of living organisms and their genetic material. So far, only India has resisted this pressure; in March of 1995, the upper house of the Indian parliament indefinitely tabled a proposal that would have brought the country's patent laws into compliance.[12]

India's resistance to patenting life is the culmination of many years of activism by Indian farmers against the corporate control of agriculture. Organized farmers in India and other third world countries are aware that, under the guise of fighting hunger, corporate agribusiness has heightened social inequality in agricultural regions and made people increasingly dependent on the corporate-dominated global economy. At the same time, industrial farming methods have lowered groundwater levels, poisoned the land with chemicals, and undermined the species diversity that has long sustained indigenous agriculture.[13]

Farmers in the southwestern Indian state of Karnataka have focused on the increasing dominance of Cargill and other transnational corporations, and the threat they pose to land, water, and regional food security. In 1992, activists entered Cargill's regional office in Bangalore, removed records and supplies of seeds, and tossed them into a bonfire, reminiscent of the bonfires stoked by British textiles during India's independence movement. The following summer, two hundred members of the state's peasant organization dismantled Cargill's regional seed storage unit and razed it to the ground.[14]

In October 1993, half a million farmers joined a day-long procession and rally in Bangalore to protest corporate control of agriculture, the patenting of seeds and other life-forms, and the new trade and patent rules required by the then-proposed GATT agreement. Their demands included a strong affirmation of the tradition of free cultivation and exchange of seeds by India's farmers.[15] S. M. Mohamed Idris of the Penang, Malaysia-based Third World Network described the farmers' predicament in these terms:

> Unlike the colonialism of the past, this new colonialism is more subtle, more invisible and therefore more dangerous. The rich countries and their corporations have already taken most of our natural resources, our minerals, our trees, our soils, as raw materials for their industries. Now that these resources are almost gone, they want to take away our rich and diverse biological materials, our seeds and our genetic resources.[16]

In 1996, one hundred farmer-activists attacked Bangalore's first Kentucky Fried Chicken outlet, breaking windows and electrical outlets to protest the fast food industry's environmental unsustainability, cruelty to animals through factory farming, human health hazards, and the resulting erosion of traditional agricultural and social values. During a protracted legal and political battle, activists in Karnataka had discovered that Kentucky Fried Chicken products contain nearly three times the level of monosodium glutamate allowed by state law, and that the company had plans to invest $40 million in building restaurants all across the country. Charges of attempted murder against the protesters and one of the movement's most prominent spokespeople were dropped in the face of international protest.[17] . . .

BLAMING THE POOR

While concerns about population growth and immigration in the United States have been exploited by those who would sever the fundamental link between ecology and social justice, third world activists face a far more extreme agenda. . . . The population issue has largely become a smoke screen to obscure the patterns of colonialism and exploitation that are primarily responsible for the destruction of the South's ecological integrity.[18] Discussions of overpopulation invariably focus on countries in Africa and southern Asia, rather than Holland, for example, which probably has the world's highest population density, or Japan, which has long imported much of its food, timber, and other necessities.[19] Population growth in the third world is certainly a matter of ecological concern, but it is a symptom, rather than a cause, of environmental and social degradation.

Western pundits, such as Robert Kaplan of the *Atlantic Monthly* in his widely quoted polemic "The Coming Anarchy," see a grim future for Africa and other "underdeveloped" regions of the world. In their apocalyptic visions, this is an inevitable consequence of nature "beginning to take its revenge."[20] This view is supported by many environmental groups working in the international sphere, such as the World Wildlife Fund, which in the late 1980s described the world's poor as the "most direct threat to wildlife and wildlands."[21] Ecological thinkers and activists in third world countries raise a very different set of questions. Is "nature" responsible for the horrifying statistics offered by Kaplan and others—African countries losing 80 percent of their rainforests, people forced to migrate to dangerous and overcrowded cities, etc.—or are such problems the consequence of an intensified neocolonialism that has stripped their land bare and turned much of the world into what author Tom Athanasiou terms "an international debtor's prison"?[22]

Even those institutions most responsible for the present state of affairs are being pressed to acknowledge who is really overconsuming the earth's resources. The World Bank, for example, has helped drive countless countries into debilitating cycles of poverty and dependency, in the name of "structural adjustment," a systematic effort to reorient the world's economies toward debt repayment, privatization of public services, and the promotion of foreign investment. Yet even the bank is compelled to admit that industrialized countries, with barely 20 percent of the world's population, consume well over 80 percent of the world's goods.[23] Between 1900 and 1990, although the world's human population tripled, fossil fuel use increased thirtyfold and industrial output increased fiftyfold.[24] A more graphic example is cited by the Malaysian activist Martin Khor, who decries the "gross inequalities in the use of natural resources epitomized by the fact that New Yorkers use more energy commuting in a week than the energy used by all Africans for all uses in a year..."[25] There is clearly nothing inevitable about the relationship between population and consumption, especially when considered in regionally specific terms.

For most mainstream commentators, in the South as well as the North, and for many mainstream environmentalists, the answer to inequality is more development, often carried out with a veneer of environmental sustainability. We have seen that development, for many third world activists, is merely the latest incarnation of the five hundred-year legacy of European colonialism. As Muto Ichiyo of the Tokyo-based Pacific-Asia Resource Center describes it, "Economic development, which was supposed to raise the world out of poverty, has so far only transformed undeveloped poverty into developed poverty, traditional poverty into modernized poverty designed to function smoothly in the world economic system."[26] It brings toxic hazards, such as Bhopal (where families of the victims of the 1984 chemical explosion are continuing to pursue legal charges against the executives of Union Carbide), and sweatshop industries that assault people's health and well-being.[27] For British political ecologists Oliver Tickell and Nicholas Hildyard, "Casting environmental problems in the language of development diverts attention from the policies, values and knowledge systems that have led to the crisis—and the interest groups that have promoted them."[28]

One expression of the current development paradigm has been the active participation of some US-based environmental groups in government-funded international development efforts. International development assistance ostensibly designed to encourage the use of environmental technologies is often used as a wedge to satisfy the needs of transnational capital. A recent report by the US Agency for International Development (AID), for example, advocated "the forging of environmental policies to favor private sector, market-based solutions . . . and supporting market-based approaches to biodiversity preservation and enhancement."[29] Technical assistance to address environmental problems is, for the most part, tied to the enactment of measures to limit foreign investors' liability for environmental damages. In 1993, $132 million in such assistance was funded by AID and channeled through the international activities of environmental organizations such as the World Wildlife Fund, Nature Conservancy, Conservation International, and World Resources Institute.[30]

As Vandana Shiva often points out, development does much more than perpetuate poverty and sustain the institutions of Northern domination. It systematically degrades the knowledge, skills, and cultural practices that have made it possible for people to thrive completely outside of a commercial context for thousands of years. In India, development turns once self-reliant farmers into "credit addicts and chemical addicts";[31] in Africa, it turns indigenous pastoralists into beggars. Even in the West, in the boreal forests of northern Quebec, for instance, it has meant the relocation of many recently intact Cree villages into prefabricated neighborhoods entirely dependent on imported consumer goods. Once relocated into the global market—physically, economically, and culturally—people invariably confront the same debilitating social ills that affect urbanized and suburban people throughout the world.

In the late 1980s and early 1990s, the concept of "sustainable development" became the accepted agenda for reconciling environmental protection with economic development. The term emerged from a series of UN studies and commissions, culminating in the oft-quoted 1988 Brundtland Commission report, *Our Common Future*, and the 1992 UN Earth Summit in Rio de Janeiro.[32] While the Rio conference made "sustainable development" a household term among mainstream environmentalists, and helped to enshrine it as the official policy of government agencies throughout the world, many activists in both the North and South see it as a fundamental contradiction in terms.

The principle of sustainability, a cornerstone of ecological thought embodying the regeneration of natural cycles and the promise of a successful weaving of human lifeways into them, has been severely distorted by the capitalist imperative of unencumbered growth. In contrast to the perpetual dynamic balance between human communities and the natural world envisioned by social ecologists, sustainable development advocates have joined forces with the global status quo within the framework of a crude, population-centered determinism. Jim McNeill, principal author of *Our Common Future*, explained it this way in a volume prepared for the Trilateral Commission on the eve of the Rio Earth Summit:

> If human numbers do double again, a five- to ten-fold increase in economic activity would be required to enable them to meet their basic needs and minimal aspirations. . . . Is there, in fact, any way to multiply economic activity a further five to ten times, without it undermining itself and compromising the future completely? *Can growth on these orders of magnitude be managed on a basis that is sustainable?* [emphasis added][33]

With a subtle and increasingly common turn of phrase—and questionable use of statistics—the project of making development environmentally sustainable has been transformed into one of sustaining development and economic growth. The earth's ecosystems cannot possibly survive a five- to ten-fold increase in economic activity—certainly not the kind of economic activity that we in the industrialized North have come to take for granted, and only recently begun to question. With third world cities already choking on pollution, and global climate change threatening to thoroughly disrupt the ecological balances that sustain life, it is difficult to imagine the consequences, for example, of providing private automobiles to each of China's three hundred million city dwellers or enough factory-raised animals to feed Kentucky Fried Chicken to the non-vegetarian portion of India's population. Two hundred years of industrial development in the North occurred largely at the expense of the lands, resources, and people of the South. Where will the emerging middle classes of the developing world's cities find the equivalent resources to appropriate in the name of development? The world may see a great deal more of the "jobless growth" that sustains an economy based increasingly on unproductive financial speculation, but what promise does this offer to the tens of millions of people who have been forced off the land and into the maelstrom of the global cash economy?

Realizing the long-range impossibility—and the immediate social and ecological consequences—of the Western model of development, ecologists and traditional peoples throughout the third world are seeking a different kind of vision for the future, a vision that embraces indigenous traditions while rejecting the mythical benefits of replacing subsistence-based economies with ones that rely on buying and selling commodified goods.[34] Campaigns to resist intrusions of the market economy against traditional lands and economic practices are rarely reported in the official international press, but such efforts have spread throughout the world, becoming more organized and more politically conscious.

Farmers from Ecuador to West Africa and the Philippines are returning to indigenous farming methods, banning the use of chemicals and modern machinery in their traditional territories. Fishing communities in India and the Philippines have established coastal zones from which mechanized commercial fishing boats are banned.[35] Activists in Malaysia forced the cancellation of a $5 billion mega-dam project, the second largest of its type in the world, which would have displaced nine thousand people. Adherents to a landless people's movement in Brazil have occupied traditional lands, rejecting the control of absentee landowners committed to raising cash crops. People throughout southern Mexico, inspired by the example of the Zapatista rebels of the state of Chiapas, are defying the corrupt political oligarchy that has dominated the country for nearly seventy years. The Zapatista rebellion of January 1994, coinciding with the enactment of NAFTA, pledged to reverse the Mexican government's recent abolition of constitutionally guaranteed communal land rights. One town in central Mexico expelled officials who supported the construction of luxury hotels, condominiums, and a golf course on indigenous lands, declaring a "free municipality" independent of the state government.[36] As the editors of the British journal *Ecologist* have explained:

> Indeed, as the structures of enclosure begin to falter and break down under the stress of economic recession, international debt, popular protest and everyday resistance to the anonymity of industrialization, new life is breathed into even the most seemingly dismal communities as people rediscover the value of coming together to resolve their problems.[37]

What role can ecological activists in the North play in furthering the movements of people in the South? As third world representatives at the 1992 Earth Summit emphasized, changes in the North's patterns of consumption are a prerequisite to meaningfully addressing the needs of the poor. As the Third World Network's Martin Khor described the situation:

> Since their basic needs are already fulfilled, the minority spend their incomes on superfluous and luxury consumption, and indeed the system requires them to do so to avoid . . . recession. The poor have basic needs but too little resources to fulfill them. Thus much of the world's finite resources are being depleted or degraded to become inputs to the production of luxuries. . . . At the international level, it should be realized that the present crisis is generated by the unsustainable economic model in the North, inappropriate development patterns in the South, and an inequitable global economic system that links the northern and southern models.[38]

As inequalities in wealth and power within the industrialized societies begin to parallel the huge disparities between the North and South, it is obvious that the

ecological crisis cannot be addressed without seriously confronting the underlying causes of poverty and inequality.

By unquestioningly adopting the "globalist" perspective of transnational
corporations, Northern environmentalists have unwittingly accepted the oppressive and antiecological mind-set of global corporate management. The false
globalism put forward by international institutions, from the United Nations to
the World Bank and International Monetary Fund, has set the stage for increasing
control by those institutions over land and resources, sometimes under the guise
of protecting the environment. "The global," explained Vandana Shiva, "does not
represent the universal human interest; it represents a particular local and
parochial interest which has been globalized through its reach and control,"
specifically, the interests of the corporate global managers.[39] Their outlook
denies the rights and sovereignty of peoples throughout the world, expropriating
and patenting their knowledge while placing conditions on access to the North's
environmental technologies. It mandates structural adjustment and the creation
of managed rainforest preserves to forestall climate change, while refusing to
limit the North's fossil fuel emissions, which are chiefly responsible for disrupting the earth's climate. The tacit acceptance, and even the celebration, of a
managerial globalism by many Northern environmentalists has become yet
another obstacle to addressing the root causes of environmental and social destruction. As Shiva has written:

> The image of planet Earth used as a visual in the discourse on global ecology
> hides the fact that, at the ethical level the global as construct does not symbolize
> planetary consciousness. The global reach by narrow and selfish interests does
> not use planetary or Gaian ethics. In fact, it excludes the planet and peoples
> from the mind, and puts global institutions in their place. The concept of the
> planet is invoked by the most rapacious and greedy institutions to destroy and
> kill the cultures which use a planetary consciousness to guide their daily
> actions.[40]

Perhaps more than ever, an ecological outlook requires that we understand
how the world looks when one steps outside the boundaries of a northern, industrial, consumerist worldview. The traditional knowledge of indigenous peoples
as well as the social and economic analyses of third world activists are helping
not only to unmask the green facade of neocolonialism, but to challenge the
political complacency of many Northern environmentalists. This challenge has
become especially urgent as mainstream environmental groups legitimate the
disingenuous adoption of environmental rhetoric by the US government, the
World Bank, and other institutions. "Given the key role they are fated to play in
the politics of an ever-shrinking world," Tom Athanasiou writes, "it is past time
for environmentalists to face their own history." Environmentalists have become
advocates, sometimes inadvertently, "merely for the comforts and aesthetics of

affluent nature lovers," he continues. Today, this is no longer tolerable. "They have no choice. History will judge Greens by whether they stand with the world's poor."[41] Adopting such a stance, and heeding the messages of third world ecologists, may ultimately help us to discover what is most sustainable in our own diverse cultures as well.

NOTES

1. While some commentators dismiss the term third world as a cold war archaism, in favor of more neutral terms like "the South," many activists in Asia, Africa, and Latin America embrace a "third world" identity. Eschewing the images of poverty and "underdevelopment" that the term evokes for many in the industrialized world, for these activists "third world" recalls the best traditions of the Non-Aligned Movement of the 1950s and 1960s, and the national independence and "third path" economic currents of the 1960s and 1970s.

2. For a comprehensive analysis of this process and its effects on traditional, land-based peoples, see "Whose Common Future," *Ecologist* 22, no. 4 (July/August 1992): 121–210 (available in book form from New Society Publishers).

3. Vandana Shiva, *Staying Alive: Women, Ecology and Development* (London: Zed Books, 1988), pp. 1–2.

4. "The Green Belt Movement in Kenya," interview with Wangari Maathai, in Steve Lerner, ed., *Beyond the Earth Summit*, 47–64 (Bolinas, CA: Common Knowledge Press, 1992).

5. Shiva, *Staying Alive*, pp. 55–89; interview with Vandana Shiva in Lerner, *Beyond the Earth Summit*, pp. 77–87.

6. Larry Lohmann, "Visitors to the Commons: Approaching Thailand's 'Environmental' Struggles from a Western Starting Point," in Bron Raymond Taylor, ed., *Ecological Resistance Movements: The Global Emergence of Radical and Popular Environmentalism*, 117 (Albany: State University of New York Press, 1995).

7. Ben Wisner, "*Luta*, Livelihood and Lifeworld in Contemporary Africa," in Taylor, *Ecological Resistance Movements*, p. 184.

8. Yash Tandon, "Grassroots Resistance to Dominant Land-Use Patterns in Southern Africa," in Taylor, *Ecological Resistance Movements*, p. 173.

9. On toxic dumping in the third world, see Mitchel Cohen, "Toxic Imperialism: Exporting Pentagonorrhea," *Z Magazine*, October 1990, 78–79; Mitchel Cohen, *Haiti and Somalia; The International Trade in Toxic Waste* (Brooklyn, NY: Red Balloon, 1995); Chin Oy Sim, "Basel Convention to Ban Waste Exports?" *Third World Resurgence*, no. 44 (April 1994): 10–12.

10. "Biopiracy Update: A Global Pandemic," *RAFI Communiqué*, September/October 1995; "The Struggle against Biopiracy" (feature section), *Third World Resurgence*, no. 63 (November 1995): 9–30.

11. Pat Mooney, "Indigenous Person from Papua New Guinea Claimed in US Government Patent," Rural Advancement Foundation International, October 1995, also reprinted in *Third World Resurgence*, no. 63: 30. See also Andrew Kimbrell, *The Human Body Shop* (San Francisco: HarperCollins, 1993), esp. chs. 14, 15.

12. Martin Khor, "A Worldwide Fight against Biopiracy and Patents on Life," *Third World Resurgence*, no. 63: 11.

13. Shiva, *Staying Alive*, ch. 5. See also Vandana Shiva, *The Violence of the Green Revolution: Third World Agriculture, Ecology and Politics* (London, England: Zed Books, 1993).

14. Vandana Shiva, "Quit India! Indian Farmers Burn Cargill Plant and Send Message to Multinationals," *Third World Resurgence*, no. 36 (August 1993): 40–41.

15. Martin Khor, "500,000 Indian Farmers Rally against GATT and Patenting of Seeds," *Third World Resurgence*, no. 39 (November 1993): 20–22.

16. S. M. Mohamed Idris, "Doublespeak and the New Biological Colonialism," *Third World Resurgence*, no. 39 (November 1993): 30–31.

17. S. M. Mohamed Idris, letter protesting the arrest of Professor M. D. Nanjundaswamy and others, February 1996; Ong Ju Lynn, "Activists Force India's First KFC Outlet to Be Closed Down," *Third World Resurgence*, no. 63: 42; Vandana Shiva, "More Than a Matter of Two Flies: Why KFC Is an Ecological Issue," *Third World Resurgence*, no. 67 (March 1996): 2–4.

18. See, for example, Amartya Sen, "Population: Delusion and Reality," *New York Review of Books*, September 22, 1994, pp. 62–71; Jonathan Ueberson, "Too Many People?" *New York Review of Books*, June 26, 1986, pp. 36–41; Murray Bookchin, "The Population Myth," in *Which Way for the Ecology Movement?* (San Francisco: AK Press, 1994); and Frances Moore Lappé and Joseph Collins, *Food First: Beyond the Myth of Scarcity* (Boston: Houghton Mifflin, 1977).

19. Canadian planners Mathis Wackernagel and William Rees have estimated that the basic food, fuel, forest, and other needs of the Netherlands consume an equivalent land area fifteen times the country's size. Mathis Wackernagel and William Rees, *Our Ecological Footprint: Reducing Human Impact on the Earth* (Gabriola Island, BC: New Society Publishers, 1996), pp. 93–95.

20. Robert D. Kaplan, 'The Coming Anarchy," *Atlantic Monthly*, February 1994, esp. pp. 46–54.

21. Quoted in Pratap Chatterjee and Matthias Finger, *The Earth Brokers: Power, Politics and World Development* (London: Routledge, 1994), p. 70.

22. Tom Athanasiou, *Divided Planet: The Ecology of Rich and Poor* (Boston: Little, Brown, 1996), p. 45.

23. The World Bank, *World Development Report, 1989*, quoted in Jim MacNeil et al., *Beyond Interdependence* (New York: Oxford University Press, 1991). The 1992 UN Development Program's Human Development Report revealed that the 20 percent of the world's population that live in the richest countries receive 82.7 percent of the world's income (David Korten, "The Limits of the Earth," *Nation* [July 15, 1996]: 16).

24. MacNeil et al., *Beyond Interdependence*, p. 3.

25. Martin Khor, "Earth Summit Ends with Disappointment and Hope," *Third World Network*, June 1992.

26. Ichiyo Muto, "For an Alliance of Hope," in Jeremy Brecher et al., eds., *Global Visions: Beyond the New World Order*, 148 (Boston: South End Press, 1993).

27. On the legacy of Bhopal, see David Denbo et al., *Abuse of Power* (New York: New Horizons Press, 1990); Peter Montague, "Things to Come," *RACHEL's Environment and Health Weekly*, no. 523, December 5, 1996.

28. Oliver Tickell and Nicholas Hildyard, "Green Dollars, Green Menace," *Ecologist* 22, no. 3 (May/June 1992): 82.

29. Quoted in Tom Barry, "Seeing Green: The AIDing of the Environment," *Covert Action Quarterly*, no. 58 (Fall 1996): 47.

30. Barry, "Seeing Green," p. 49.

31. Vandana Shiva, "The Politics of Diversity," *Permaculture Activist* 7, no. 1 (Spring 1991): 6.

32. World Commission on Environment and Development, *Our Common Future* (New York: Oxford University Press, 1988). For an analysis of the outcome of the Rio summit and third world activists' response, see Brian Tokar, "After the 'Earth Summit,'" *Z Magazine*, September 1992, pp. 8–14. On the co-optation of international environmentalists at Rio, see Chatterjee and Finger, *The Earth Brokers.*

33. MacNeil et al., *Beyond Interdependence*, p. 5.

34. Maria Mies, "The Need for a New Vision: The Subsistence Perspective," in Maria Mies and Vandana Shiva, *Ecofeminism* (London: Zed Books, 1993), pp. 297–324.

35. *Ecologist*, "Whose Common Future?" pp. 197–200.

36. Jo Bedingfield, "Golf Course Tees Off Mexican Village," *San Francisco Chronicle* (November 12, 1995); Alberto Ruz Buenfil, "The Revolution in Tepotztlán," *Revista ArcoRedes* (April 1996). For background on the Zapatista movement, see John Ross, *Rebellion from the Roots* (Monroe, ME: Common Courage Press, 1995).

37. *Ecologist*, "Whose Common Future?" p. 204.

38. Martin Khor, "Sustainable Development and Sustainability: Ten Points to Clarify the Concepts," *Earth Summit Briefings*, no. 4 (Third World Network, 1992).

39. Vandana Shiva, "The Greening of the Global Reach," in Brecher, *Global Visions*, pp. 53–54.

40. Ibid. See also Vandana Shiva, *Monocultures of the Mind* (Penang, Malaysia: Third World Network, 1993).

41. Athanasiou, *Divided Planet*, p. 304.

9

POPULATION AND POVERTY

BARRY COMMONER

One of the virtues of the environmental point of view is that we see the planet as a harmonious whole, a global system of water, soil, and living things bounded by the thin skin of air. However, when we look at the planet with an eye on human manifestations—the technosphere and the social systems that create it—it is split in two. The Northern hemisphere contains most of the modern technosphere—its factories, power plants, automotive vehicles, and petrochemical plants—and the wealth that it generates. The Southern hemisphere contains most of the people, nearly all of them desperately poor.

The result of this division is a painful global irony: the poor countries of the South, while deprived of an equitable share of the world's wealth, suffer the environmental hazards generated by the creation of that wealth in the North. The developing countries of the South will not only experience the impact of global warming and ozone depletion, which are now chiefly due to the industrialized countries, but are also victimized by the North's toxic exports. For example, as bans have been imposed on particularly dangerous pesticides in industrialized countries, manufacturers have marketed them in developing countries instead. There, poorly regulated, they have created in the bodies of local populations the world's highest concentrations of pesticides. Similarly, as environmental concerns have limited disposal sites for trash and the toxic trash-burning incinerators in the United States, efforts have been made to get rid of these pollutants—not always successfully—in developing countries.

Yet the gravest threat of the environmental crisis to developing countries comes not from pollutants so generously imposed on them by their wealthy plan-

From: *Making Peace with the Planet* (New York: Pantheon, 1990), pp. 141–43, 155–60, 161–64, 168.

etary neighbors, but from a more subtle source. This threat arises from a serious, frequently voiced misconception about the origin of the environmental crisis. In this and earlier analyses, I have argued that the environmental crisis originates not in the natural ecosphere, but in the man-made technosphere. The data about both the development of the post-1950 assault on the environment and the effort since 1970 to reduce it support this conclusion. There is, however, another view of the environmental crisis that turns these relationships upside down. This view holds that the problem is ecological; that environmental degradation originates in an imbalance between the earth's limited resources and the rapidly growing human population, which stresses the environment and also causes social problems such as poverty and hunger.

This position had a popular following in the early days of the environmental movement, based on the unequivocal assertions by some well-known environmentalists. In a widely quoted article, "The Tragedy of the Commons," Garrett Hardin put it this way:

> The pollution problem is a consequence of population. It did not matter much how a lonely American frontiersman disposed of his waste. . . . But as population became denser, the natural chemical and biological recycling processes became overloaded. . . . Freedom to breed will bring ruin to all.

Paul Ehrlich's best seller, *The Population Bomb*, was even more explicit about the origin of the environmental crisis:

> The causal chain of the deterioration [of the environment] is easily followed to its source. Too many cars, too many factories, too much detergent, too much pesticide, multiplying contrails, inadequate sewage treatment plants, too little water, too much carbon dioxide—all can be traced easily to *too many people*.

. . . Like all living things, people have an inherent tendency to multiply geometrically—that is, the more people there are, the more people they tend to produce. In contrast, the supply of food rises more slowly, for unlike people it does not increase in proportion to the existing rate of food production. This is, of course, the familiar relationship described by Malthus that led him to conclude that the population will eventually outgrow the food supply (and other needed resources), leading to famine and mass death. The problem is whether other countervailing forces will intervene to limit population growth and to increase food production.

When we turn from merely stating the problem to analyzing and attempting to solve it, the issue becomes much more complex. The simple statement that there is a limit to the growth of the human population, imposed on it by the limited availability of the necessary resources, is a useful but abstract idea. In order to reduce it to the level of reality in which the problem must be solved, we need to analyze the actual relationship between population growth and resources. Current views on this question are neither simple nor unanimous.

One view is that the cause of the population problems is uncontrolled fertility, the countervailing force—the death rate—having been weakened by medical advances. According to this view, given the freedom to do so, people will inevitably produce children faster than the goods needed to support them. It follows then, that the birthrate must be deliberately reduced to the point of "zero population growth."

The methods that have been proposed to achieve this kind of direct reduction in birthrate vary considerably. One method is family planning: providing people with effective contraception and access to abortion facilities and educating them about the value of having fewer children. Another suggestion, sometimes called the "lifeboat ethic," is to withhold from the people of starving developing countries, which, having failed to limit their birthrate sufficiently, are deemed to be too far gone or too unworthy to be saved. The author of this so-called ethic, Garrett Hardin, stated it this way:

> So long as we nations multiply at different rates, survival requires that we adopt the ethic of the lifeboat. A lifeboat can hold only so many people. There are more than two billion wretched people in the world—ten times as many as in the United States. It is literally beyond our ability to save them all. . . . Both international granaries and lax immigration policies must be rejected if we are to save something for our grandchildren.

But there is another view of population that is much more complex. It is based on the evidence, amassed by demographers, that the birthrate is not only affected by biological factors, such as fertility and contraception, but also by equally powerful social and economic influences. Demographers have delineated a complex network of interactions among the various biological and social factors. It shows that population growth is not the consequence of a simple arithmetic relationship between birthrate and death rate. Instead, there are circular relationships in which, as in an ecological cycle, every step is connected to several others.

Thus, while a reduced death rate does, of course, increase the rate of population growth, it can have also the opposite effect, since families usually respond to a reduced rate of infant mortality by opting for fewer children. This negative feedback modulates the effect of a decreased death rate on population size. Similarly, although a rising population increases the demand on resources, it also stimulates economic activity, which in turn improves educational levels. This tends to raise the average age at marriage and to facilitate contraceptive practices, leading to a reduced birthrate, which mitigates the pressure on resources.

In these processes, there is a powerful social force that reduces the death rate (thereby stimulating population growth) and leads people voluntarily to restrict the production of children (thereby reducing population growth). That force, simply stated, is the quality of life: a high standard of living; a sense of well-being; security in the future. When and how the two opposite effects of this force are felt differs with the stages in a country's economic development. In a pre-modern society, such as England before the Industrial Revolution or India before the advent of the English, both death rates and birthrates were high. But they were in balance and population size was stable. Then, as agricultural and industrial production began to increase and living conditions improved, the death rate began to fall. With the birthrate remaining high, the population grew rapidly. However, some thirty to forty years later, as living standards continued to improve, the decline in the death rate persisted, but the birthrate began to decline as well, reducing the rate of population growth.

Swedish demographic data, which are particularly detailed, provide a good example of this process. In around 1800, Sweden had a high birthrate, about 33 per 1,000 population, but since the death rate was equally high, the population was in balance. Then as agriculture and, later, industrial production advanced, the death rate dropped until, by the mid-nineteenth century, it stood at about 20 per 1,000. Since the birthrate remained virtually constant during that period, there was a large excess of births over deaths and the population increased rapidly—an early version of the "population explosion." Then the birthrate began to drop, until in the mid-twentieth century it reached about 14 per 1,000, when the death rate was about 10 per 1,000. Thus, under the influence of a constantly rising standard of living, the population moved, with time, from a position of balance at high birth and death rates to a new position of near balance at low birth and death rates. But in between, the population increased considerably.

This process, the demographic transition, has been characteristic of all industrialized countries. In these countries, the death rate began to decline in the mid-eighteenth century, reaching an average of 30 per 1,000 in 1850, 24 per 1,000 in 1900, 16 per 1,000 in 1950, and 9 per 1,000 in 1985. In contrast, the birthrate remained constant at about 40 per 1,000 until 1850, then dropping rapidly, reaching 32 per 1,000 in 1900, 23 per 1,000 in 1950, and 14 per 1,000 in 1985. As a result, populations grew considerably, especially in the nineteenth century, then slowed to the present net rate of growth of 0.4 percent per year.

The same process has been under way in developing countries, but with a longer time lag between the declines in death rate and birthrate. In developing countries, the average death rate was more or less constant, at about 38 per 1,000 until 1850, then declined to 33 per 1,000 in 1900, 23 per 1,000 in 1950, and 10 per 1,000 in 1985. The average birthrate, on the other hand, remained at a constant high level, 43 per 1,000 until about 1925; it has since declined at an increasing rate, reaching 37 per 1,000 in 1950, and 30 per 1,000 in 1985. As a result, the increase in population of the developing countries that began around 1850 has started to slow down and those country's populations are now growing at an average rate of 1.74 percent annually. It is important to note that the *death rates* of developed and developing countries are now nearly the same and, given the inherent biological limits, are not likely to decline much further. Thus, in developing countries the progressively rapid drop in birthrate will accelerate progress toward populations that, like those of developed countries, are approximately in balance.

One indicator of the quality of life—infant mortality—is especially decisive in this process. Couples respond to a low rate of infant mortality by realizing that they no longer need to have more children to replace the ones that die. Birth control is, of course, an essential part of this process; but it can succeed—barring compulsion—only in the presence of a rising standard of living which generates the necessary motivation. There is a critical point in the rate of infant mortality below which the birthrate begins to drop sharply, creating the conditions for a balanced population. This process appears to be just as characteristic of developing countries as of developed ones. Thus, where infant mortality is particularly high, as in African countries, the birthrate is also very high. Infant mortality is always very responsive to improved living conditions, especially with respect to nutrition. Consequently, there is a kind of critical standard of living, which, if achieved, can lead to a rapid reduction in birthrate and an approach to a balanced population.

Thus, in human societies, there is a built-in process that regulates population size: if the standard of living, which initiates the rise in population, continues to increase, the reason that populations in developing countries have not yet leveled off is that this basic condition has not yet been met. . . .

Given this background, what can be said about the various alternative methods of achieving a balanced world population? In India, there has been an interesting, if partially inadvertent, comparative test of two of the possible approaches: family-planning programs and efforts (also on a family basis) to elevate the living standard. The results of this test show that while the family-planning effort itself failed to reduce the birthrate, improved living standards succeeded.

In 1954, a Harvard team undertook the first major field study of birth control in India. The population of a number of test villages was provided with contraceptives and suitable educational programs. Over a six-year period, 1954–60, birthrates, death rates, and health status in this population were compared with the rates found in an equivalent population in villages not provided with the birth control program. A follow-up in 1969 showed that the population control program had failed. Although in the test population the birthrate dropped from 40 per 1,000 in 1957 to 35 per 1,000 in 1968, a similar reduction also occurred in the comparison population.

We now know why the study failed, thanks to a remarkable book by Mahmood Mamdani, *The Myth of Population Control.* He investigated in detail the impact of the study on one of the test villages, Manupur. What Mamdani discovered confirms the view that population control in a country like India depends on the *economic* factors that indirectly limit fertility. Talking with the Manupur villagers he discovered why, despite the study's statistics regarding ready "acceptance" of the offered contraceptives, the birthrate was not affected:

One such "acceptance" case was Asa Singh, a sometime land laborer who is now a watchman at the village high school. I questioned him as to whether he used the tablets or not: "Certainly I did. You can read it in their books. From 1957 to 1960, I never failed." Asa Singh, however, had a son who had been born sometime in "late 1958 or 1959." At our third meeting I pointed this out to him. ... Finally he looked at me and responded "Babuji, someday you'll understand. It is sometimes better to lie. It stops you from hurting people, does no harm, and might even help them." The next day Asa Singh took me to a friend's house ... and I saw small rectangular boxes and bottles, one piled on top of the other, all arranged as a tiny sculpture in a corner of the room. This man had made a sculpture of birth control devices. Asa Singh said: "Most of us threw the tablets away. But my brother here, he makes use of everything."

Such stories have been reported before and are often taken to indicate how much "ignorance" has to be overcome before birth control can be effective in countries like India. But Mamdani takes us much further into the problem, by finding out why the villagers preferred not to use the contraceptives. In one interview after another he discovered a simple, decisive fact: that in order to advance their economic condition, to take advantage of the opportunities newly created by the development of independent India, *children were essential.* Mamdani makes this very explicit:

To begin with, most families have either little or no savings, and they can earn too little to be able to finance the education of *any* children, even through high school. Another source of income must be found, and the only solution is, as one tailor told me, "to have enough children so that there are at least three or four sons in the family." Then each son can finish high school by spending part of the afternoon working. ... After high school, one son is sent on to college while the others work to save and pay the necessary fees. ... Once his education is completed, he will use his increased earnings to put his brother through college. He will not marry until the second brother has finished his college education and can carry the burden of educating the third brother.

Mamdani points out that "it was the rise in the age of marriage—from 17.5 years in 1956 to 20 in 1969—and not the birth control program that was responsible for the decrease in the birthrate in the village from 40 per 1,000 in 1957 to 35 per 1,000 in 1968. While the birth control program was a failure, the net result of the technological and social change in Manupur was to bring down the birthrate."

Here, then, in the simple realities of an Indian village are the principles of the demographic transition at work. There is a way to control the rapid growth of populations in developing countries. It is to help them develop, and to achieve more rapidly, the level of welfare that everywhere in the world is the real motivation for reducing the birthrate.

Against this conclusion it will be argued, to quote Hardin, that "it is literally beyond our ability to save them all." This reflects the view that there is simply insufficient food and other resources in the world to support the present world population at the standard of living required to motivate the demographic transition. It is sometimes pointed out, for example, that the United States consumes about one-third of the world's resources to support only 6 percent of the world's population, the inference being that there are simply not enough resources in the world to permit the rest of the world to achieve the standard of living and low birthrate characteristic of the United States.

The fault in this reasoning is readily apparent from the actual relationship between the birthrates and living standards of different countries. The only available comparative measure of standard of living is gross national product (GNP) per capita. Neglecting for this purpose the faults inherent in GNP as a measure of the quality of life, a plot of birthrate against GNP per capita is very revealing. For example, in 1984 in the United States GNP per capita was $15,541 and the birthrate was 16 per 1,000. In the poorest countries (GNP per capita less than $500 per year), the birthrates were 32 to 55 per 1,000. In those countries where GNP per capita was $4,000 to $5,000 (for example, Greece), the birthrate ranged from 15 to 19 per 1,000. Thus, in order to bring the birthrates of the poor countries down to the low levels characteristic of the richer ones, the poor countries do not need to become as affluent (at least as measured by GNP per capita) as the United States. By achieving a per capita GNP only, let us say, one-third of that of the United States, these countries could reach birthrates almost as low as those of the European and North American countries.

In a sense the demographic transition is a means of translating the availability of a decent level of resources, especially food, into a voluntary reduction in birthrate. The per capita cost, in GNP, of increasing the standard of living of developing countries to the point that would motivate a voluntary reduction in birthrate is small, compared to the wealth of the rich, developed countries—a much-neglected, global bargain. . . .

My own purely personal conclusion is, like all of these, not scientific but political: the world population crisis, which is the ultimate outcome of the exploitation of poor nations by rich ones, ought to be remedied by returning to the poor countries enough of the wealth taken from them to give their peoples both the reason and the resources voluntarily to limit their own fertility.

In sum, I believe that if the root cause of the world population crisis is poverty, then to end it we must abolish poverty. And if the cause of poverty is the grossly unequal distribution of world's wealth, then to end poverty, and with it the population crisis, we must redistribute that wealth, among nations and within them.

10

NATURAL CAPITALISM

PAUL HAWKEN, AMORY LOVINS, AND L. HUNTER LOVINS

Natural capital can be viewed as the sum total of the ecological systems that support life, different from human-made capital in that natural capital cannot be produced by human activity. It is easy to overlook because it is the pond in which we swim, and, like fish, we are not aware we're in the water. One can live perfectly well without ever giving a thought to the sulfur cycle, mycorrhizal formation, alleles, wetland functions, or why giant sequoia trees can't reproduce without chattering squirrels. We need not know that 80 percent of the 1,330 cultivated species of plants that supply our food are pollinated by wild or semi-wild pollinators,[1] but we should be aware that we are losing many of those pollinators including half of our honeybee colonies in the past fifty years in the United States, one-fourth since 1990. As biologists Gary Paul Nabhan and Steven Buchmann write in their book *Forgotten Pollinators*, "Nature's most productive workers [are] slowly being put out of business."[2]

Only when the services provided by ecosystem functions are unmistakably disrupted do we step back and reconsider. Virtually every fish caught and consumed in the Great Lakes region comes with some amount of industrially produced contamination. When rain disappears and soil blows away in the Midwest, when towns are flooded downstream by clear-cutting upstream, the absence of natural capital services becomes more apparent. Sometimes we mourn the loss much later. Kelp has become an increasingly valuable commodity, producing a wide range of products from food additives to nutritional supplements and pharmaceuticals. But Russian trappers critically injured Pacific Coast kelp beds in the

From: *Natural Capitalism: Creating the Next Industrial Revolution* (Boston: Little, Brown, 1999), pp. 150–59.

eighteenth and nineteenth centuries, when sea otters from Alaska to Baja were hunted to near extinction. The otters ate urchins that eat kelp. Without the otters, the urchin population soared, and the beds, described by early explorers as vast underwater forests, were decimated. The Russians wanted the otter because after the invention of the samovar, Russian appetite for Chinese tea soared and otter furs were the only currency the Chinese would accept. Worth as much as precious metals, the fur was desired as trim for ornate robes.[3]

Compared to the rest of the world, North Americans have been fortunate in not having suffered debilitating degradation of their ecosystem services. Many countries and regions, more densely and historically populated, face far more severe effects of natural capital depletion. Yet American ecosystems cannot long endure without the health of their counterparts around the world. The atmosphere does not distinguish whether CO_2 comes from US oil or Chinese coal, nor do the record-breaking 240 mph winds recorded in Guam in 1997 lose force if you don't happen to believe in climate change.[4]

SUBSTITUTES OR COMPLEMENTS?

Many economists continue to insist that natural and manufactured capital are interchangeable, that one can replace the other. While they may acknowledge some loss of living systems, they contend that market forces will combine with human ingenuity to bring about the necessary technological adaptations to compensate for that loss. The effort of creating substitutes, they argue, will drive research, promote spending, increase jobs, and create more economic prosperity. Hydroponics, for example, could theoretically replace farms, creating potential benefits. There are substitutes for many resource commodities, as is the case with copper, coal, and metals. And there may be other beneficial substitutes on the drawing boards or not yet invented. Nevertheless, look at this very human-oriented list and try to imagine the technologies that could replace these services:

- production of oxygen
- maintenance of biological and genetic diversity
- purification of water and air
- storage, cycling, and global distribution of freshwater
- regulation of the chemical composition of the atmosphere
- maintenance of migration and nursery habitats for wildlife
- decomposition of organic wastes
- sequestration and detoxification of human and industrial waste
- natural pest and disease control by insects, birds, bats, and other organisms
- production of genetic library for food, fibers, pharmaceuticals, and materials
- fixation of solar energy and conversion into raw materials
- management of soil erosion and sediment control

- flood prevention and regulation of runoff
- protection against harmful cosmic radiation
- regulation of the chemical composition of the oceans
- regulation of the local and global climate
- formation of topsoil and maintenance of soil fertility
- production of grasslands, fertilizers, and food
- storage and recycling of nutrients[5]

Thus far there are precious few if any substitutes for the services that natural capital invisibly provides. If it took a $200 million investment to minimally keep eight people alive for two years in Biosphere 2, how much would it cost to replicate functions in the preceding list?

In 1997 a group of highly respected scientists, primarily biologists, wrote a consensus paper on ecosystem services in an attempt to raise public awareness of their concern about this issue. Published in the spring 1997 *Issues in Ecology*, it noted:

Based on available scientific evidence, we are certain that:

- Ecosystem services are essential to civilization.
- Ecosystem services operate on such a grand scale and in such intricate and little-explored ways that most could not be replaced by technology.
- Human activities are already impairing the flow of ecosystem services on a large scale.
- If current trends continue, humanity will dramatically alter or destroy virtually all of Earth's remaining natural ecosystems within a few decades.

That the public does not understand the economic implications of declining ecosystem services has been frustrating to scientists. But in 1994, a group of Pew Scholars gathered in Arizona. Out of this meeting came the book *Nature's Services*, edited by Gretchen Daily, and a paper, whose lead author was economist Robert Costanza, entitled "The Value of the World's Ecosystem Services and Natural Capital," published in the British journal *Nature* on May 15, 1997. Both publications occasioned headlines, press conferences, and follow-up stories. The issues finally received proper attention because the scientists shrewdly put a price tag on the annual value of seventeen ecosystem services: $36 trillion on average, with a high estimate of $58 trillion (1998 dollars). Given that in 1998 the gross world product was $39 trillion, the figures were surprising.[6]

Most of the ecosystem values the scientists identified had never been economically measured. They included $1.3 trillion a year for atmospheric regulation of gases, $2.3 trillion for the assimilation and processing of waste, $17 trillion for nutrient flows, and $2.8 trillion for the storage and purification of water. The greatest contribution, $20.9 trillion, was from marine systems, especially coastal environments. Terrestrial systems added $12.3 trillion, with forests and wetlands each responsible for about $4.7 trillion. The value of all terrestrial systems averaged just over $466 per acre per year. Marine systems were lower, averaging $234 per acre, but more highly concentrated in coastal environments, including the continental shelf, where the yield was $1,640 per acre. The highest annual value per acre recorded was for estuaries, at $9,240. The primary value of coastal estuaries is not as a food source but in their capacity to provide nutrient recycling services for 40 trillion cubic meters of river water every year. On land, the highest-valued environments were wetlands and floodplains, at $7,924 per acre per year. The greatest benefits derived from these systems are flood control, storm protection, waste treatment and recycling, and water storage.

At first glance, these numbers may seem unduly high. After all, many farmers have much more modest incomes per acre; US annual gross farm income averages about $200 per acre per year. But bear in mind that the values measured do not simply record resources extracted and sold. An acre of ocean or chaparral can't be conventionally monetized according to the standard economic point of view, which counts only what's taken away to market, not the service of supporting life itself.

In the United States, the decline in ecosystem services can be gauged in part by the loss of major ecosystems. These habitats or ecological communities, and many more, are all unique and are all under threat of destruction:

- California wetlands and riparian communities
- tallgrass prairies (which once nurtured nearly 100 million buffalo, elk, and antelope)
- Hawaiian dry forests
- longleaf pine forests and savannas
- forest wetlands in the South
- ancient ponderosa pine forests
- ancient eastern deciduous forests
- California native grasslands
- southern Appalachian spruce-fir forests
- Midwestern wetlands
- marine coastal communities in all lower forty-eight states and Hawaii
- ancient redwood forests
- ancient cedar forests of the Northwest
- pine forests of the Great Lakes
- eastern grasslands and savannas
- Southern California coastal sage scrub[7]

If we capitalized the annual income of $36 trillion for ecosystem services, using the going rate for US treasuries, it would mean that nature is roughly worth a little more than $500 trillion—an absurdly low figure, as it is comparable to the next thirteen years of economic output.[8] What prices can do, however, is to illustrate vividly and concretely a relationship that is breaking down. Establishing values for natural capital stocks and flows, as rough as they may be, or—as natural capitalism does—*behaving as if* we were doing so, is a first step toward incorporating the value of ecosystem services into planning, policy, and public behavior. When a Philippine fisherman tosses a stick of dynamite into coral reefs, harvesting stunned fish for local markets and broken pieces of coral for the pharmaceutical industry, he pockets cash at market prices. He does not pay for the loss of the coral reef, but it should be obvious that the net present value of the coral reef habitat as a future home of fish far outweighs the few pesos garnered by its destruction. Nevertheless, governments from developed and developing nations still use accounting methods that register the fish and coral harvest as net gains rather than net losses.

If the services provided by natural capital provide in effect annual "subsidies" to production worth tens of trillions of dollars, and these subsidies are declining while affluence and population growth are accelerating their depletion, at what point will civilization be affected? How will businesses all reliant on natural capital, and some especially so, prosper in the future? Given that all of the biomes studied in the *Nature* article are declining in area, viability, and productivity, perhaps a revision in economics is overdue. A reassessment of national and international balance sheets is needed in which the stock and flow of services from natural capital are at least partially if not fully valued.

Biologist Peter Raven, director of the Missouri Botanical Garden and one of the world's foremost experts in biodiversity, writes that ecosystem services are not merely "a series of factors lying on the side of industrial processes, which added up could cause trouble, but rather an expression of the functioning of a healthy Earth. . . . [W]e're disrupting that functioning to an incredible degree." The cash estimate of their value commodifies the living world and says

> nothing about our real place in nature, morality, or the simple joy of living in a richly diverse, interesting, living world. As a biologist, I always think about such broad subjects in the way the world functions, as if there were no people there; and then I think about the flow of energy from the Sun, and the activities of all the photosynthetic organisms, the food chains and communities that regulate the flow of the stored energy here on Earth, and the ways in which human beings impact or break that flow, or divert it for their own purposes—what are the actual biological limits. For me, it is always the centrality of those functions, within which we evolved and which are so essential to our continued existence, that keeps looming so large.[9]

LIMITING FACTORS

Former World Bank economist Herman Daly believes that humankind is facing a historic juncture: For the first time, the limits to increased prosperity are due to the lack not of human-made capital but rather of natural capital.

Historically, economic development has periodically faced one or another limiting factor, including the availability of labor, energy resources, and financial capital. A limiting factor is one that prevents a system from surviving or growing if it is absent. If marooned in a mountain snowstorm, you need water, food, and warmth to survive; the resource in shortest supply limits your ability to survive. One factor does not compensate for the lack of another. Drinking more water will not make up for lack of clothing if you are freezing, just as having more clothing will not satisfy hunger. Because limiting factors in a complementary system cannot be substituted one for the other, the complement in shortest supply is what must be increased if the enterprise is to continue. Increasingly, the limiting factor for humanity is the decline of the living systems, quintessentially complements. Remove any of the ecosystem services listed previously, and others start to break down and eventually disappear.

The knowledge that shortages of ecosystem services will not lead to substitutions causes a different kind of anguish on both sides of the environmental debate. Eminent scientists and economists including Peter Raven, Herman Daly, J. Peterson Myers, Paul Ehrlich, Norman Myers, Gretchen Daily, Robert Costanza, Jane Lubchenco, and thousands more are trying to reach business, academic, and political audiences with this message. On the other hand, business acts as if scientists have either been unduly pessimistic or simply wrong in the past, and, in the case of climate change, will buy full-page ads in the *Wall Street Journal* arguing for, ironically, more studies and science, little of which they offer to fund. In the meantime, the loss of living systems is accelerating worldwide, despite huge capital spending on environmental cleanup by industrial nations and responsible corporations. The gap in understanding would be comical were it not potentially tragic. It's as if you are intent on cleaning your house, which is situated on a floodplain whose river is rising. Cleaning house is an admirable activity, but it's not an appropriate response to the immediate problem.

Whenever the economy has faced limiting factors to development in the past, industrial countries were able to continue to grow by maximizing the productivity or increasing the supply of the limiting factor. These measures sometimes came at a high cost to society. "From this foul drain the greatest stream of human industry flows out to fertilize the whole world," as de Tocqueville wrote.[10] Labor shortages were "satisfied" shamefully by slavery, as well as by immigration and high birthrates. Labor-saving machinery was supplied by the industrial revolution. New energy sources came from the discovery and extraction of coal, oil, and gas. Tinkerers and inventors created steam engines, spinning jennies, cotton gins, and telegraphy. Financial capital became universally accessible through central banks, credit, stock exchanges, and currency exchange mechanisms. Typically, whenever new limiting factors emerged, a profound restructuring of the economy was the response. Herman Daly believes we are once more in such a period of restructuring, because the relationship between natural and human-made capital is changing so rapidly.

As natural capital becomes a limiting factor, we ought to take into consideration what we mean by the concept of "income." In 1946, economist J. R. Hicks defined income as the "maximum amount that a community can consume over some time period and still be as well off at the end of the period as at the beginning."[11] Being well-off at the end of a given year requires that some part of the capital stock is used to produce income, whether that capital is a soybean farm, semiconductor factory, or truck fleet. In order to continue to allow people to be well off, year after year, that capital must either increase or remain in place. In the past, this definition of income was applied only to human-made capital, because natural capital was abundant. Today, the same definition should also apply to natural capital. This means that in order to keep our levels of income stable, much less increase them, we must sustain the original stocks of both types of capital. The less able we are to substitute artificial for natural capital, the more *both* forms of capital must be safeguarded from liquidation.

To maintain income, we need not only to maintain our stock of natural capital but to increase it dramatically in preparation for the possible doubling of population that may occur in the next century. This fourth principle of natural capitalism, investing in natural capital, is a matter of common sense. The only way to maximize natural capital's productivity in the near term is by changing consumption and production patterns. Since today 80 percent of the world receives only about 20 percent of the resource flow, it is obvious that this majority will require more consumption, not less. The industrialized world will need radically improved resource productivity, both at home and abroad, and then begin to reverse the loss of natural capital and increase its supply. This is the only way to improve the quality of life everywhere in the world at once, rather than merely redistributing scarcity.

As economist Herman Daly explains, "[W]hen the limiting factor changes, then behavior that used to be economic becomes uneconomic. Economic logic remains the same, but the pattern of scarcity in the world changes, with the result that behavior must change if it is to remain economic."[12] This proposition explains the despair and excitement on both sides of the issue of resource management. On the environmental side, scientists are frustrated that many businesspeople do not yet understand the basic dynamics involved in the degradation of biological systems. For business, it seems unthinkable if not ludicrous that you shouldn't be able to create the future by using the same methods that have been successful in the present and past. In this transitional phase, however, business is gradually coming to realize that economic activities that were once lucrative may no longer lead to a prosperous future. That realization is already fueling the next industrial revolution.

NOTES

1. J. N. Abramovitz, "Putting a Value on Nature's 'Free' Services," *World Watch* 11, no. 1 (January/February 1998): 10.

2. Gary Paul Nabhan and Steven Buchmann have formed the Forgotten Pollinators Campaign to educate the public about the growing threat human activities place on domestic crops and wild plants. Forgotten pollinators include the 4,000–5,000 wild bees native to North America but also hummingbirds, butterflies, beetles, moths, bats, and even certain species of flies.

3. P. McHugh, "Rare Otters Surface in SF Bay," *San Francisco Chronicle*, January 17, 1998.

4. S. Newman, "Super Typhoon," Earthweek, *San Francisco Chronicle*, December 12, 1997.

5. R. S. de Groot, "Environmental Functions and the Economic Value of Natural Ecosystems," in *Investing in Natural Capital*, ed. A. Jansson et al., 151 (Washington, DC, 1994). (This is an amended and reworded rendition of de Groot's list of Regulation, Carrier, Production, and Information Functions). Cairns 1997 contains a very useful framework for understanding conditions for sustaining natural capital, based in part on the work of Dr. Karl-Henrik Robèrt, founder of the Natural Step.

6. The analysis published in *Nature* included sixteen specific biomes—geographical regions containing specific communities of flora and fauna—and identified the value of seventeen ecosystem services by economic activity within each biome. The biomes included marine and terrestrial environments: open ocean (33.2 million hectares), estuaries (180 million hectares), seaweed and algae beds (200 million hectares), coral reefs (62 million hectares), continental shelves (2.6 million hectares), lakes and rivers (200 million hectares), tropical forests (1,900 million hectares), temperate forests (2,955 million hectares), grasslands and rangelands (3,898 million hectares), tidal marshes and mangroves (165 million hectares), and swamps and floodplains (165 million hectares). The economic values included net income, replacement cost, market value, resource production, real estate value in the case of cultural services, damage prevention, shadow prices, external costs mitigated, direct or estimated revenues in the case of recreation, avoided costs and damages, lost income in the case of erosion, restoration costs in cases of erosion control, option value, rents, opportunity costs, dockside prices for marine products, flood damage control, and energy flow analyses. One hundred seventeen prior studies, surveys, and papers were used as primary data and valuation sources for the paper. The study did not include the value of nonrenewable fuels and minerals or of the atmosphere itself.

7. R. F. Noss, E. T. LaRoe, and J. M. Scott, *Endangered Ecosystems of the United States: A Status Report and Plan for Action* (Washington, DC: Defenders of Wildlife, 1995); R. F. Noss and R. L. Peters, *Endangered Ecosystems of the United States: A Preliminary Assessment of Loss and Degradation*, biological report 28, USDI National Biological Service, Washington, DC, 1995.

8. We are indebted to Susan Meeker-Lowry, who made this point eloquently in a letter to the editor regarding Jane Abramovitz's article in *World Watch* (cited in n1): "Nature doesn't provide us with services, like a waitress or mechanic or doctor. The diverse species and processes that make up ecosystems do what they do naturally.... Putting a price tag on it is terribly misleading, because in the end we cannot buy nature because we cannot create it."

9. Peter Raven, private correspondence, February 15, 1999.

10. As quoted in E. Hobsbawm, *The Age of Revolution* (New York: Vintage, 1996).

11. H. E. Daly, "Operationalizing Sustainable Development by Investing in Natural Capital," in *Investing in Natural Capital*, ed. A. Jansson et al., ch. 22 (Washington, DC: Island Press, 1994).

12. Ibid.

PART III

DEEP, SOCIAL, AND SOCIALIST ECOLOGY

11

DEEP ECOLOGY

ARNE NAESS

Ecologically responsible policies are concerned only in part with pollution and resource depletion. There are deeper concerns which touch upon principles of diversity, complexity, autonomy, decentralization, symbiosis, egalitarianism, and classlessness.

T he emergence of ecologists from their former relative obscurity marks a turning point in our scientific communities. But their message is twisted and misused. A shallow, but presently rather powerful movement, and a deep, but less-influential movement, compete for our attention. I shall make an effort to characterize the two.

1. THE SHALLOW ECOLOGY MOVEMENT:

Fight against pollution and resource depletion. Central objective: the health and affluence of people in the developed countries.

2. THE DEEP ECOLOGY MOVEMENT:

1. Rejection of the man-in-environment image in favor of the *relational, total-field image*. Organisms as knots in the biospherical net or field of intrinsic relations. An intrinsic relation between two things *A* and *B* is such that the rela-

---

From: "The Shallow and the Deep, Long-Range Ecology Movement: A Summary," *Inquiry* 16 (1973): 95–100.

tion belongs to the definitions or basic constitutions of A and B, so that without the relation, A and B are no longer the same things. The total-field model dissolves not only the man-in-environment concept, but every compact thing-in-milieu concept—except when talking at a superficial or preliminary level of communication.

2. *Biospherical egalitarianism*—in principle. The "in principle" clause is inserted because any realistic praxis necessitates some killing, exploitation, and suppression. The ecological field worker acquires a deep-seated respect, or even veneration, for ways and forms of life. He reaches an understanding from within, a kind of understanding that others reserve for fellow men and for a narrow section of ways and forms of life. To the ecological field worker, *the equal right to live and blossom* is an intuitively clear and obvious value axiom. Its restriction to humans is an anthropocentrism with detrimental effects upon the life quality of humans themselves. This quality depends in part upon the deep pleasure and satisfaction we receive from close partnership with other forms of life. The attempt to ignore our dependence and to establish a master-slave role has contributed to the alienation of man from himself.

Ecological egalitarianism implies the reinterpretation of the future-research variable, "level of crowding," so that general mammalian crowding and loss of life-equality is taken seriously, not only human crowding. (Research on the high requirements of free space of certain mammals has, incidentally, suggested that theorists of human urbanism have largely underestimated human life-space requirements. Behavioral crowding symptoms [neuroses, aggressiveness, loss of traditions . . .] are largely the same among mammals.)

3. *Principles of diversity and of symbiosis.* Diversity enhances the potentialities of survival, the chances of new modes of life, the richness of forms. And the so-called struggle of life, and survival of the fittest, should be interpreted in the sense of ability to coexist and cooperate in complex relationships, rather than ability to kill, exploit, and suppress. "Live and let live" is a more powerful ecological principle than "Either you or me."

The latter tends to reduce the multiplicity of kinds of forms of life, and also to create destruction within the communities of the same species. Ecologically inspired attitudes therefore favor diversity of human ways of life, of cultures, of occupations, of economies. They support the fight against economic and cultural, as much as military, invasion and domination, and they are opposed to the annihilation of seals and whales as much as to that of human tribes or cultures.

4. *Anticlass posture.* Diversity of human ways of life is in part due to (intended or unintended) exploitation and suppression on the part of certain groups. The exploiter lives differently from the exploited, but both are adversely affected in their potentialities of self-realization. The principle of diversity does not cover differences due merely to certain attitudes or behaviors forcibly blocked or restrained. The principles of ecological egalitarianism and of symbiosis support the same anticlass posture. The ecological attitude favors the

extension of all three principles to any group conflicts, including those of today between developing and developed nations. The three principles also favor extreme caution toward any overall plans for the future, except those consistent with wide and widening classless diversity.

5. Fight against *pollution and resource depletion*. In this fight ecologists have found powerful supporters, but sometimes to the detriment of their total stand. This happens when attention is focused on pollution and resource depletion rather than on the other points, or when projects are implemented which reduce pollution but increase evils of the other kinds. Thus, if prices of life necessities increase because of the installation of antipollution devices, class differences increase too. An ethics of responsibility implies that ecologists do not serve the shallow, but the deep ecological movement. That is, not only point (5), but all seven points must be considered together.

Ecologists are irreplaceable informants in any society, whatever their political color. If well organized, they have the power to reject jobs in which they submit themselves to institutions or to planners with limited ecological perspectives. As it is now, ecologists sometimes serve masters who deliberately ignore the wider perspectives.

6. *Complexity, not complication*. The theory of ecosystems contains an important distinction between what is complicated without any gestalt or unifying principles—we may think of finding our way through a chaotic city—and what is complex. A multiplicity of more or less lawful, interacting factors may operate together to form a unity, a system. We make a shoe or use a map or integrate a variety of activities into a workaday pattern. Organisms, ways of life, and interactions in the biosphere in general, exhibit complexity of such an astoundingly high level as to color the general outlook of ecologists. Such complexity makes thinking in terms of vast systems inevitable. It also makes for a keen, steady perception of the profound *human ignorance* of biospherical relationships and therefore of the effect of disturbances.

Applied to humans, the complexity-not-complication principle favors division of labor, *not fragmentation of labor*. It favors integrated actions in which the whole person is active, not mere reactions. It favors complex economies, an integrated variety of means of living. (Combinations of industrial and agricultural activity, of intellectual and manual work, of specialized and nonspecialized occupations, of urban and nonurban activity, of work in city and recreation in nature with recreation in city and work in nature . . .)

It favors soft technique and "soft future-research," less prognosis, more clarification of possibilities. More sensitivity toward continuity and live traditions, and—most important—toward our state of ignorance.

The implementation of ecologically responsible policies requires in this century an exponential growth of technical skill and invention—but in new directions, directions which today are not consistently and liberally supported by the research policy organs of our nation-states.

7. *Local autonomy and decentralization.* The vulnerability of a form of life is roughly proportional to the weight of influences from afar, from outside the local region in which that form has obtained an ecological equilibrium. This lends support to our efforts to strengthen local self-government and material and mental self-sufficiency. But these efforts presuppose an impetus toward decentralization. Pollution problems, including those of thermal pollution and recirculation of materials, also lead us in this direction, because increased local autonomy, if we are able to keep other factors constant, reduces energy consumption. (Compare an approximately self-sufficient locality with one requiring the importation of foodstuff, materials for house construction, fuel and skilled labor from other continents. The former may use only 5 percent of the energy used by the latter.) Local autonomy is strengthened by a reduction in the number of links in the hierarchical chains of decision. (For example, a chain consisting of local board, municipal council, highest subnational decision maker, a statewide institution in a state federation, a federal national government institution, a coalition of nations, and of institutions; for example, EEC top levels, and a global institution can be reduced to one made up of local board, nationwide institution, and global institution.) Even if a decision follows majority rules at each step, many local interests may be dropped along the line, if it is too long.

Summing up, then, it should, first of all, be borne in mind that the norms and tendencies of the Deep Ecology movement are not derived from ecology by logic or induction. Ecological knowledge and the lifestyle of the ecological field worker have *suggested, inspired, and fortified* the perspectives of the Deep Ecology movement. Many of the formulations in the above seven-point survey are rather vague generalizations, only tenable if made more precise in certain directions. But all over the world the inspiration from ecology has shown remarkable convergencies. The survey does not pretend to be more than one of the possible condensed codifications of these convergencies.

Secondly, it should be fully appreciated that the significant tenets of the Deep Ecology movement are clearly and forcefully *normative.* They express a value priority system only in part based on results (or lack of results, cf. point [6]) of scientific research. Today, ecologists try to influence policy-making bodies largely through threats, through predictions concerning pollutants and resource depletion, knowing that policy makers accept at least certain minimum norms concerning health and just distribution. But it is clear that there is a vast number of people in all countries, and even a considerable number of people in power, who accept as valid the wider norms and values characteristic of the Deep Ecology movement. There are political potentials in this movement which should not be overlooked and which have little to do with pollution and resource depletion. In plotting possible futures, the norms should be freely used and elaborated.

Thirdly, insofar as ecology movements deserve our attention, they are *ecophilosophical* rather than ecological. Ecology is a *limited* science which makes *use* of scientific methods. Philosophy is the most general forum of debate on fundamentals, descriptive as well as prescriptive and political philosophy is one of its subsections. By an *ecosophy* I mean a philosophy of ecological harmony or equilibrium. A philosophy as a kind of *sofia* wisdom, is openly normative, it contains *both* norms, rules, postulates, value priority announcements *and* hypotheses concerning the state of affairs in our universe. Wisdom is policy wisdom, prescription, not only scientific description and prediction.

The details of an ecosophy will show many variations due to significant differences concerning not only "facts" of pollution, resources, population, etc., but also value priorities. Today, however, the seven points listed provide one unified framework for ecosophical systems.

In general system theory, systems are mostly conceived in terms of causally or functionally interacting or interrelated items. An ecosophy, however, is more like a system of the kind constructed by Aristotle or Spinoza. It is expressed verbally as a set of sentences with a variety of functions, descriptive and prescriptive. The basic relation is that between subsets of premises and subsets of conclusions, that is, the relation of derivability. The relevant notions of derivability may be classed according to rigor, with logical and mathematical deductions topping the list, but also according to how much is implicitly taken for granted. An exposition of an ecosophy must necessarily be only moderately precise considering the vast scope of relevant ecological and normative (social, political, ethical) material. At the moment, ecosophy might profitably use models of systems, rough approximations of global systematizations. It is the global character, not preciseness in detail, which distinguishes an ecosophy. It articulates and integrates the efforts of an ideal ecological team, a team comprising not only scientists from an extreme variety of disciplines, but also students of politics and active policy makers.

Under the name of *ecologism*, various deviations from the deep movement have been championed—primarily with a one-sided stress on pollution and resource depletion, but also with a neglect of the great differences between under- and overdeveloped countries in favor of a vague global approach. The global approach is essential, but regional differences must largely determine policies in the coming years.

12

THE DEEP ECOLOGY MOVEMENT

BILL DEVALL

There are two great streams of environmentalism in the latter half of the twentieth century. One stream is reformist, attempting to control some of the worst of the air and water pollution and inefficient land-use practices in industrialized nations and to save a few of the remaining pieces of wildlands as "designated wilderness areas." The other stream supports many of the reformist goals but is revolutionary, seeking a new metaphysics, epistemology, cosmology, and environmental ethics of person/planet. This chapter is an intellectual archeology of the second of these streams of environmentalism, which I will call "Deep Ecology."

There are several other phrases that some writers are using for the perspective I am describing in this paper. Some call it "eco-philosophy" or "foundational ecology" or the "new natural philosophy." I use "Deep Ecology" as the shortest label. Although I am convinced that Deep Ecology is radically different from the perspective of the dominant social paradigm, I do not use the phrase "radical ecology" or "revolutionary ecology" because I think those labels have such a burden of emotive associations that many people would not hear what is being said about Deep Ecology because of their projection of other meanings of "revolution" onto the perspective of Deep Ecology.

I contend that both streams of environmentalism are reactions to the successes and excesses of the implementation of the dominant social paradigm. Although reformist environmentalism treats some of the symptoms of the environmental crisis and challenges some of the assumptions of the dominant social paradigm (such as growth of the economy at any cost), Deep Ecology questions the fundamental premises of the dominant social paradigm. In the future, as the limits of

From: *Natural Resources Journal* 20 (April 1980): 299–313.

reform are reached and environmental problems become more serious, the reform environmental movement will have to come to terms with Deep Ecology.

The analysis in the present paper was inspired by Arne Naess's paper on "shallow and deep, long-range" environmetalism.[1] The methods used are patterned after John Rodman's seminal critique of the resources conservation and development movement in the United States.[2] The data are the writings of a diverse group of thinkers who have been developing a theory of Deep Ecology, especially during the last quarter of a century. Relatively few of these writings have appeared in popular journals or in books published by mainstream publishers. I have searched these writings for common threads or themes much as Max Weber searched the sermons of Protestant ministers for themes which reflected from and back to the intellectual and social crisis of the emerging Protestant ethic and the spirit of capitalism.[3] Several questions are addressed in this paper: What are the sources of Deep Ecology? How do the premises of Deep Ecology differ from those of the dominant social paradigm? What are the areas of disagreement between reformist environmentalism and Deep Ecology? What is the likely future role of the Deep Ecology movement?

THE DOMINANT PARADIGM

A paradigm is a shorthand description of the worldview, the collection of values, beliefs, habits, and norms which form the frame of reference of a collectivity of people—those who share a nation, a religion, a social class. According to one writer, a *dominant* social paradigm is the mental image of social reality that guides expectations in a society.

The dominant paradigm in North America includes the belief that "economic growth," as measured by the gross national product, is a measure of progress, the belief that the primary goal of the governments of nation-states, after national defense, should be to create conditions that will increase production of commodities and satisfy material wants of citizens, and the belief that "technology can solve our problems." Nature, in this paradigm, is only a storehouse of resources which should be "developed" to satisfy ever-increasing numbers of humans and ever-increasing demands of humans. Science is wedded to technology, the development of *techniques* for control of natural processes (such as weather modification). Change ("planned obsolescence") is an end in itself. The new is valued over the old and the present over future generations. The goal of persons is personal satisfaction of wants and a higher standard of living as measured by possession of commodities (houses, autos, recreation vehicles, etc.).[4] Whatever its origin, this paradigm continues to be dominant, to be preached through publicity (i.e., advertising), and to be part of the worldview of most citizens in North America.[5]

For some writers, the dominant social paradigm derives from Judeo-

Christian origins.[6] For others, the excesses of air and water pollution, the demand for more and more centralization of political and economic power and the disregard for future generations, and the unwise use of natural resources derive from the ideology and structure of capitalism or from the Lockean view that property must be "improved" to make it valuable to the "owner" and to society.[7] For others, the dominant social paradigm derives from the "scientism" of the modern West (Europe and North America) as applied to the technique of domination.[8]

Following Thomas Kuhn's theory of the dominance of paradigms in modern science and the operation of scientists doing what he calls normal science within a paradigm, it can be argued that (1) those who subscribe to a given paradigm share a definition of what problems are and their priorities; (2) the general heuristics, or rules of the game, for approaching problems is widely agreed upon; (3) there is a definitive, underlying confidence among believers of the paradigm that solutions within the paradigm do exist; and (4) those who believe the assumptions of the paradigm may argue about the validity of data, but rarely are their debates about the definition of what the problem is or whether there are solutions or not. Proposed solutions to problems arising from following the assumptions of the paradigms are evaluated as "reasonable," "realistic," or "valid" in terms of the agreed upon "rules of the game." When the data is difficult to fit to the paradigm, frequently there is dissonance disavowal, an attempt to explain away the inconsistency.[9]

It is possible for a paradigm shift to occur when a group of persons finds in comparing its data with generally accepted theory that the conclusions become "weird" when compared with expectations. In terms of the shared views of the goals, rules, and perceptions of reality in a nation, a tribe, or a religious group, for example, a charismatic leader, a social movement, or a formation of social networks of persons exploring a new social paradigm may be at the vanguard of a paradigm shift.

Reformist environmentalism in this paper refers to several social movements which are related in that the goal of all of them is to change society for "better living" without attacking the premises of the dominant social paradigm. These reform movements each defined a problem—such as need for more open space—and voluntary organizations were formed to agitate for social changes. There has also been considerable coalition building between different voluntary organizations espousing reform environmentalism. Several reformist environmental movements, including at least the following, have been active during the last century: (1) the movement to establish urban parks, designated wilderness areas, and national parks;[10] (2) the movement to mitigate the health and public safety hazards created by the technology which was applied to create the so-called Industrial Revolution.[11] The Union of Concerned Scientists, for example, has brought to the attention of the general public some of the hazards to public health and safety of the use of nuclear power to generate electricity; (3) the movement to develop "proper" land-use planning. This includes the city beautiful movement of the late nineteenth century and the movement to zone and plan land use such as the currently controversial attempts to zone uses along the coastal zones;[12] (4) the resources conservation and development movement symbolized by the philosophy of multiple use of Gifford Pinchot and the US Forest Service;[13] (5) the "back to the land" movement of the 1960s and 1970s and the "organic farming" ideology; (6) the concern with exponential growth of human population and formation of such groups as Zero Population Growth;[14] (7) the "humane" and "animal liberation" movement directed at changing the attitudes and behavior of humans toward some other aspects of animals;[15] and (8) the "limits to growth" movement, which emphasizes we should control human population and move toward a "steady-state" or "conserver society" as rapidly as possible.[16]

SOURCES OF DEEP ECOLOGY

What I call Deep Ecology in this paper is premised on a gestalt of person-in-nature. The person is not above or outside of nature. The person is part of creation ongoing. The person cares for and about nature, shows reverence toward and respect for nonhuman nature, loves and lives with nonhuman nature, is a person in the "earth household" and "lets being be," lets nonhuman nature follow separate evolutionary destinies. Deep Ecology, unlike reform environmentalism, is not just a pragmatic, short-term social movement with a goal like stopping nuclear power or cleaning up the waterways. Deep Ecology first attempts to question and present alternatives to conventional ways of thinking in the modern West. Deep Ecology understands that some of the "solutions" of reform environmentalism are counterproductive. Deep Ecology seeks transformation of values and social organization.

The historian Lynn White Jr., in his influential 1967 article, "The Historical Roots of Our Ecologic Crisis," provided one impetus for the current upwelling of interest in Deep Ecology by criticizing what he saw as the dominant Judeo-Christian view of man versus nature, or man at war with nature. But there are other writers, coming from diverse intellectual and spiritual disciplines, who have provided, in the cumulative impact of their work, a profound critique of the dominant social paradigm and the "single vision" of science in the modern (post-1500) West.[17]

One major stream of thought influencing the development of Deep Ecology has been the influx of Eastern spiritual traditions into the West, which began in the 1950s with the writings of such people as Alan Watts[18] and Daisetz Suzuki.[19] Eastern traditions provided a radically different man/nature vision than that of the dominant social paradigm of the West. During the 1950s the so-called beat poets such as Alan Ginsberg seemed to be groping for a way through Eastern philosophy to cope with the violence, insanity, and alienation of people from people and people from nature they experienced in North America. Except for Gary

Snyder, who developed into one of the most influential ecophilosophers of the 1970s, these beat poets, from the perspective of the 1970s, were naive in their understanding of Eastern philosophy, ecology, and the philosophical traditions of the West.

During the late 1960s and 1970s, however, philosophers, scientists, and social critics have begun to compare Eastern and Western philosophic traditions as they relate to science, technology, and man/nature relations. Fritjof Capra's *Tao of Physics*, for example, emphasizes the parallels between Eastern philosophies and the theories of twentieth-century physics.[20] Joseph Needham's massive work, *Science and Civilization in China*, brought to the consciousness of the West the incredibly high level of science, technology, and civilization achieved in the East for millennia and made available to Western readers an alternative approach to science and human values.[21] More recently, Needham has suggested that modern Westerners take the philosophies of the East as a spiritual and ethical basis for modern science.[22] Works by Huston Smith, among others, have also contributed to this resurgent interest in relating the environmental crisis to the values expressed in the dominant Western paradigm. Smith and others have looked to the Eastern philosophies for spiritual-religious guidance.[23]

Several social philosophers have written brilliant critiques of Western societies but have not presented a new metaphysical basis for their philosophy nor attempted to incorporate Eastern philosophy into their analyses. Jacques Ellul wrote on *technique* and the technological society.[24] Paul Goodman discussed the question "can there be a humane technology?"[25] Herbert Marcuse analyzed "one-dimensional man" as the prototypical "modern" urbanite.[26] The works of Theodore Roszak have also had considerable impact on those thinkers interested in understanding the malaise and contradictions of modern societies by examining the premises of the dominant social paradigm.[27]

A second stream of thought contributing to Deep Ecology has been the reevaluation of Native Americans (and other preliterate peoples) during the 1960s and 1970s. This is not a revival of the romantic view of Native Americans as "noble savages" but rather an attempt to evaluate traditional religions, philosophies, and social organizations of Native Americans in objective, comparative, analytic, and critical ways.

A number of questions have been asked. How did different tribes at different times cope with changes in their natural environment (such as prolonged drought) and with technological innovation? What were the "separate realities" of Native Americans and can modern Western man understand and know, in a phenomenological sense, these "separate realities"? The experiences of Carlos Castaneda, for example, indicate it may be very difficult for modern man to develop such understanding since this requires a major perceptual shift of man/nature. Robert Ornstein concludes, "Castaneda's experience demonstrates primarily that the Western-trained intellectual, even a 'seeker,' is by his culture almost completely unprepared to understand esoteric traditions."[28]

From the many sources on Native Americans which have become available during the 1970s, I quote a statement by Luther Standing Bear, an Oglala Sioux, from *Touch the Earth* to illustrate the contrast with the modern paradigm of the West:

> We did not think of the great open plains, the beautiful rolling hills, and winding streams with tangled growth, as "wild." Only to the white man was nature a "wilderness" and only to him was the land "infested" with "wild" animals and "savage" people. To us it was tame. Earth was bountiful and we were sur-rounded with the blessings of the Great Mystery. Not until the hairy man from the east came and with brutal frenzy heaped injustices upon us and the families we loved was it "wild" for us. When the very animals of the forest began fleeing from his approach, then it was that for us the "wild west" began.[29]

A third source of Deep Ecology is found in the "minority tradition" of Western religious and philosophical traditions. The philosopher George Sessions has claimed that

> in the civilized West, a tenuous thread can be drawn through the Presocratics, Theophrastus, Lucretius, St. Francis, Bruno and other neo-Platonic mystics, Spinoza, Thoreau, John Muir, Santayana, Robinson Jeffers, Aldo Leopold, Loren Eiseley, Gary Snyder, Paul Shepard, Arne Naess, and maybe that desert rat, Edward Abbey. This minority tradition, despite differences, could have pro-vided the West with a healthy basis for a realistic portrayal of the balance and interconnectedness of three artificially separable components (God/Nature/Man) of an untimely seamless and inseparable Whole.[30]

Sessions, together with Arne Naess and Stuart Hampshire, has seen the philoso-pher Spinoza as providing a unique fusion of an integrated man/nature metaphysic with modern European science.[31] Spinoza's ethics is most naturally interpreted as implying biospheric egalitarianism, and science is endorsed by Spinoza as valu-able primarily for contemplation of a pantheistic, sacred universe and for spiritual discipline and development. Spinoza stands out in a unique way in opposition to other seventeenth-century philosophers—for example, Bacon, Descartes, and Leibniz—who were at that time laying the foundations for the technocratic-industrial social paradigm and the fulfillment of the Christian imperative that man *must* dominate and control all nature. It has been claimed by several writers that the poet-philosopher Robinson Jeffers, who lived most of his life on the California coastline at Big Sur, was Spinoza's twentieth-century "evangelist" and that Jeffers gave Spinoza's philosophy an explicitly ecological interpretation.[32]

Among contemporary European philosophers, the two most influential have been Alfred North Whitehead and Martin Heidegger.[33] In particular, more Amer-ican philosophers, both those with an interest in ecological consciousness and those interested in contemporary philosophers, are discussing Heidegger's critique of Western philosophy and contemporary Western societies. Because Heidegger's approach to philosophy and language is so different from the language we are accustomed to in American academia, any summary of his ideas would distort the theory he is presenting. The reader is referred to the books and articles on Heidegger cited below.[34]

A fourth source of reference for the Deep Ecology movement has been the scientific discipline of *ecology*. For some ecology is a science of the "home," of the "relationships between," while for others ecology is a perspective. The difference is important, for ecology as a science is open for co-optation by the engineers, the "technological fixers" who want to "enhance," "manage," or "humanize" the biosphere. At the beginning of the "environmental decade" of the 1970s, two ecologists issued a warning against this approach:

> Even if we dispense with the idea that ecologists are some sort of environmental engineers and compare them to the pure physicists who provide scientific rules for engineers, do the tentative understandings we have outlined (in their article) provide a sound basis for action by those who would manage the environment? It is self-evident that they do not. We submit that ecology as such probably cannot do what many people expect it to do; it cannot provide a set of "rules" of the kind needed to manage the environment.[35]

Donald Worster, at the conclusion of his scholarly and brilliant history of ecological thinking in the West, is of the same opinion.[36]

But ecologists do have an important task in the Deep Ecology movement. They can be *subversive* in their perspective. For human ecologist Paul Shepard, "the ideological status of ecology is that of a resistance movement" because its intellectual leaders such as Aldo Leopold challenge the major premises of the dominant social paradigm.[37] As Worster in his history of ecology points out:

> All science, though primarily concerned with the "Is," becomes implicated at some point in the "Ought." Conversely, spinning out moral visions without reference to the material world may ultimately be an empty enterprise; when . . . moral values depart too far from nature's ways. . . . Scientist and moralist might together explore a potential union of their concerns; might seek a set of empirical facts with ethical meaning, a set of moral truths. The ecological ethic of interdependence may be the outcome of just such a dialectical relation.[38]

A final source of inspiration for the deep, long-range ecology movement is those artists who have tried to maintain a sense of place in their work.[39] Some artists, standing against the tide of midcentury pop art, minimalist art, and conceptual art have shown remarkable clarity and objectivity in their perception of nature. This spiritual-mystical objectivism is found, for example, in the photographs of Ansel Adams.[40] For these artists, including Morris Graves, who introduced concepts of Eastern thought (including Zen Buddhism) into his art, and Larry Gray, who reveals the eloquent light of revelation of nature in his skyscapes, men reaffirm their spiritual kinship with the eternity of God in nature through art.[41]

THEMES OF DEEP ECOLOGY

I indicated in preceding pages that many thinkers are questioning some of the premises of the dominant social paradigm of the modern societies. They are attempting to extend on an appropriate metaphysics, epistemology, and ethics for what I call an "ecological consciousness." Some of these writers are very supportive of reformist environmental social movements, but they feel reform, while necessary, is not sufficient. They suggest a new paradigm is required and a new utopian vision of "right livelihood" and the "good society." Utopia stimulates our thinking concerning alternatives to present society.[42] Some persons, such as Aldo Leopold, have suggested that we begin our thinking on utopia not with a statement of "human nature" or "needs of humans" but by trying to "think like a mountain." This profound extending, "thinking like a mountain," is part and parcel of the phenomenology of ecological consciousness.[43] Deep Ecology begins with unity rather than dualism, which has been the dominant theme of Western philosophy.[44]

Philosopher Henryk Skolimowski, who has written several papers on the options for the ecology movements, asserts:

> We are in a period of ferment and turmoil, in which we have to challenge the limits of the analytical and empiricist comprehension of the world as we must work out a new conceptual and philosophical framework in which the multitude of new social, ethical, ecological, epistemological, and ontological problems can be accommodated and fruitfully tackled. The need for a new philosophical framework is felt by nearly everybody. It would be lamentable if professional philosophers were among the last to recognize this.[45]

Numerous other writers on Deep Ecology, including William Ophuls, E. F. Schumacher, George Sessions, Theodore Roszak, Paul Shepard, Gary Snyder, and Arne Naess, have in one way or another called for a new social paradigm or a new environmental ethic. We must "think like a mountain," according to Aldo Leopold. And Roberick Nash says:

> Do rocks have rights? If the times comes when to any considerable group of us such a question is no longer ridiculous, we may be on the verge of a change of value structures that will make possible measures to cope with the growing ecologic crisis. One hopes there is enough time left.[46]

Any attempt to create artificially a "new ecological ethics" or a "new ontology of man's place in nature" out of the diverse strands of thought which make up the

Deep Ecology movement is likely to be forced and futile. However, by explicating some of the major themes embodied in and presupposed by the intellectual movement I am calling Deep Ecology, some groundwork can be laid for further discussion and clarification.[47] Following the general outline of perennial philosophy, the order of the following statements summarizing Deep Ecology's basic principles are metaphysical-religious, psychological-epistemological, ethical, and social-economic-political. These concerns of Deep Ecology encompass most of reformist environmentalism's concerns but subsume them in its fundamental critique of the dominant paradigm.

According to Deep Ecology:

1. A new cosmic/ecological metaphysics which stresses the identity (I/thou) of humans with nonhuman nature is a necessary condition for a viable approach to building an ecophilosophy. In Deep Ecology, the wholeness and integrity of person/planet together with the principle of what Arne Naess calls "biological equalitarianism" are the most important ideas. Man is an integral part of nature, not over or apart from nature. Man is a "plain citizen" of the biosphere, not its conqueror or manager. There should be a "democracy of all God's creatures" according to St. Francis; or as Spinoza said, man is a "temporary and dependent mode of the whole of God/Nature." Man flows with the system of nature rather than attempting to control all of the rest of nature. The hand of man lies lightly on the land. Man does not perfect nature, nor is man's primary duty to make nature more efficient.[48]

2. An objective approach to nature is required. This approach is found, for example, in Spinoza and in the works of Spinoza's disciple, Robinson Jeffers. Jeffers describes his orientation as a philosophy of "inhumanism" to draw a sharp and shocking contrast with the subjective anthropocentrism of the prevailing humanistic philosophy, art, and culture of the twentieth-century West.[49]

3. A new psychology is needed to integrate the metaphysics in the mind field of postindustrial society. A major paradigm shift results from psychological changes of perception. The new paradigm requires rejection of subject/object, man/nature dualisms and will require a pervasive awareness of total intermingling of the planet Earth. Psychotherapy seen as adjustment to ego-oriented society is replaced by a new ideal of psychotherapy as spiritual development.[50] The new metaphysics and psychology leads logically to posture of biospheric egalitarianism and liberation in the sense of autonomy psychological/emotional freedom of the individual, spiritual development for *Homo sapiens*, and the right of other species to pursue their own evolutionary destinies.[51]

4. There is an objective basis for environmentalism, but objective science in the new paradigm is different from the narrow, analytic conception of the "scientific method" currently popular. Based on "ancient wisdom," science should be both objective and participatory without modern science's subject/object dualism. The main *value* of science is seen in its ancient perspective as contemplation of the cosmos and the enhancement of understanding of self and creation.[52]

5. There is wisdom in the stability of natural processes unchanged by human intervention. Massive human-induced disruptions of ecosystems will be unethical and harmful to men. Design for human settlement should be with nature, not against nature.[53]

6. The quality and human existence and human welfare should not be measured only by quantity of products. Technology is returned to its ancient place as an appropriate tool for human welfare, not an end in itself.[54]

7. Optimal human carrying capacity should be determined for the planet as a biosphere and for specific islands, valleys, and continents. A drastic reduction of the rate of growth of population of *Homo sapiens* through humane birth-control programs is required.[55]

8. Treating the symptoms of man/nature conflict, such as air or water pollution, may divert attention from more important issues and thus be counterproductive to "solving" the problems. Economics must be subordinate to ecological-ethical criteria. Economics is to be treated as a small subbranch of ecology and will assume a rightfully minor role in the new paradigm.[56]

9. A new philosophical anthropology will draw on data of hunting/gathering societies for principles of healthy, ecologically viable societies. Industrial society is not the end toward which all societies should aim or try to aim.[57] Therefore, the notion of "reinhabiting the land" with hunting-gathering, and gardening as a goal and standard for postindustrial society should be seriously considered.[58]

10. Diversity is inherently desirable both culturally and as a principle of health and stability of ecosystems.[59]

11. There should be a rapid movement toward "soft" energy paths and "appropriate technology" and toward life-styles which will result in a drastic decrease in per capita energy consumption in advanced industrial societies while increasing appropriate energy in decentralized villages in so-called third world nations.[60] Deep ecologists are committed to rapid movement to a "steady-state" or "conservor society" both from ethical principles of harmonious integration of humans with nature and from appreciation of ecological realities.[61] Integration of sophisticated, elegant, unobtrusive, ecologically sound, appropriate technology with greatly scaled down, diversified, organic, labor-intensive agriculture, hunting, and gathering is another goal.[62]

12. Education should have as its goal encouraging the spiritual development and personhood development of the members of a community, not just training them in occupations appropriate for oligarchic bureaucracies and for consumerism in advanced industrial societies.[63]

13. More leisure as contemplation in art, dance, music, and physical skills will return play to its place as the nursery of individual fulfillment and cultural achievement.[64]

14. Local autonomy and decentralization of power is preferred over centralized political control through oligarchic bureaucracies. Even if bureaucratic modes of organization are more "efficient," other modes of organization for

small-scale human communities are more "effective" in terms of the principles of Deep Ecology.[65]

15. In the interim, before the steady-state economy and radically changed social structure are instituted, vast areas of the planet biospheres will be zoned "off limits" to further industrial exploitation and large-scale human settlement; these should be protected by defensive groups of people. . . .

Deep Ecology is liberating ecological consciousness. The writers I have cited in this paper provide radical critiques of modern society and of the dominant values of this society. They also provide, or some of them do, a profound utopian alternative. The elaborating of or deepening of ecological consciousness is a continuing process. The goal is to have action and consciousness as one. But the development of ecological consciousness is seen as prior to ecological resistance in many of the writings cited. This ecological consciousness may not be very well articulated except by intellectuals who are in the business of verbalizing. But they, as much as anyone, realize the limitation of just verbalizing. Consciousness is *knowing*. From the perspective of Deep Ecology, ecological resistance will naturally flow from and with a developing ecological consciousness.

NOTES

1. Arne Naess, "The Shallow and the Deep, Long-Range Ecology Movement," *Inquiry* 16 (1973): 95–100.

2. John Rodman, "Four Forms of Ecological Consciousness: Beyond Economics, Resources, and Conservation" (1977) Pitzer College. Revised version published as "Four Forms of Ecological Consciousness Reconsidered," in *Ethics and the Environment*, ed. Donald Scherer and Thomas Attig (Englewood Cliffs, NJ: Prentice-Hall, 1983).

3. Max Weber, *The Protestant Ethic and the Spirit of Capitalism* (New York: Scribner's, 1930).

4. Dennis C. Pirages and Paul Ehrlich, *Ark II: Social Response to Environmental Imperatives* (New York: Viking, 1974), p. 43. See also *The Future of the Great Plains*, H.R. Doc. 144, 75th Cong., 1st sess. (1937).

5. On the history of the paradigm, see Victor Ferkiss, *The Future of Technological Civilization* (New York: George Braziller, 1974). For a critique of the "me-now" consumerism of the 1970s, see Christopher Lasch, *The Culture of Narcissism: American Life in an Age of Diminishing Expectations* (New York: Warner Books, 1979). See also Manager's Journal, "Monitoring America, Values of Americans," *Wall Street Journal*, October 2, 1978.

6. Lynn White Jr., "The Historical Roots of Our Ecologic Crisis," *Science* 155 (1967): 1203.

7. Barry Weisberg, *Beyond Repair: The Ecology of Capitalism* (Boston: Beacon, 1971); England and Bluestone, "Ecology and Class Conflict," *Review of Radical Political Economics* 3 (1971): 21. On Locke's view of "property," see Ferkiss, *Future of Technological Civilization*.

8. Leo Marx, *The Machine in the Garden: Technology and the Pastoral Ideal in America* (New York: Oxford, 1967); Lewis Mumford, *The Myth of the Machine*, vol. 2, *The Pentagon of Power* (New York: Harcourt Brace, 1970).

9. Thomas Kuhn, *The Structure of Scientific Revolutions*, 2nd ed. (Chicago: University of Chicago Press, 1970). For criticism of Kuhn, see Imre Lakatos and Alan Musgrave, eds., *Criticism and the Growth of Knowledge* (Cambridge: Cambridge University Press, 1970).

10. Roderick Nash, *Wilderness and the American Mind*, 2nd ed. (New Haven, CT: Yale University Press, 1973); Joseph Sax, "America's National Parks: Their Principles, Purposes and their Prospects," *Natural History* 35 (1976): 57.

11. Barry Commoner, *The Closing Circle* (New York: Knopf, 1971); James Ridgeway, *The Politics of Ecology* (New York: Dutton, 1970).

12. Natural Resources Defense Council, *Land Use Controls in the United States: A Handbook of Legal Rights of Citizens* (New York: Dial Press, 1977); Ian McHarg, *Design with Nature* (Garden City, NY: Doubleday, 1971).

13. Rodman, "Four Forms of Ecological Consciousness"; Samuel Hays, *Conservation and the Gospel of Efficiency* (Cambridge, MA: Harvard University Press, 1959); Gifford Pinchot, *Breaking New Ground* (New York: Harcourt Brace, 1947).

14. Paul Ehrlich, *The Population Bomb* (New York: Ballantine, 1968). See, generally, publications of the organization Zero Population Growth. Also, United States Commission on Population and the American Future, *The Report of the Commission on Population Growth and the American Future* (Washington, DC: Government Printing Office, 1972).

15. Tom Regan and Paul Singer, *Animal Rights and Human Obligations* (Englewood Cliffs, NJ: Prentice-Hall, 1976); Peter Singer, *Animal Liberation* (New York: Random House, 1977).

16. Donella H. Meadows, Dennis L. Meadows, Jorgen Randers, and William W. Behrens III, *The Limits to Growth*, 2nd ed. (New York: Universe Books, 1974); Mihajlo Mesarovic and Edward Pestel, *Mankind at the Turning Point: Second Report of the Club of Rome* (New York: Dutton, 1974); Donella H. Meadows, *Alternatives to Growth* (Cambridge, MA: Ballinger, 1977). For a critique of the limits to growth model, see H. S. D. Cole, Christopher Freeman, Marie Johoda, and K. L. R. Pavitt, eds., *Models of Doom: A Critique of the Limits to Growth* (New York: Universe, 1973); Herman Daly, ed., *Toward a Steady-State Economy* (San Francisco: W. H. Freeman, 1973).

17. White, "Historical Roots."

18. Alan Watts, *Psychotherapy East and West* (New York: Pantheon, 1975); Alan Watts, *Nature, Man and Woman* (New York: Pantheon, 1970); Alan Watts, *The Spirit of Zen: A Way of Life, Work and Art in the Far East* (London: J. Murray, 1936); Alan Watts, *The Essence of Alan Watts* (Millbrae, CA: Celestial Arts, 1977).

19. Daisetz Suzuki, *Essays in Zen Buddhism* (New York: Grove, 1961).

20. Fritjof Capra, *The Tao of Physics* (Berkeley, CA: Shambala, 1975).

21. Joseph Needham, *Science and Civilization in China*, 6 vols. (New York: Cambridge University Press, 1954–1976).

22. Joseph Needham, "History and Human Values: A Chinese Perspective for World Science and Technology," *Centennial Review* 20 (1976): 1.

23. Houston Smith, *Forgotten Truth: The Perennial Tradition* (New York: Harper

and Row, 1976); H. Smith, "Tao Now: An Ecological Testament," in *Earth Might Be Fair*, ed. Ian Barbour (Englewood Cliffs, NJ: Prentice-Hall, 1972).

24. Jacques Ellul, *The Technological Society* (New York: Vintage, 1964).

25. Paul Goodman, "Can Technology Be Humane?" in *Western Man and Environmental Ethics*, ed. Ian Barbour, 225–42 (Reading, MA: Addison-Wesley, 1973).

26. Herbert Marcuse, *One-Dimensional Man* (Boston: Beacon Press, 1964).

27. Theodore Roszak, *The Making of a Counter-Culture* (Garden City, NY: Doubleday, 1969); T. Roszak, *Where the Wasteland Ends: Politics and Transcendence in Post-Industrial Society* (Garden City, NY: Doubleday, 1972); T. Roszak, *Unfinished Animal: The Aquarian Frontier and the Evolution of Consciousness* (New York: Harper and Row, 1975); T. Roszak, *Person/Planet* (Garden City, NY: Doubleday, 1978).

28. Robert Ornstein, *The Mind Field* (New York: Octagon Press, 1976), p. 105. The works of Carlos Castaneda have been influential. They include Castaneda, *The Teachings of Don Juan* (Berkeley and Los Angeles: University of California Press, 1968); *A Separate Reality* (New York: Simon and Schuster, 1971); *Journey to Ixtlan* (New York: Simon and Schuster, 1972); *Tales of Power* (New York: Simon and Schuster, 1974).

29. Luther Standing Bear in *Touch the Earth*, ed. T. C. McLuhan (New York: Simon and Schuster, 1971). Among the most significant and original theories of Native Americans and nonhuman nature are: Vine Deloria, *God Is Red* (New York: Grosset and Dunlap, 1975); Calvin Martin, *Keepers of the Game* (Berkeley and Los Angeles: University of California Press, 1978); Stan Steiner, *The Vanishing White Man* (New York: Harper and Row, 1976).

30. George Sessions, "Spinoza and Jeffers on Man in Nature," *Inquiry* 20 (1977): 481. See also George Sessions, *Spinoza, Perennial Philosophy and Deep Ecology* (unpublished MS, Sierra College, 1979).

31. Stuart Hampshire, *Two Theories of Morality* (Oxford: Oxford University Press, 1977); Stuart Hampshire, *Spinoza* (Harmondsworth, UK: Penguin Books, 1962); Arne Naess, "Spinoza and Ecology," *Philosophia* 7 (1977): 45.

32. Sessions, "Spinoza and Jeffers on Man in Nature." See also Arthur Coffin, *Robinson Jeffers: Poet of Inhumanism* (Madison: University of Wisconsin Press, 1971); Bill Hotchkiss, *Jeffers: The Siviastic Vision* (Auburn, CA: Blue Oak Press, 1975); R. Brophy, "Robinson Jeffers, Metaphysician of the West" (unpublished MS, Department of English, Long Beach State University, Long Beach, CA).

33. John Cobb Jr., *Is It Too Late? A Theology of Ecology* (1972); Charles Hartshorne, *Beyond Humanism: Essays in the Philosophy of Nature* (Chicago: Willett Clark, 1937); Alfred North Whitehead, *Science and the Modern World* (New York: Macmillan, 1925), chs. 5, 13; Griffin, "Whitehead's Contribution to the Theology of Nature," *Bucknell Review* 20 (1972): 95.

34. Martin Heidegger, *The Question concerning Technology and Other Essays*, trans. William Lovitt (New York: Harper and Row, 1977); George Steiner, *Martin Heidegger* (Hassocks, UK: Harvester Press, 1978); Vincent Vycinas, *Earth and Gods: An Introduction to the Philosophy of Martin Heidegger* (The Hague: M. Nijhoff, 1961). On the approach taken by Heidegger and the contemporary ecological consciousness, see Delores LaChapelle, *Earth Wisdom* (Los Angeles: Guild of Tutors Press, 1978), ch. 9. The writings of Michael Zimmerman on Heidegger are also useful, including his "Beyond 'Humanism': Heidegger's Understanding of Technology," *Listening* 12 (1977): 74. See

also M. Zimmerman, "Marx and Heidegger on the Technological Domination of Nature," *Philosophy Today* 12 (Summer 1979): 99.

35. William Murdoch and Joseph Connell, "All about Ecology," in *Western Man and Environmental Ethics*, ed. Ian Barbour, 156–70 (Reading, MA: Addison-Wesley, 1973).

36. Donald Worster, "Epilogue," in *Nature's Economy* (San Francisco: Sierra Club Books, 1977). The importance of the thinking of ecologist Aldo Leopold should be emphasized. There are many articles interpreting Leopold's message. See, for example, Hwa Yol and Petee Jung, "The Splendor of the Wild: Zen and Aldo Leopold," *Atlantic Naturalist* 5 (1974): 29.

37. Paul Shepard, "Introduction: Man and Ecology," in *The Subversive Science*, ed. Paul Shepard and Daniel McKinley, 1 (Boston: Houghton Mifflin, 1969). See also Neil Everndon, "Beyond Ecology," *North American Review* 263 (1978): 16–20.

38. Worster, *Nature's Economy*, 338.

39. Hans Huth, "Wilderness and Art," in *Wilderness, America's Living Heritage*, ed. David Brower, 60 (San Francisco: Sierra Club Books, 1961); Paul Shepard, "A Sense of Place," *North American Review* 262 (1977): 22.

40. Ansel Adams, "The Artist and the Ideals of Wilderness," in *Wilderness, America's Living Heritage*, ed. David Brower, 49 (San Francisco: Sierra Club Books, 1961).

41. Morris Graves, *The Drawings of Morris Graves*, ed. Ida E. Rubin (Boston: New York Graphic Society, 1974).

42. Mulford Quichert Sibley, *Nature and Civilization: Some Implications for Politics* (Ithaca, IL: F. E. Peacock, 1977), ch. 7, p. 251. Sibley makes a case for more utopian visions from contemporary intellectuals. Although many people have been revulsed by the visions of Marxism and fascist dictatorships, "The student of politics has an obligation not only to explain and criticize but also to propose and explicate ideals. We need more utopian visions, not fewer. For if politics be that activity through which man seeks consciously and deliberately to order and control his collective life, then one of the salient questions in all politics must be: Order and control for what ends? Without utopian visions these ends cannot be stated as wholes; and even a discussion of means and strategies will be clouded unless ends are at least relatively clear" (p. 47).

43. Hwa Yol and Petee Jung, "To Save the Earth," *Philosophy Today* 8 (1975): 108.

44. Sessions, "Spinoza and Jeffers on Man and Nature."

45. Henryk Skolimowski, "The Ecology Movement Re-examined," *Ecologist* 6 (1976): 298; Skolimowski, "Options for the Ecology Movement," *Ecologist* 7 (1977): 318.

46. Roderick Nash, "Do Rocks Have Rights?" *Center Magazine* 10 (1977): 2.

47. Skolimowski, "Ecology Movement Re-examined."

48. Jacob Needleman, *A Sense of the Cosmos* (Garden City, NY: Doubleday, 1975), pp. 76–77, 100–102.

49. Sessions, "Spinoza and Jeffers on Man and Nature." Spinoza is one of the important philosophers for Deep Ecology. The new translations of Spinoza's work are absolutely essential for understanding his thought. See Paul Weinpahl, *The Radical Spinoza* (New York: New York University Press, 1979).

50. Ornstein, *Mind Field*.

51. Sessions, "Spinoza and Jeffers on Man and Nature"; Gary Snyder, *The Old Ways* (San Francisco: City Lights Books, 1977).

52. Capra, *Tao of Physics*; Sessions, "Spinoza and Jeffers on Man and Nature"; Needleman, *Sense of the Cosmos*.

53. Commoner, *Closing Circle*; McHarg, *Design with Nature*.

54. Needleman, *Sense of the Cosmos*; Sessions, "Spinoza and Jeffers on Man and Nature."

55. E. F. Schumacher, *Small Is Beautiful: Economics as If People Mattered* (New York: Harper and Row, 1973). As an example of this argument, see Paul Ehrlich, Anne Ehrlich, and John Holdren, *Ecoscience* (San Francisco: W. H. Freeman, 1977).

56. William Ophuls, *Ecology and the Politics of Scarcity* (San Francisco: W. H. Freeman, 1977). Schumacher, *Small Is Beautiful*.

57. Steiner, *Vanishing White Man*; Snyder, *Old Ways*; Paul Shepard, *The Tender Carnivore and the Sacred Game* (New York: Scribner's, 1974).

58. Peter Berg and Raymond Dasmann, *Reinhabiting A Separate Country* (San Francisco: Planet Drum, 1978).

59. Raymond Dasmann, *A Different Kind of Country* (New York: John Wiley, 1968); Norman Myers, *The Sinking Ark: A New Look at the Problems of Disappearing Species* (New York: Pergamon, 1979).

60. Raymond Dasmann, *Ecological Principles for Economic Development* (New York: John Wiley, 1973).

61. Ophuls, *Politics of Scarcity*. On the "steady-state," see Daly, *Toward a Steady-State Economy*.

62. Schumacher, *Small Is Beautiful*.

63. Roszak, *Making of a Counter-Culture*.

64. Johan Huizinga, *Homo Ludens: A Study of the Play Element in Culture*, trans. R. F. C. Hull (London: Routledge, 1949). John Collier, "The Fullness of Life through Leisure," in *The Subversive Science*, ed. Paul Shepard and Daniel McKinley, 416–36 (Boston: Houghton Mifflin, 1969).

65. Ophuls, *Politics of Scarcity*; Dennis C. Pirages, ed., *The Sustainable Society* (New York: Praeger, 1977); Berg and Dasmann, *Reinhabiting a Separate Country*.

13

ECOCENTRISM AND THE ANTHROPOCENTRIC DETOUR

GEORGE SESSIONS

Warwick Fox points out that philosophical Deep Ecology has two main tasks: "A positive or constructive task of encouraging an egalitarian [or ecocentric] attitude on the part of humans towards all entities in the ecosphere, [and] a negative or critical task of dismantling anthropocentrism."[1] In hopes of helping contribute to these tasks, I offer a brief summary of the history of the twists and turns of ecocentrism and anthropocentrism in the West. This summary discusses only the main highlights; no attempt is made to be all inclusive.

I. ECOCENTRISM AND PRIMAL CULTURES

Although this issue has been hotly debated in the literature over the last twenty years, it seems accurate to say that the cultures of most primal (hunting/ gathering) societies throughout the world were permeated with nature-oriented religions that expressed the ecocentric perspective. These cosmologies, involving a sacred sense of the earth and all its inhabitants, helped order their lives and determine their values. For example, anthropologist Stan Steiner describes the traditional American Indian philosophy of the sacred "circle of life": "In the Circle of Life, every being is no more, or less, than any other. We are all sisters and brothers. Life is shared with the bird, bear, insects, plants, mountains, clouds, stars, sun."[2] Countless expressions of this kind can be cited from primal cultures all over the globe.[3] Given that the vast majority of humans who have lived on Earth over the millennia have been hunter/gatherers, it is clear

From: *ReVISION* 13, no. 3 (Winter 1991): 109–15.

that ecocentrism has been the dominant human religious/philosophical perspective through time.

With the beginning of agriculture, most ecocentric cultures (and religions) were gradually replaced or driven off into remote corners of the earth by pastoral and, eventually, "civilized" cultures (Latin: *civitas*, cities).[4] It seems likely that one of the functions of the Garden of Eden story, for instance, was to provide a moral justification for this process. Thus, the environmental crisis, in Paul Shepard's view, has been ten thousand years in the making: "As agriculture replaced hunting and gathering it was accompanied by radical changes in the way men saw and responded to their natural surroundings. . . . [Agriculturalists] all shared the aim of completely humanizing the earth's surface, replacing wild with domestic, and creating landscapes from habitat."[5] Whereas Taoism and certain other Eastern religions retained elements of the ancient shamanistic nature religions, the Western religious tradition radically distanced itself from wild nature and, in the process, became increasingly anthropocentric. Henri Frankfort claims, for example, that Judaism sacrificed "the greatest good ancient Near East religion could bestow—the harmonious integration of man's life with the life of nature—Man remained outside nature, exploiting it for a livelihood . . . using its imagery for the expression of his moods, but never sharing its mysterious life."[6] This overall analysis is also supported by Loren Eiseley, who points out that

> primitive man existed in close interdependence with his first world . . . he was still inside that world; he had not turned her into an instrument or a mere source of materials. Christian man in the West strove to escape this lingering illusion the primitives had projected upon nature. Intent upon the destiny of his own soul, and increasingly urban, man drew back from too great an intimacy with the natural. . . . If the new religion was to survive, Pan had to be driven from his hillside or rendered powerless by incorporating him into Christianity.[7]

Similar conclusions were arrived at by D. H. Lawrence in his extraordinary 1924 essay, "Pan in America": "Gradually men moved into cities. And they loved the display of people better than the display of a tree. They liked the glory they got out of overpowering one another in war. And, above all, they loved the vainglory of their own words, the pomp of argument and the vanity of ideas. . . . Til at last the old Pan died and was turned into the devil of the Christians."[8]

The intellectual Greek strand in Western culture also exhibits a similar development from early ecocentric nature religions, and the nature-oriented cosmological speculations of the pre-Socratics, to the anthropocentrism of the classical Athenian philosophers. The Milesian cosmologists, according to Karl Popper, "envisaged the world as a kind of a house, the home of all creatures—our home."[9] Beginning with Socrates, however, philosophical speculation was characterized by "an undue emphasis upon man as compared with the universe," as Bertrand Russell and certain other historians of Western philosophy have observed.

With the culmination of Athenian philosophy in Aristotle, an anthropocentric system of philosophy and science was set in place that was to play a major role in shaping Western thought until the seventeenth century. Aristotle rejected the pre-Socratic ideas of an infinite universe, cosmological and biological evolution, and heliocentrism, and proposed instead an Earth-centered finite universe wherein humans were differentiated from, and seen as superior to, animals and plants by virtue of their rationality. Although Aristotle's philosophy was biologically inspired, nevertheless he arrived at a hierarchical concept of the "great chain of being" in which nature *made* plants for the use of animals, and animals were *made* for the sake of humans (*Politics* 1).

In the great medieval Christian synthesis of Saint Thomas Aquinas, there were problems reconciling Aristotle's naturalism with Christian otherworldliness (including the idea of an immaterial soul), but Aristotle's anthropocentric cosmology turned out to be quite compatible with Judeo-Christian anthropocentrism. In the Christian version of the "great chain of being," the hierarchical ladder led from a transcendent God, angels, men, women, and children, down to animals, plants, and the inanimate realm.[10] By way of summarizing this medieval culmination and synthesis of Greek and Christian thought, philosopher Kurt Baier remarked: "The medieval Christian world picture assigned to man [humans] a highly significant, indeed the central part in the grand scheme of things. The universe was made for the express purpose of providing a stage on which to enact a drama starring Man in the title role."[11] The West has had several decisive historical opportunities to leave the path of the narrow and ecologically destructive "anthropocentric detour" and return to ecocentrism, but the dominant culture has not done so. In his classic paper of the 1960s that linked anthropocentrism with the environmental crisis, historian Lynn White Jr. claimed that, in the thirteenth century, Saint Francis of Assisi tried to undermine the Christian views of human dominance over, and separation from, the rest of nature. According to White: "Francis tried to depose man from his monarchy over creation and set up a democracy of all God's creatures. . . . The greatest spiritual revolutionary in Western history, Saint Francis, proposed what he thought was an alternative Christian view of nature and man's relation to it: he tried to substitute the idea of the equality of all creatures, including man, for the idea of man's limitless rule of creation. He failed."[12]

II. THE RISE OF THE ANTHROPOCENTRIC MODERN WORLD

The second opportunity to abandon the "anthropocentric detour" came in the seventeenth century with the development of the nonanthropocentric philosophical system of the Dutch philosopher Baruch Spinoza. But even as the West underwent the major intellectual/social paradigm shift from the medieval to the modern world in the seventeenth and eighteenth centuries, the anthropocentrism of the Greek and Judeo-Christian traditions continued to dominate the major theorists of this period. For example, the leading philosophical spokesmen for the scientific revolution (Francis Bacon, René Descartes, and Leibniz) were all strongly influenced by Christian anthropocentric theology. Bacon claimed that modern science would allow humans to regain a command over nature, which had been lost with Adam's fall in the garden. Descartes, considered to be the "father" of modern Western philosophy, argued that the new science would make humans the "masters and possessors of nature." Also in keeping with his Christian background, Descartes' famous "mind-body dualism" resulted in the view that only humans had minds (or souls); all other creatures were merely bodies (machines); they had no sentience (mental life) and, as a result, could feel no pain. And so, in the middle of the nineteenth century, Darwin had to argue, against prevailing opinion, that at least the great apes experienced various feelings and emotions! Descartes set Western academic philosophy on a subjectivist, "inside-out" epistemological path that, among other things, made the reality of the world, apart from human consciousness, derivative and thus highly suspect. It was certainly not possible, in this view, to recognize that nature itself had intrinsic value or was sacred.[13]

These Christian anthropocentric attitudes combined with, and reinforced, Renaissance anthropocentric humanism, which arose prior to the scientific revolution (from classical Greek and Roman sources), in, for example, the fifteenth-century pronouncements of Pico della Mirandola and continued with the Enlightenment philosophers into the twentieth century with Karl Marx, John Dewey, and the humanistic existentialism of Jean-Paul Sartre. Like medieval Christianity, Renaissance humanism portrayed humans as the central fact in the universe while also supporting the exalted view that humans had unlimited powers, potential, and freedom (what ecophilosopher Peter Gunter refers to as "man-infinite").[14]

Modern science, however, turned out to be a two-edged sword. As we have seen, seventeenth-century science, on the one hand, was conceived within a Christian matrix with the avowed anthropocentric purpose of conquering and dominating nature. This led Lynn White to claim that both modern science and the more recent scientifically based technology "is permeated with Christian arrogance toward nature."[15]

On the other hand, the development of modern theoretical science over the last three hundred years has resulted in the replacement of the Aristotelian anthropocentric cosmology with essentially the original nonanthropocentric cosmological worldview of the pre-Socratics, first in astronomy with heliocentrism, the infinity of the universe, and cosmic evolution, then in biology with Darwinian evolution. Ecology, as the "subversive science," has, even more than Darwinian biology, stepped across the anthropocentric threshold, so to speak, and implied an ecocentric orientation to the world. Thus, as the modern West has tried to cling to its anthropocentric illusions, each major theoretical scientific development since the seventeenth century has served to "decentralize" humans from their preeminent place in the Aristotelian-Christian cosmology. Lynn White also recognized this aspect of the development of modern science when he pointed out: "Despite Copernicus, all the cosmos rotates around our little globe. Despite Darwin, we are not, in our hearts, part of the natural process. We are superior to nature, contemptuous of it, willing to use it for our slightest whim."[16]

Carrying this a step further, the physicist Fritjof Capra claims that twentieth-century developments in Einsteinian relativity and quantum physics have undermined the atomistic "discrete entity" mechanistic paradigm of seventeenth-century science and now point instead to a process metaphysics of interrelatedness more typical of Eastern cosmologies such as Hinduism. Capra finds the parallels between the "new physics," Eastern cosmologies, and the ecological sense of interrelatedness so striking that he has said: "I think what physics can do is help generate ecological awareness."[17] Thomas Berry, and the physicist Brian Swimme, claim that the unfolding of twentieth-century science actually provides us with a new, nonanthropocentric "cosmic story" within which humans can now metaphysically find their place and values in an evolutionary/ecological context.[18]

From the vantage point of the late twentieth century and the deepening environmental crisis, it is now possible to see that the transition from the medieval to the modern world was not as radical as many Enlightenment-inspired historians have thought. The key was not so much a shift to reason, proper empirical scientific method, or the theories of the new sciences (which in many cases were a rediscovery of the theories of the pre-Socratics). Rather, the transition amounted to an increasingly unfettered playing out of the full implications of a radical anthropocentrism with roots deep in ancient Western culture: the discovery of new worlds to plunder by a resource-depleted, ecologically damaged Europe; the development of a new science, science-driven technology, and industrialization capable of inflicting great damage on the earth; the exponential growth of human populations; the crowding out and extinction of other species through habitat degradation and loss; but ultimately the desanctification of Nature and the concept of private property giving rise to economic systems designed to treat the earth exclusively as a human resource and as a commodity, together with the reconceptualization and reinforcement of humans as basically Hobbesian, selfish exploiters and consumers. All this was to be the new freedom under democratic regimes for humans.

III. SPINOZA'S INFLUENCE UPON MODERN ECOCENTRISM

Spinoza's seventeenth-century philosophical system provided the second intellectual opportunity for Western culture to abandon the anthropocentric detour. Drawing upon ancient Jewish pantheistic roots, Spinoza seemed to foresee the nonanthropocentric development of modern science. Spinoza attempted to resanctify the world by identifying God with nature, which was conceived of as each and every existing entity—human and nonhuman. Unlike Descartes, mind (or the mental attribute) for Spinoza was found throughout nature. Through criticism of Hobbes and Descartes, Spinoza developed a philosophy that would channel the new scientific understanding of nature primarily into spiritual human self-realization and into an appreciation of God/nature, rather than into the misguided attempt to dominate and control nature. Some contemporary philosophers believe that Spinoza's system is the most sophisticated ever developed in the West and that it provides a fruitful guide to ecological understanding and self-realization. Others point out that it is the Western system most similar to Eastern thought, especially Zen Buddhism.[19]

Spinoza's pantheistic vision did not derail the dominant Western philosophic and religious anthropocentrism and the dream of the human conquest of nature in the seventeenth century, but it influenced many who questioned and resisted this trend. Some of the leading figures of the eighteenth-century European romantic movement (the main Western countercultural force speaking on behalf of nature and against the uncritical and unbridled enthusiasm of the scientific and Industrial Revolutions) were inspired by Spinoza's pantheism.

Two of the leading intellects of the early twentieth century, Bertrand Russell and Albert Einstein, were deeply influenced by Spinoza. Einstein called himself a "disciple of Spinoza," expressed his admiration as well for Saint Francis, and held that "cosmic religious feeling" was the highest form of religious life.[20]

Recent scholarship has revealed that Russell's cosmology, ethics, and religious orientation were essentially Spinozistic. At the end of World War II and long before there was any public awareness of a pending environmental crisis, Russell was a lone philosopher (with the possible exception of Heidegger) speaking out against anthropocentrism. The philosophies of Marx and Dewey are anthropocentric, Russell claimed; they place humans in the center of things and "have been inspired by scientific *technique*." Further, they are "power philosophies, and tend to regard everything non-human as mere raw material." Prophetically, Russell warned that the desire of Dewey (and Marx) for social power over nature "contributes to the increasing danger of vast social disaster."[21]

Einstein and Russell were drawn to Spinoza largely from the perspective of cosmology and astronomy (although Russell was beginning to see the implications of Spinoza's pantheism for the human exploitation of the earth).[22] An explicitly ecological expression of Spinoza's pantheism had to await the California poet Robinson Jeffers and the Norwegian philosopher Arne Naess.

From his perch on the Hawk Tower on the Carmel coast beginning in the 1920s, Jeffers developed a pantheistic philosophy, expressed through his poetry, which he called "Inhumanism," as a counterpoint to Western anthropocentrism. One commentator claimed that Jeffers was "Spinoza's twentieth century evangelist." And Jeffers's ecocentric philosophy and poetry has inspired David Brower, Ansel Adams, Nancy Newhall, and other prominent contemporary environmentalists.[23]

Arne Naess lived in wild mountain and coastal areas of Norway as a young boy and, at the age of seventeen, Spinoza became his favorite philosopher. Two years later, while a young student in Paris, Naess became "hooked" on Gandhi as well, as a result of Gandhi's 1931 "salt march." As the age of ecology dawned in the 1960s, Naess was influenced by Rachel Carson's great efforts and began to see the relevance of the philosophies of Spinoza and Gandhi for ecological understanding and action. He formed Spinoza study groups at the University of Oslo that produced important and original Spinoza scholarship. Naess has claimed that "no great philosopher has so much to offer in the way of clarification and articulation of basic ecological attitudes as Baruch Spinoza."[24]

IV. THE ECOCENTRISM OF SANTAYANA AND MUIR

While Spinoza's path to ecocentrism was not taken in the seventeenth and eighteenth centuries by the dominant Western culture, yet a third opportunity occurred in America at the beginning of the twentieth century. Echoing the spirit of Thoreau, Harvard University philosopher George Santayana had grown increasingly disillusioned with his anthropocentric, pragmatist, and idealist colleagues, with the direction of urbanization, and with the economic-technological domination-over-nature path of late nineteenth-century America. Upon his retirement, Santayana came to the University of California, Berkeley, in 1911 to deliver a parting shot at the prevailing anthropocentric American philosophy and religion. In his lecture, Santayana pointed out that

> a Californian whom I had recently the pleasure of meeting observed that, if the philosophers had lived among your mountains their systems would have been different . . . from what those systems are which the European genteel tradition has handed down since Socrates; for these systems are egotistical; directly or indirectly they are anthropocentric, and inspired by the conceited notion that man, or human reason, is the center and pivot of the universe. That is what the mountains and the woods should make you at last ashamed to assert.[25]

Santayana claimed that only one American writer, Walt Whitman, had escaped anthropocentrism by extending the democratic principle "to the animals, to inanimate nature, to the cosmos as a whole. Whitman was a pantheist; but his pantheism, unlike that of Spinoza, was unintellectual, lazy, and self-indulgent." Santayana looked forward to an ecocentric revolution in philosophy.[26]

It is not known whether Santayana was aware of John Muir. But Muir, still alive in California at the time of Santayana's Berkeley lecture, was the one American who (with the possible exception of Thoreau) preeminently exemplified Santayana's ecocentric revolution. Muir's personal papers were unavailable to scholars until the mid-1970s, and so the full extent of his philosophic ecocentric achievement was unknown until studies on Muir began to appear in the

1980s.[27] As a result of his deep personal experiences in nature (and independently of the Transcendentalists), Muir had rejected the anthropocentrism of his strict Calvinist upbringing ("Lord Man"), arriving at a basically ecocentric perspective by the age of twenty-nine during his one-thousand-mile walk to the Gulf of Mexico in 1867. For the next ten years, Muir wandered through Yosemite and the High Sierra, climbing mountains, studying glaciers, and further developing his ecological consciousness. He arrived at the major generalizations of ecology by direct observation and intuitive experiencing of nature. In 1892, Muir was drafted to be the first president of the Sierra Club, a position he held until his death in 1914. Only now is he belatedly recognized by historians as the founder of the American conservation (environmental) movement.[28]

The turning point for early twentieth-century ecocentrism occurred in the confrontation between Muir, Gifford Pinchot, and President Theodore Roosevelt. Muir camped with Roosevelt in Yosemite and tried to influence his philosophical outlook but, by 1908, Roosevelt had turned away from Muir's ecocentrism and adopted Pinchot's anthropocentric policies of the scientific-technological management and development of nature as a human resource and commodity. A unique opportunity to set America on an ecocentric ecological path was lost, and Santayana's lecture had largely been in vain.[29]

V. ECOCENTRISM AND THE FUTURE

While America and the rest of Western culture continued with the anthropocentric detour, now under the resource conservation and development policies of Pinchot, ecocentrism remained alive during this period in the writings of D. H. Lawrence, Robinson Jeffers, Joseph Wood Krutch, and various professional ecologists, including Aldo Leopold. Ecocentrism was to reemerge as an intellectual force in American life during the age of ecology of the 1960s and 1970s (the transition "from conservation to ecology") in the writings of Aldous Huxley, Paul Sears, Loren Eiseley, Rachel Carson, Lynn White Jr., Paul Shepard, and in Gary Snyder, Edward Abbey, Arne Naess, and the Deep Ecology movement. John Muir's philosophy again came to the fore as a guiding beacon.

In the 1980s, as the philosophical differences between ecocentrism and anthropocentrism loomed more dramatically into view, and as Earth First! appeared and green political parties tried to sort out their philosophical underpinnings, debates broke out among the supporters of the traditions of Muir, Pinchot, and Marx.[30] From an ecological standpoint, which constitutes the genuinely radical tradition? Scholars are now in agreement that Marx's environmental views are anthropocentric and most closely resemble Pinchot's.[31] As environmental historian Stephen Fox assesses the situation: "The conservation movement, the most successful exercise in anti-modernism, corresponded to the Russian Revolution. Muir was its Lenin. Pinchot was its Stalin."[32]

As professional ecologists have been documenting since the 1960s, humans are seriously out of balance with the rest of the earth. A human population of one to two billion living lightly on the planet would probably be sustainable given the ecological requirements of carrying capacity for all the earth's species. As things now stand, according to Michael Soulé and the new field of conservation biology, almost all currently protected wilderness areas and wildlife preserves around the world are too small, and the boundaries are not ecologically drawn. As a result, natural evolution and continued speciation for many species on the planet have ground to a halt, and massive species extinctions and wild ecosystem degradation are inevitable. To correct this situation, Arne Naess once suggested that an ideal ecological balance on the planet might consist of one-third wilderness (wild species habitat), one-third "free nature" (where there are mixed communities of human and wild species living in largely nondomesticated ecosystems), and one-third cities, roads, agriculture, etc., for intensive human inhabitation of the planet.[33]

As the "social justice" movement (with its anthropocentric lineage) attempts to join forces with the ecocentric ecological movement, what will be the outcome?[34] Will the very real need to secure human livelihood and equality, a toxin-free human environment, and "jobs" in the short run in a seriously overcrowded world overshadow the necessity to protect and restore the long-range ecological integrity of the planet? Have most humans now become so thoroughly domesticated in their urban environments that they are unable to perceive the need for a wild planet, wild species and ecosystems, and the wild in themselves? Murray Bookchin and the "social ecologists" have, for the most part, demonstrated an unwillingness or inability to transcend a narrow anthropocentric perspective and consider the necessity for human population stabilization and reduction, and the high priority of protection and restoration of wild species and ecosystems.[35] The recent upheaval in Earth First! has resulted in the ouster of ecocentrically oriented leaders and their replacement by less-imaginative leaders with a "social justice" background who want to emphasize urban pollution problems and deemphasize wild ecosystem protection.

It appears that the age of ecology stands today at a major crossroads. One path follows the crucial philosophical and ecological insights of Thoreau and Muir and Naess, while incorporating, to the extent possible in this radically out-of-balance world, the concerns of social justice. It involves what Gary Snyder calls "the practice of the wild."[36] It means recapturing for humanity what philosopher Max Oelschlaeger refers to as "Paleolithic consciousness."[37] All of the ecological evidence and wisdom suggest that the other path (a continuation of the "anthropocentric detour") will lead inexorably, perhaps with the best of intentions, to an accelerating decline of the earth and all its inhabitants.

NOTES

1. Warwick Fox, "The Deep Ecology–Ecofeminism Debate and Its Parallels," *Environmental Ethics* 11, no. 1 (1989): 5–25, quotation on p. 5.

2. Stan Steiner, *The Vanishing White Man* (New York: Harper and Row, 1976), p. 113.

3. See also J. Baird Callicott, "Traditional American Indian and Western European Attitudes toward Nature," *Environmental Ethics* 4 (Winter 1982): 293; J. Donald Hughes, *American Indian Ecology* (El Paso: Texas Western Press, 1982); Max Oelschlaeger, *The Idea of Wilderness: From Prehistory to the Age of Ecology* (New Haven, CT: Yale University Press, 1991), pp. 1–30.

4. See also Oelschlaeger, *Idea of Wilderness*.

5. Paul Shepard, *The Tender Carnivore and the Sacred Game* (New York: Scribner's, 1973), p. 237.

6. H. Frankfort, *Kingship and the Gods* (Chicago: University of Chicago Press, 1948), p. 342.

7. Loren Eiseley, "The Last Magician," in *The Invisible Pyramid*, ed. L. Eiseley (New York: Scribner's, 1970), p. 154.

8. D. H. Lawrence, "Pan in America" (1926), reprinted in *The Everlasting Universe*, ed. L. Forstner and J. Todd (Lexington, MA: D. C. Heath, 1971), p. 221. For discussions of Lawrence's ecological views, see Del Ivan Janik, "Environmental Consciousness in Modern Literature," in *Ecological Consciousness*, ed. J. Hughes and R. Schwartz (Washington, DC: University Press of America, 1981); and Dolores LaChapelle, "D. H. Lawrence and Deep Ecology," *Trumpeter* 7 (Spring 1990): 26–30.

9. Karl Popper, "Back to the Presocratics," in *Conjectures and Refutations,* ed. Karl Popper, 141 (London: Routledge and Kegan Paul, 1965).

10. For a critical discussion of the anthropocentric Christian "great chain of being" from an ecofeminist perspective, see Elizabeth Gray, *Green Paradise Lost* (Wellesley, MA: Roundtable Press, 1982).

11. K. Baler, "The Meaning of Life: Christianity versus Science," in *Philosophy for a New Generation*, ed. A. Bierman and J. Gould, 596 (New York: Macmillan, 1973).

12. L. White Jr., "Historical Roots of Our Ecologic Crisis," in Forstner, and Todd, *Everlasting Universe*, p. 16.

13. For an ecological critique of Bacon, Descartes, and Leibniz, see Clarence J. Glacken, "Man against Nature: An Outmoded Concept," in *The Environmental Crisis*, ed. H. Helfrich Jr. (New Haven, CT: Yale University Press, 1970); William Leiss, *The Domination of Nature* (New York: Braziller, 1970). For critiques of the anthropocentric development of Western academic philosophy, see John B. Cobb Jr., "The Population Explosion and the Rights of the Subhuman World," in *Environment and Society: A Book of Readings on Environmental Policy, Attitudes, and Values*, ed. Robert T. Roelofs (Englewood Cliffs, NJ: Prentice-Hall, 1974); George Sessions, "Anthropocentrism and the Environmental Crisis," *Humboldt Journal of Social Relations* 2 (Spring 1974): 71–81; Eugene Hargrove, *Foundation of Environmental Ethics* (Englewood Cliffs, NJ: Prentice-Hall, 1989), pp. 14–47.

14. For ecological critiques of humanism, see Roderick French, "Is Ecological Humanism a Contradiction in Terms?: The Philosophical Foundations of the Humanities

under Attack," in Hughes and Schwartz, *Ecological Consciousness*, 43–66; David Ehrenfeld, *The Arrogance of Humanism* (Oxford: Oxford University Press, 1978).

15. White, "Historical Roots," p. 15.

16. Ibid.

17. Fritjof Capra, *The Tao of Physics* (Berkeley, CA: Shambala, 1975). See also J. Baird Callicott, "Intrinsic Value, Quantum Theory, and Environmental Ethics," *Environmental Ethics* 7 (Fall 1985): 257; Michael Zimmerman, "Quantum Theory, Intrinsic Value, and Panentheism," *Environmental Ethics* 10 (Spring 1988): 3; Andrew McLaughlin, "Images and Ethics of Nature," *Environmental Ethics* 7 (Winter 1985): 293.

18. Thomas Berry, *The Dream of the Earth* (San Francisco: Sierra Club Books, 1989); Thomas Berry and Brian Swimme, *The Universe Story* (Clinton, WA: New Story Productions, 1989).

19. George Sessions, "Western Process Metaphysics: Heraclitus, Whitehead, and Spinoza," in *Deep Ecology: Living as if Nature Mattered*, ed. Bill Devall and George Sessions (Salt Lake City: Peregrine Smith, 1985), pp. 236–42; Paul Wienpahl, *The Radical Spinoza* (New York: New York University Press, 1979); Arne Naess, "Through Spinoza to Mahayana Buddhism, or through Mahayana Buddhism to Spinoza?" in *Spinoza's Philosophy of Man*, ed. Jon Wetlesen (Oslo: University Press, 1978).

20. Albert Einstein, *The World as I See It* (New York: Philosophical Library, 1942), p. 14.

21. Bertrand Russell, *A History of Western Philosophy* (New York: Simon and Schuster, 1945), p. 494.

22. B. Hoffman and H. Dukas, *Albert Einstein: Creator and Rebel* (New York: New American Library, 1972), pp. 94–95; Arne Naess, "Einstein, Spinoza, and God," in *Old and New Questions in Physics, Cosmology, Philosophy and Theoretical Biology*, ed. Alwyn van der Merwe, 683–87 (Holland: Plenum, 1983); Kenneth Blackwell, *The Spinozistic Ethics of Bertrand Russell* (London: Allen and Unwin, 1985).

23. David Brower, ed., *Not Man Apart: Lines from Robinson Jeffers* (San Francisco: Sierra Club Books, 1965); Ansel Adams, *Ansel Adams: An Autobiography* (Boston: Little, Brown, 1985), pp. 84–87; George Sessions, "Spinoza and Jeffers on Man in Nature," *Inquiry* 20 (1977): 481; for "Jeffers as Spinoza's Evangelist," see Arthur Coffin, *Robinson Jeffers: Poet of Inhumanism* (Madison: University of Wisconsin Press, 1971), p. 255.

24. Arne Naess, *Freedom, Emotion, and Self-Subsistence: The Structure of a Central Part of Spinoza's Ethics* (Oslo: University Press, 1975), pp. 118–19; Arne Naess, "Spinoza and Ecology," in *Speculum Spinozanum*, ed. S. Hessing (Boston: Routledge and Kegan Paul, 1978); Arne Naess, *Ghandi and Group Conflict* (Oslo: University Press, 1974).

25. George Santayana, "The Genteel Tradition in American Philosophy," in *Winds of Doctrine* (New York: Scribner's, 1926).

26. Ibid., p. 203. For a discussion of Santayana's address in an ecological context, see William Everson, *Archetype West* (Berkeley, CA: Oyez Press, 1976), pp. 54–60.

27. For the new Muir scholarship and discussions of his ecocentric philosophy, see Frederick Turner, *Rediscovering America: John Muir in His Time and Ours* (San Francisco: Sierra Club Books, 1985); Stephen Fox, *John Muir and His Legacy: The American Conservation Movement* (Boston: Little, Brown, 1981), pp. 43–53, 59, 79–81, 289–91, 350–55, 361; and esp. Michael Cohen, *The Pathless Way: John Muir and American Wilderness* (Madison: University of Wisconsin Press, 1984), chs. 1, 6, 7.

28. See Fox, *John Muir and His Legacy*; Michael Cohen, *The History of the Sierra Club 1892–1970* (San Francisco: Sierra Club Books, 1988).

29. For the split between Muir, Pinchot, and Roosevelt, see Roderick Nash, *Wilderness and the American Mind*, 3rd ed. (New Haven, CT: Yale University Press, 1982), pp. 129–40; Fox, *John Muir and His Legacy*, pp. 109–30; Cohen, *Pathless Way*, pp. 160–61, 292ff; Bill Devall and George Sessions, "The Development of Natural Resources and the Integrity of Nature," *Environmental Ethics* 6 (Winter 1984): 293.

30. See, for example, Robyn Eckersley, "The Road to Ecotopia? Socialism versus Environmentalism," *Trumpeter* 5 (Summer 1988): 60; Francis Moore Lappé and J. Baird Callicott, "Marx Meets Muir: Toward a Synthesis of the Progressive Political and Ecological Visions," *Tikkun* 2 (Winter 1987): 16.

31. Howard L. Parsons, ed., *Marx and Engels on Ecology* (New Haven, CT: Greenwood Press, 1977); Val Routley [Plumwood], "On Karl Marx as an Environmental Hero," *Environmental Ethics* 3 (Spring 1981): 237; Donald Lee, "Toward a Marxian Ecological Ethic," *Environmental Ethics* 4 (Winter 1982): 339–43; John Clark, "Marx's Inorganic Body," *Environmental Ethics* 11 (Fall 1989): 243.

32. Fox, *John Muir and His Legacy*.

33. For a discussion of the findings of Soulé and the conservation biologists, see Christopher Manes, *Green Rage: Radical Environmentalism and the Unmaking of Civilization* (Boston: Little, Brown, 1990), pp. 34–35; George Sessions, "Ecocentrism, Wilderness and Global Ecosystem Protection," in *The Wilderness Condition: Essays on Environment and Civilization*, ed. Max Oelschlaeger (San Francisco: Sierra Club Books, 1991); Arne Naess, "Ecosophy, Population, and Free Nature," *Trumpeter* 5 (Fall 1988): 113.

34. Lappé and Callicott, "Marx Meets Muir."

35. See Murray Bookchin, *The Philosophy of Social Ecology: Essays on Dialectical Naturalism* (New York: Black Rose Books, 1990).

36. Gary Snyder, *The Practice of the Wild* (San Francisco: North Point Press, 1990).

37. Oelschlaeger, *Idea of Wilderness*.

14

THE CONCEPT
OF SOCIAL ECOLOGY

MURRAY BOOKCHIN

The reconstructive and destructive tendencies in our time are too much at odds with each other to admit of reconciliation. The social horizon presents the starkly conflicting prospects of a harmonized world and an ecological sensibility based on a rich commitment to community, mutual aid, and new technologies on the one hand, and the terrifying prospect of some sort of thermonuclear disaster on the other. Our world, it would appear, will either undergo revolutionary changes, so far-reaching in character that humanity will totally transform its social relations and its very conception of life, or it will suffer an apocalypse that may well end humanity's tenure on the planet.

The tension between these two prospects has already subverted the morale of the traditional social order. We have entered an era that consists no longer of institutional stabilization but of institutional decay. A widespread alienation is developing toward the forms, the aspirations, the demands, and above all the institutions of the established order. The most exuberant, in fact, theatrical evidence of this alienation occurred in the sixties, when the "youth revolt" exploded into what seemed to be a counterculture. Considerably more than protest and adolescent nihilism marked the period. Almost intuitively, new values of sensuousness, new forms of communal lifestyle, changes in dress, language, and music, all borne on the wave of a deep sense of impending social change, rolled over a sizable section of an entire generation. We still do not know in what sense this wave began to ebb: whether as a historic retreat or as a transformation into a serious project for inner and social development. That the symbols of this movement eventually became the artifacts for a new culture industry does not

From: *CoEvolution Quarterly* (Winter 1981): 15–22.

alter the movement's far-reaching effects. Western society will never be the same again. . . .

Crucial as this decay of institutions and values may be, it by no means exhausts the problems that confront the existing society. Intertwined with the social crisis is a crisis that has emerged directly from man's exploitation of the planet.[1] Established society is faced with a breakdown not only of its values and institutions, but also of its natural environment. This problem is not unique to our times: the desiccated wastelands of the Near East, the areas where the arts of agriculture and urbanism had their beginnings, are evidence of ancient human despoliation. But this example pales before the massive destruction of the environment that has occurred since the days of the Industrial Revolution, and especially since the end of the Second World War. The damage inflicted on the environment by contemporary society encompasses the entire earth. The exploitation and pollution of the earth have damaged not only the integrity of the atmosphere, climate, water resources, soil, flora and fauna of specific regions, but also the basic natural cycles on which all living things are dependent.

Yet modern man's capacity for destruction is quixotic evidence of his capacity for reconstruction. The powerful technological agents we have unleashed against the environment include many of the very agents we require for its reconstruction. What we crucially lack is the consciousness and sensibility that will help us achieve such eminently desirable goals—a consciousness and sensibility that is far broader than we customarily mean by these terms. Our definitions must include not only the ability to reason logically and respond emotionally in a humanistic fashion; they must also include a fresh awareness of the relatedness between things and an imaginative insight into the possible. On this score, Marx was entirely correct to emphasize that the revolution required by our time must draw its poetry not from the past but from the future, from the humanistic potentialities that lie on the horizons of social life.

The new consciousness and sensibility cannot be poetic alone; they must also be scientific. Indeed, there is a level at which our consciousness must be neither poetry nor science, but a transcendence of both into a new realm of theory and practice, an artfulness that combines fancy with reason, imagination with logic, vision with technique. We cannot shed our scientific heritage without returning to a rudimentary technology, with its shackles of material insecurity, toil, and renunciation. By the same token, we cannot allow ourselves to be imprisoned within a mechanistic outlook and a dehumanizing technology—with its shackles of alienation, competition, and brute denial of humanity's potentialities. Poetry and imagination must be integrated with science and technology, for we have evolved beyond an innocence that can be nourished exclusively by myths and dreams.

Is there a scientific discipline that allows for the indiscipline of fancy, imagination, and artfulness? Can it encompass problems created by the social and environmental crises of our time? Can it integrate critique with reconstruction, theory with practice, vision with technique? In view of the enormous dislocations that now confront us, our own era raises the need for a more sweeping and insightful body of knowledge—scientific as well as social—to deal with our problems. Without renouncing the gains of earlier scientific and social theories, we are obliged to develop a more rounded critical analysis of our relationship with the natural world. We must seek the foundations for a more reconstructive approach to the grave problems posed by the apparent "contradictions" between nature and society. We can no longer afford to remain captives to the tendency of the more traditional sciences to dissect phenomena and examine their fragments. We must combine them, relate them, and see them in their totality as well as their specificity.

In response to these needs, we have formulated a discipline unique to our age: "social ecology." The more well-known term "ecology" was coined by Ernst Haeckel a century ago to denote the investigation of the interrelationships between animals, plants, and their inorganic environment. Since Haeckel's day, the term has been expanded to include ecologies of cities, of health, and of the mind. This proliferation of a word into widely disparate areas may seem particularly desirable to an age that fervently seeks some kind of intellectual coherence and unity of perception. But it can also prove to be extremely treacherous. Like such newly arrived words as "holism," "decentralization," and "dialectics," the term "ecology" runs the peril of merely hanging in the air without any roots, context, or texture. Often it is used as a metaphor, an alluring catchword, that loses the potentially compelling internal logic of its premises.

Accordingly, the radical thrust of these words is easily neutralized. "Holism" evaporates into a mystical sigh, a rhetorical expression for ecological fellowship and community that ends with such in-group greetings and salutations as "holistically yours." What was once a serious philosophical stance has been reduced to environmentalist kitsch. "Decentralization" commonly means logistical alternatives to gigantism, not the human scale that would make an intimate and direct democracy possible. "Ecology" fares even worse. All too often it becomes a metaphor, like the word "dialectics," for any kind of integration and development. Perhaps even more troubling, the word in recent years has been identified with a very crude form of natural engineering which might well be called "environmentalism."

I am mindful that many ecologically oriented individuals use "ecology" and "environmentalism" interchangeably. Here, I would like to draw a semantically convenient distinction. By "environmentalism" I propose to designate a mechanistic, instrumental outlook that sees nature as a passive habitat composed of "objects" such as animals, plants, minerals, and the like which must merely be rendered more serviceable for human use. Given my use of the term, "environmentalism" tends to reduce nature to a storage bin of "natural resources" or "raw materials." Within this context, very little of a social nature is spared from the environmentalist's vocabulary: cities become "urban resources" and their inhabitants "human resources." If the word "resources" leaps out so frequently from environmentalistic discussions of nature, cities, and people, an issue more important than mere wordplay is at stake. Environmentalism, as I use this term, tends to view the ecological project for attaining a harmonious relationship between humanity and nature as a truce rather than a lasting equilibrium. The harmony of the environmentalist centers around the development of new techniques for plundering the natural world with a minimal disruption of the human habitat. Environmentalism does not question the most basic premise of the present society, notably, that humanity must dominate nature; rather, it seeks to *facilitate* that notion by developing techniques for diminishing the hazards caused by the reckless despoliation of the environment.

To distinguish ecology from environmentalism and from abstract, often obfuscatory definitions of the term, I must return to its original usage and explore its direct relevance to society. Put quite simply: ecology deals with the dynamic balance of nature, with the interdependence of living and nonliving things. Since nature also includes human beings, the science must include humanity's role in the natural world—specifically, the character, form, and structure of humanity's relationship with other species and with the inorganic substrate of the biotic environment. From a critical viewpoint, ecology opens to wide purview the vast disequilibrium that has emerged from humanity's split with the natural world. One of nature's very unique species, *Homo sapiens*, has slowly and painstakingly developed from the natural world into a unique social world of its own. As both worlds interact with each other through highly complex phases of evolution, it has become as important to speak of a *social ecology* as to speak of a natural ecology.

Let me emphasize that the failure to explore these phases of human evolution—which have yielded a succession of hierarchies, classes, cities, and finally states—is to make a mockery of the term "social ecology." Unfortunately, the discipline has been beleaguered by self-professed acolytes who continually try to collapse all the phases of natural and human development into a universal "oneness" (not wholeness), a yawning "night in which all cows are black," to apply one of Hegel's caustic phrases to a widely accepted pop mysticism that clothes itself in ecological verbiage. If nothing else, our common use of the word "species" to denote the wealth of life around us should alert us to the fact of *specificity*, of *particularity*—the rich abundance of *differentiated* beings and things that enter into the very subject matter of natural ecology. To explore these differentia, to examine the phases and interfaces that enter into their making and into humanity's long development from animality to society—a development latent with problems and possibilities—is to make social ecology one of the most powerful disciplines from which to draw our critique of the present social order.

But social ecology not only provides a critique of the split between humanity and nature; it also poses the need to heal them. Indeed, it poses the need to radically transcend them. As E. A. Gutkind pointed out, "The goal of Social Ecology is wholeness, and not mere adding together of innumerable details collected at random and interpreted subjectively and insufficiently." The science deals with social and natural relationships in communities or "ecosystems."[2] In conceiving them holistically, that is to say, in terms of their mutual interdependence, social ecology seeks to unravel the *forms* and *patterns* of interrelationships that give intelligibility to a community, be it natural or social. Holism, here, is the result of a conscious effort to discern how the particulars of a community are arranged, how its geometry (as the Greeks might have put it) makes the whole more than the sum of its parts. Hence, the wholeness to which Gutkind refers is not to be mistaken for a spectral oneness that yields cosmic dissolution in a structureless nirvana; it is a richly articulated structure that has a history and an internal logic of its own.

History, in fact, is as important as form or structure. To a large extent, the history of a phenomenon *is* the phenomenon itself. We are, in a real sense, everything that existed before us and, in turn, we can eventually become vastly more than we are. Surprisingly, *very* little in the evolution of life-forms has been lost in natural and social evolution, as the embryonic development in our very bodies attests. Evolution lies within us (as well as around us) as parts of the very nature of our beings. For the present, it suffices to point out that wholeness is not a bleak undifferentiated universality that involves the reduction of a phenomenon to what it has in common with everything else. Nor is it a celestial, omnipresent energy that replaces the vast material differentia of which the natural and social realms are composed. To the contrary, wholeness comprises the variegated structures, articulations, and mediations that impart to the whole a rich variety of forms and thereby add unique qualitative properties to what a strictly analytic mind often reduces to "innumerable" and "random" details.

Terms like "wholeness," "totality," and even "community" have perilous nuances for a generation that has known fascism and other totalitarian ideologies. The words evoke images of a "wholeness" achieved through homogenization, standardization, and a repressive coordination of human beings. These fears are reinforced by a "wholeness" that seems to provide an inexorable finality to the course of human history—one that implies a suprahuman, narrowly teleological concept of "social law" which denies the ability of human will and individual choice to shape the course of social events. Such notions of social law and teleology have been used to achieve a ruthless subjugation of the individual to suprahuman forces beyond human control. Our century has been afflicted by a plethora of totalitarian ideologies that, placing human beings in the service of "history," have denied them a place in the service of their own humanity.

Actually, such a totalitarian concept of wholeness stands sharply at odds with what ecologists denote by the term. In addition to comprehending its heightened awareness of form and structure, we now come to a very important tenet of ecology: ecological wholeness means not an immutable homogeneity but rather the very opposite—a dynamic *unity of diversity*. In nature, balance and harmony are achieved by ever-changing differentiation, by ever-expanding diversity. Ecological stability, in effect, is a function not of simplicity and homogeneity but of complexity and variety. The capacity of an ecosystem to retain its integrity depends not upon the uniformity of the environment but upon its diversity.

If we assume that the thrust of natural evolution has been toward increasing complexity, that the colonization of the planet by life has been possible only as result of biotic variety, a prudent rescaling of man's hubris should call for caution in disturbing natural processes. To assume that science commands life's vast nexus of organic and inorganic interrelationships in all its details is worse than arrogance: it is sheer stupidity. If unity in diversity forms one of the cardinal tenets of ecology, the wealth of biota that exists in a single acre of soil leads us to still another basic ecological tenet: the need to allow for a high degree of natural spontaneity. The compelling dictum "respect for nature" has concrete implications.

Thus a considerable amount of leeway must be permitted for natural spontaneity, for the diverse biological forces that yield a variegated ecological situation. "Working with nature" largely means that we must foster the biotic variety that emerges from a spontaneous development of natural phenomena. I hardly mean that we must surrender ourselves to a mythical nature that is beyond all human comprehension and intervention, a nature that demands human awe and subservience. Perhaps the most obvious conclusion we can draw from these ecological tenets is Charles Elton's sensitive observation: "The world's future has to be managed, but this management would not be just like a game of chess—[but] more like steering a boat." What ecology, both natural and social, can hope to teach us is the way to find the current and understand the direction of the stream.

What ultimately distinguishes an ecological outlook as uniquely liberatory is the challenge it raises to conventional notions of hierarchy. Let me emphasize, however, that this challenge is implicit: it must be painstakingly elicited from the discipline, which is permeated by conventional scientistic biases. Ecologists are rarely aware that their science provides strong philosophical underpinnings to a nonhierarchical view of reality. Like many natural scientists, they resist philosophical generalizations as alien to their research and conclusions—a prejudice that is itself a philosophy rooted in the Anglo-American empirical tradition. Moreover, they follow their colleagues in other disciplines who model their notions of science on physics. This prejudice, which goes back to Galileo's day, has led to a widespread acceptance of systems theory in ecological circles. While systems theory has its place in the repertoire of science, it can easily become an all-encompassing, quantitative, reductionist theory of energetics if it acquires preeminence over *qualitative* descriptions of ecosystems, that is, descriptions rooted in organic evolution, variety, and holism. Whatever the merits of systems theory as an account of energy flow through an ecosystem, the primacy it gives to this quantitative aspect of ecosystem analysis fails to take adequate account of life-forms as more than consumers and producers of calories.

If we recognize that every ecosystem can also be viewed as a food web, we can think of it as a circular, interlacing nexus of plant-animal relationships (rather than a stratified pyramid with man at its apex) that includes such widely varying creatures as microorganisms and large mammals. What ordinarily puzzles anyone who sees food-web diagrams for the first time is the impossibility of discerning a point of entry into the nexus. The web can be entered at any point and leads back to its point of departure without any apparent exit. Aside from the energy provided by sunlight (and dissipated by radiation), the system to all appearances is closed. Each species, be it a form of bacteria or deer, is knitted together in a network of interdependence with each other, however indirect the links may be. A predator in the web is also prey, even if the "lowliest" of organisms merely makes it ill or helps to consume it after death.

Nor is predation the sole link that unites one species with another. A resplendent literature now exists that reveals the enormous extent to which symbiotic mutualism is a major factor in fostering ecological stability and organic evolution. That plants and animals continually adapt to unwittingly aid each other, be it by an exchange of biochemical functions that are mutually beneficial or even dramatic instances of physical assistance and succor, has opened a whole new perspective on the nature of ecosystem stability and development.

We must not get caught up in direct comparisons between plants, animals, and human beings or in direct comparisons between plant-animal ecosystems and human communities. None of these is completely congruent with another. It is not in the *particulars* of differentiation that plant-animal communities are ecologically united with human communities but rather in their logic of differentiation. Wholeness, in fact, is completeness. The dynamic stability of the whole derives from a visible level of completeness in human communities as in climax ecosystems. What unites these modes of wholeness and completeness, however different they are in their specificity and their qualitative distinctness, is the *logic of development itself.* A climax forest is whole and complete as a result of the same unifying process—the same *dialectic* that makes a particular social form whole and complete.

When wholeness and completeness are viewed as the result of an immanent dialectic within phenomena, we do no more violence to the uniqueness of these phenomena than the principle of gravity does violence to the uniqueness of objects that fall within its "lawfulness." In this sense, the ideal of human roundedness, a product of the rounded community, is the legitimate heir to the ideal of a stabilized nature, a product of the rounded natural environment. Marx tried to root humanity's identity and self-discovery in its productive interaction with nature. But I must add that not only does humanity place its imprint on the natural world and transform it, but also nature places its imprint on the human world and transforms it. To use the language of hierarchy against itself: it is not only we who "tame" nature but also nature that "tames" us.

These turns of phrase should be taken as more than metaphors. Lest it seem that I have rarefied the concept of wholeness into an abstract dialectical principle, let me note that natural ecosystems and human communities interact with each other in very existential ways. Our animal nature is never so distant from our social nature that we can remove ourselves from the organic world outside us and from the one within us. From our embryonic development to our layered brain, we recapitulate the totality of our natural evolution. We are not so remote from our primate ancestry that we can ignore its physical legacy in our stereoscopic vision, acuity of intelligence, and grasping fingers. We phase into society as individuals in the same way that society, phasing out of nature, comes into itself.

These continuities, to be sure, are obvious enough. What is often less obvious is the extent to which nature itself is a realm of potentiality for the emergence of *social* differentia. Nature is as much a precondition for the *development* of society—not merely its emergence—as technics, labor, language, and mind. And it is a precondition not merely in William Petty's sense—that if labor is the "father" of wealth, nature is its "mother." This formula, so dear to Marx, actually slights nature by imparting to it the patriarchal notion of feminine passivity. The affinities between nature and society are more active than we care to admit. Very specific forms of nature, that is to say, very specific *ecosystems*, constitute the ground for very specific forms of society. At the risk of using a highly embattled phrase, I might say that a "historical materialism" of natural development could be written that would transform "passive nature"—the object of human labor—into "active nature," the creator of human labor. Labor's metabolism with nature cuts both ways, so that nature interacts *with* humanity to yield the actualization of their common potentialities in the natural and social worlds.

An interaction of this kind, in which terms like "father" and "mother" strike a false note, can be stated very concretely. The recent emphasis on bioregions as frameworks for various human communities provides a strong case for the need to readapt technics and work styles to accord with the requirements and possibilities of particular ecological areas. Bioregional requirements and possibilities place a heavy burden on humanity's claims of "sovereignty" over nature and "autonomy" from its needs. If it is true, as Marx wrote, that men make history but not under conditions of their own choosing, it is no less true that history makes society but not under condition of its own choosing. The hidden dimension that lurks in this wordplay with Marx's famous formula is the natural history that enters into the making of social history—active, concrete, existential nature that emerges from stage to stage of its own evermore complex development in the form of equally complex and dynamic ecosystems. Our ecosystems, in turn, are interlinked into highly dynamic and complex bioregions. How concrete the hidden dimension of social development is—and how much humanity's claims to "sovereignty" must defer to it—has only recently become evident from our need to design an alternative technology that is as adaptive to a bioregion as it is productive to society. Hence our concept of wholeness is not a finished tapestry of natural and social relations that we can exhibit to the hungry eyes of sociologists. It is a fecund natural history, ever-active and ever-changing—the way childhood presses toward and is absorbed into youth, and youth into adulthood.

Within this highly complex context of ideas we must now try to transpose the nonhierarchical character of natural ecosystems to society. Sociobiology has made this project deceptively simple and crudely mechanistic. To refute the notion that savanna baboons are hierarchical, for example, we take refuge in the complete exclusivity of society—its immunity to natural principles. If nature is hierarchical, so the argument goes, need this be true of a human community guided by reason, love, and mutualism?

What renders social ecology so important is that it offers no case whatsoever for hierarchy in nature and society; it decisively challenges the very function of hierarchy as a stabilizing or "ordering" principle in *both* realms. The association of *order as such* with hierarchy is ruptured. And this association is ruptured without rupturing the association of nature with society—as sociology, in its well-meaning opposition to sociobiology, has been wont to do. In contrast to sociologists, we do not have to render the social world so supremely autonomous over nature that we are obliged to dissolve the continuum that phases nature into society. In short, we do not have to accept the brute tenets of sociobiology that link us crudely to nature at one extreme or the naive tenets of sociology that cleave us sharply from nature at the other extreme. Of course, the fact that hierarchy does exist in present-day society does not mean that it has to remain so—irrespective of its lack of meaning or reality for nature. But the case against hierarchy is not contingent on its uniqueness as a social phenomenon. That hierarchy threatens the existence of social life today certainly *does* mean that it cannot remain a social fact. And that it threatens the integrity of organic nature means that it *will* not remain so, given the harsh verdict of "mute" and "blind" nature. . . .

In concrete terms, what tantalizing issues does social ecology raise for our time and our future? In restoring a new and more-advanced interface with nature than we have had previously, will it be possible to achieve a new balance between humanity and nature by sensitively tailoring our agricultural practices, urban areas, and technologies to the natural requirements of a region and the ecosystems of which it is composed? Can we hope to manage the natural environment by a drastic decentralization of agriculture which makes it possible to cultivate land as though it were a garden, balanced by a diversified fauna and flora? Will these changes require the decentralization of our cities into moderate-sized communities, creating a new balance between town and country? What technology will be required to achieve these goals—indeed, to avoid the further pollution of Earth? What institutions will be required to create a new public sphere, what social relations to foster a new ecological sensibility, what forms of work to render human practice playful and creative, what sizes and populations of communities to scale life to human dimensions controllable by all? What kind of poetry? Concrete questions—ecological, social, political, and behavioral—rush in upon us like a flood that heretofore has been dammed up by the constraints of traditional ideologies and habits of thought.

Let there be no mistake about it: the answers we provide to these questions have a direct bearing on whether or not humanity will be able to survive on the planet. The trends in our time are visibly directed against ecological diversity; in fact, they point toward a brute simplification of the entire biosphere. Complex food chains in the soil and on the earth's surface are being ruthlessly undermined by the fatuous application of industrial techniques to agriculture, with the result that soil has been reduced in many areas to a mere sponge for absorbing simple chemical nutrients. The cultivation of single crops over vast stretches of land is effacing natural, agricultural, and even physiographic variety. Immense urban belts are encroaching unrelentingly on the countryside, replacing flora and fauna with concrete, metals, and glass, and enveloping large regions with a haze of atmospheric pollutants. In this mass urban world, human experience itself becomes crude and elemental, subject to brute noisy stimuli and crass bureaucratic manipulation. A national division of labor, standardized along industrial lines, is replacing regional and local variety, reducing entire continents to immense smoking factories and cities to garish, plastic supermarkets.

Modern society, in effect, is disassembling the biotic complexity achieved by eons of organic evolution. The great movement of life from fairly simple to increasingly complex forms and relations is being ruthlessly reversed in the direction of an environment that will be able to support only simpler living things. To continue this reversal of biological evolution, to undermine the biotic food webs on which humanity depends for its means of life, places in question the very survival of the human species. If the reversal of the evolutionary process continues, there is good reason to believe—all control of other toxic agents aside—that the preconditions for complex forms of life will be irreparably destroyed and the earth will be incapable of supporting us as a viable species.

In this confluence of social and ecological crises, we can no longer afford to be unimaginative; we can no longer afford to do without utopian thinking. The crises are too serious and the possibilities too sweeping to be resolved by customary modes of thought—the very sensibilities that produced these crises in the first place. Years ago, the French students in the May-June uprising of 1968 expressed this sharp contrast of alternatives magnificently in their slogan: "Be practical! Do the impossible!" To this demand the generation that faces the next century can add the more solemn injunction: "If we don't do the impossible, we shall be faced with the unthinkable!"

NOTES

1. I use the word "man," here, advisedly. The split between humanity and nature has been precisely the work of the male, who, in the memorable lines of Theodor Adorno and Max Horkheimer, "dreamed of acquiring absolute mastery over nature, of converting the cosmos into one immense hunting ground" (*Dialectic of Enlightenment* [New York: Herder and Herder, 1972], p. 248). For the words "one immense hunting ground," I would be disposed to substitute "one immense killing ground" to describe the male-oriented "civilization" of our era.

2. The term "ecosystem"—or ecological system—is often used very loosely in many ecological works. Here I employ it, as in natural ecology, to mean a fairly demarcatable animal-plant community and the abiotic or nonliving factors needed to sustain it. I also use it in social ecology to mean a distinct human and natural community, the social as well as organic factors that interrelate with each other to provide the basis for an ecologically rounded and balanced community.

15

SOCIALISM AND ECOLOGY

JAMES O'CONNOR

The premise of red-green political action is that there is a global ecological and economic crisis; that the ecological crisis cannot be resolved without a radical transformation of capitalist production relationships; and that the economic crisis cannot be resolved without an equally radical transformation of capitalist productive forces. This means that solutions to the ecological crisis presuppose solutions to the economic crisis and vice versa. Another a priori of red-green politics is that both sets of solutions presuppose an ecological socialism.

The problem is that socialism in theory and practice has been declared "dead on arrival." In theory, post-Marxist theorists of radical democracy are completing what they think is the final autopsy of socialism. In practice, in the North, socialism has been banalized into a species of welfare capitalism. In Eastern Europe, the moment for democratic socialism seems to have been missed over twenty years ago and socialism is being overthrown. In the South, most socialist countries are introducing market incentives, reforming their tax structures, and taking other measures that they hope will enable them to find their niches in the world market. Everywhere market economy and liberal democratic ideas on the right, and radical democratic ideas on the left, seem to be defeating socialism and socialist ideas.

Meanwhile, a powerful new force in world politics has appeared, an ecology or green movement that puts the earth first and takes the preservation of the ecological integrity of the planet as the primary issue. The simultaneous rise of the

From: *Capitalism Nature, Socialism* 2, no. 3 (October 1991): 1–12. Reprinted in *Our Generation* (Canada), *Il Manifesto* (Italy), *Ecologia Politica* (Spain), *Nature and Society* (Greece), *Making Sense* (Ireland), and *Salud y Cambio* (Chile). A revised version appears in James O'Connor, *Natural Causes: Essays in Ecological Marxism* (New York: Guilford Press, 1998).

free market and the Greens together with the decline of socialism suggests that capitalism has an ally in its war against socialism. This turns out to be the case. Many or most Greens dismiss socialism as irrelevant. Some or many Greens attack it as dangerous. Especially are they quick to condemn those who they accuse of trying to appropriate ecology for Marxism.[1] The famous green slogan, "Neither left nor right, but out front," speaks for itself.[2]

But most Greens are not friends of capitalism, either, as the Green slogan makes clear. The question then arises, who or what are the greens allied with? The crude answer is, the small farmers and independent business, that is, those who used to be called the "peasantry" and "petty bourgeoisie"; "liveable cities" visionaries and planners; "small is beautiful" technocrats; and artisans, cooperatives, and others engaged in ecologically friendly production. In the South, Greens typically support decentralized production organized within village communal politics; in the North, Greens are identified with municipal and local politics of all types.

By the way of contrast, mainstream environmentalists might be called "fictitious Greens."[3] These environmentalists support environmental regulations consistent with profitability and the expansion of global capitalism, for example, resource conservation for long-run profitability and profit-oriented regulation or abolition of pollution. They are typically allied with national and international interests. In the United States, they are environmental reformers, lobbyists, lawyers, and others associated with the famous "Group of Ten."

As for ecology, everywhere it is at least tinged with populism, a politics of resentment against not only big corporations and the national state and central planning but also against environmentalism.

Ecology (in the present usage) is thus associated with "localism," which has always been opposed to the centralizing powers of capitalism. If we put two and two together, we can conclude that ecology and localism in all of their rich varieties have combined to oppose both capitalism and socialism. Localism uses the medium or vehicle of ecology and vice versa. They are both the content and context of one another. Decentralism is an expression of a certain type of social relationship, a certain social relation of production historically associated with small-scale enterprise. Ecology is an expression of a certain type of relationship between human beings and nature—a relationship which stresses the integrity of local and regional ecosystems. Together ecology and localism constitute the most visible political and economic critique of capitalism (and state socialism) today.

Besides the fact that both ecology and localism oppose capital and the national state, there are two main reasons why they appear to be natural allies. First, ecology stresses the site specificity of the interchange between human material activity and nature, hence opposes both the abstract valuation of nature made by capital and also the idea of central planning of production, and centralist approaches to global issues generally.[4] The concepts of site specificity of ecology, local subsistence or semi-autarkic economy, communal self-help principles, and direct forms of democracy all seem to be highly congruent.

Second, the socialist concept of the "masses" has been deconstructed and replaced by a new "politics of identity" in which cultural factors are given the place of honor. The idea of the specificity of cultural identities seems to meld easily with the site specificity of ecology in the context of a concept of social labor defined in narrow, geographic terms. The most dramatic examples today are the struggles of indigenous peoples to keep both their cultures and subsistence-type economies intact. In this case, the struggle to save local cultures and local ecosystems turns out to be two different sides of the same fight.

For their part, most of the traditional Left, as well as the unions, remain focused on enhanced productivity, growth, and international competitiveness, that is, jobs and wages, or more wage labor—not to abolish exploitation but to be exploited less. This part of the Left does not want to be caught any more defending any policies which can be identified with "economic austerity" or policies which labor leaders and others think would endanger past economic gains won by the working class (although union and worker struggles for healthy and safe conditions inside and outside of the workplace obviously connect in positive ways with broader ecological struggles). Most of those who oppose more growth and development are mainstream environmentalists from the urban middle classes who have the consumer goods that they want and also have the time and knowledge to oppose ecologically dangerous policies and practices. It would appear, therefore, that any effort to find a place for the working class in this equation, that is, any attempt to marry socialism and ecology, is doomed from the start.

But just because something has never happened does not mean that it cannot happen, or that it is not happening in various ways right now. In the developed capitalist countries, one can mention the green caucuses within Canada's NDP; the work of Barry Commoner, who calls for source reduction, the "social governance of technology," and economic planning based on a "deep scientific understanding of nature"; the antitoxic and worker and community health and safety movements which bring together labor, community, and ecological issues; various red-green third world solidarity movements, such as the Third World Network and Environmental Project on Central America; and the new emphasis on fighting ecological racism. One thinks of the Socialist Party's struggle for control of the Upper House of the Diet against the long-entrenched liberal Democrats, which reflects rising concern about both ecological and social issues in Japan. In Europe, we can see the greening of Labor, Social Democratic, and Communist parties, even if reluctantly and hesitatingly, as well as the rise of the Green parties, some of which (as in Germany) are to the left of these parties with respect to some traditional demands of the labor movement. And in the subimperialist powers, which are taking the brunt of the world capitalist crisis, for example, Brazil, Mexico, and Argentina in Latin America, India, and perhaps Nigeria, Korea, and Taiwan, there are new ecological movements in which the traditional working class is engaged. And we cannot forget the Nicaraguan experiment which combined policies aimed at deep environmental reforms with socialism and populism.

There are good reasons to believe that these and other ecosocialist tendencies are no flash in the pan, which permits us to propose that ecology and socialism is not a contradiction in terms. Or, to put the point differently, there are good reasons to believe that world capitalism itself has created the conditions for an ecological socialist movement. These reasons can be collected under two general headings. The first pertains to the causes and effects of the world economic and ecological crisis from the mid-1970s to the present. The second pertains to the nature of the key ecological issues, most of which are national and international, as well as local, issues.

First, the vitality of Western capitalism since World War II has been based on the massive externalization of social and ecological costs of production. Since the slowdown of world economic growth in the mid-1970s, the concerns of both socialism and ecology have become more pressing than ever before in history. The accumulation of global capital through the modern crisis has produced even more devastating effects not only on wealth and income distribution, norms of social justice, and treatment of minorities, but also on the environment. An "accelerated imbalance of (humanized) nature" is a phrase that neatly sums this up. Socially, the crisis has led to more wrenching poverty and violence, rising misery in all parts of the world, especially the South, and, environmentally, to toxification of whole regions, the production of drought, the thinning of the ozone layer, the greenhouse effect, and the withering away of rain forests and wildlife. The issues of economic and social justice and ecological justice have surfaced as in no other period in history. It is increasingly clear that they are, in fact, two sides of the same historical process.

Given the relatively slow rate of growth of worldwide market demand since the mid-1970s, capitalist enterprises have been less able to defend or restore profits by expanding their markets and selling more commodities in booming markets. Instead, global capitalism has attempted to rescue itself from its deepening crisis by cutting costs, by raising the rate of exploitation of labor, and by depleting and exhausting resources. This "economic restructuring" is a two-sided process.

Cost cutting has led big and small capitals alike to externalize more social and environmental costs, or to pay less attention to the global environment, pollution, depletion of resources, worker health and safety, and product safety (meanwhile, increasing efficiency in energy and raw-material use in the factories). The modern ecological crisis is aggravated and deepened as a result of the way that capitalism has reorganized itself to get through its latest economic crisis.

In addition, new and deeper inequalities in the distribution of wealth and income are the result of a worldwide increase in the rate of exploitation of labor. In the United States during the 1980s, for example, property income increased three times as fast as wage and salary income. Higher rates of exploitation have also depended upon the ability to abuse undocumented workers and set back labor unions, Social Democratic parties, and struggles for social justice generally, especially in the South. It is no accident that in those parts of the world where ecological degradation is greatest—Central America, for example—there is greater poverty and heightened class struggles. The feminization of poverty is also a part of this trend of ecological destruction. It is the working class, oppressed minorities, women, and the rural and urban poor worldwide who suffer most from both economic and ecological exploitation. The burden of ecological destruction falls disproportionately on these groups.

Crisis-ridden and crisis-dependent capitalism has forced the traditional issues of socialism and the relatively new issues ("new" in terms of public awareness) of ecology to the top of the political agenda. Capitalism itself turns out to be a kind of marriage broker between socialism and ecology, or, to be more cautious, if there is not yet a prospect for marriage, there are at least openings for an engagement.

Second, the vast majority of economic and social and ecological problems worldwide cannot be adequately addressed at the local level. It is true that the degradation of local ecological systems often do have local solutions in terms of prevention and de-linking (although less so in terms of social transformation). Hence it comes as no surprise to find strong connections between the revival of municipal and village politics and local ecological destruction. But most ecological problems, as well as the economic problems which are both cause and effect of the ecological problems, cannot be solved at the local level alone. Regional, national, and international planning are also necessary. The heart of ecology is, after all, the interdependence of specific sites and the need to situate local responses in regional, national, and international contexts, that is, to sublate the "local" and the "central" into new political forms.

National and international priorities are needed to deal with the problem of energy supplies, and supplies of nonrenewable resources in general, not just for the present generation but especially for future generations. The availability of other natural resources, for example, water, is mainly a regional issue, but in many parts of the globe it is a national or international issue. The same is true of the destruction of forests. Or take the problem of soil depletion, which seems to be local or site specific. Insofar as there are problems of soil quantity and quality, or water quantity or quality, in the big food-exporting countries, for example, the United States, the food-importing countries are also affected. Further, industrial and agricultural pollution of all kinds spills over local, regional, and national boundaries. North Sea pollution, acid rain, ozone depletion, and global warming are obvious examples.

Furthermore, if we broaden the concept of ecology to include urban environments, or what Marx called "general, communal conditions of production," problems of urban transport and congestion, high rents and housing, and drugs, which appear to be local issues amenable to local solutions, turn out to be global issues pertaining to the way that money capital is allocated worldwide; the loss of foreign markets for raw materials and foodstuffs in drug-producing countries; and the absence of regional, national, and international planning of infrastructures.

If we broaden the concept of ecology even more to include the relationship between human health and well-being and environmental factors (or what Marx called the "personal condition of production"), given the increased mobility of labor nationally and internationally, and greater emigration and immigration, partly thanks to the way capital has restructured itself to pull out of the economic crisis, we are also talking about problems with only or mainly national and international solutions.

Finally, if we address the question of technology and its transfer, and the relationship between new technologies and local, regional, and global ecologies, given that technology and its transfer are more or less monopolized by international corporations and nation-states, we have another national and international issue.

In sum, we have good reasons to believe that both the causes and consequences of, and also the solutions to, most ecological problems are national and international, hence that far from being incompatible, socialism and ecology presuppose one another. Socialism needs ecology because the latter stresses site specificity and reciprocity, as well as the central importance of the material interchanges within nature and between society and nature. Ecology needs socialism because the latter stresses democratic planning, and the key role of the social interchanges between human beings. By contrast, popular movements confined to the community, municipality, or village cannot by themselves deal effectively with most of both the economic and ecological aspects of the general destructiveness of global capitalism, not to speak of the destructive dialectic between economic and ecological crisis.

If we assume that ecology and socialism presuppose one another, the logical question is, why haven't they gotten together before now? Why is Marxism especially regarded as unfriendly to ecology and vice versa? To put the question another way, where did socialism go wrong, ecologically speaking?

The standard, and in my opinion correct, view is that socialism defined itself as a movement which would complete the historical tasks of fulfilling the promises of capitalism. This meant two things. First, socialism would put real social and political content into the formal claims of capitalism of equality, liberty, and fraternity. Second, socialism would realize the promise of material abundance, which crisis-ridden capitalism was incapable of doing. The first pertains to the ethical and political meanings of socialism; the second, to the economic meaning.

It has been clear for a long time to almost everyone that this construction of socialism failed on two counts. First, instead of an ethical, political society, in which the state is subordinated to civil society, we have the party bureaucratic state; and thus the post-Marxist attempt to reconcile social justice demands with liberalism.

Second, and related to the first point, in place of material abundance, we have the economic crisis of socialism; and thus the post-Marxist attempt to reconcile not only social justice demands and liberalism but also both of these with markets and market incentives.

However, putting the focus on these obvious failures obscures two other issues that have moved into the center of political debates in the past decade or two. The first is that the ethical and political construction of socialism borrowed from bourgeois society ruled out any ethical or political practice that is not more or less thoroughly human-centered, as well as downplaying or ignoring reciprocity and "discursive truth." The second is that the economic construction of abundance borrowed with only small modifications from capitalism ruled out any material practice that did not advance the productive forces, even when these practices were blind to nature's economy. Stalin's plan to green Siberia, which fortunately was never implemented, is perhaps the most grotesque example.

These two issues, or failures, one pertaining to politics and ethics, the other to the relationship between human economy and nature's economy, are connected to the failure of historical materialism itself. Hence they need to be addressed in methodological as well as theoretical and practical terms.

Historical materialism is flawed in two big ways. Marx tended to abstract his discussions of social labor, that is, the divisions of labor, from both culture and nature. A rich concept of social labor which includes both society's culture and nature's economy cannot be found in Marx or traditional historical materialism.

The first flaw is that the traditional conception of the productive forces ignores or plays down the fact that these forces are social in nature, and include the mode of cooperation, which is deeply inscribed by particular cultural norms and values.

The second flaw is that the traditional conception of the productive force also plays down or ignores the fact that these forces are natural as well as social in character.

It is worth recalling that Engels himself called Marxism the "materialist conception of history," where "history" is the noun and "materialist" is the modifier. Marxists know the expression "in material life social relations between people are produced and reproduced" by heart, and much less well the expression "in social life the material relations between people and nature are produced and reproduced." Marxists are very familiar with the "labor process" in which human beings are active agents, and much less familiar with the "waiting process" or "tending process" characteristic of agriculture, forestry, and other nature-based activities in which human beings are more passive partners and, more generally, where both parties are "active" in complex, interactive ways.

Marx constantly hammered away on the theme that the material activity of human beings is two-sided, that is, a social relationship as well as a material relationship; in other words, that capitalist production produced and reproduced a specific mode of cooperation and exploitation and a particular class structure as well as the material basis of society. But in his determination to show that material life is also social life, Marx tended to neglect the opposite and equally important fact that social life is also material life. To put the same point differently, in the formulation "material life determines consciousness," Marx stressed that since material life is socially organized, the social relationships of production determine consciousness. He played down the equally true fact that since material life is also the interchange between human beings and nature, that these material or natural relationships also determine consciousness. These points have been made in weak and strong ways by a number of people, although they have never been integrated and developed into a revised version of the materialist conception of history.

It has also been suggested *why* Marx played up history (albeit to the exclusion of culture) and played down nature. The reason is that the problem facing Marx in his time was to show that capitalist property relationships were historical, not natural. But so intent was Marx to criticize those who naturalized hence reified capitalist production relationships, competition, the world market, etc., that he forgot or downplayed the fact that the development of human-made forms of "second nature" does not make nature any less natural. This was the price he paid for inverting Feuerbach's passive materialism and Hegel's active idealism into his own brand of active materialism. As Kate Soper has written, "The fact is that in its zeal to escape the charge of biological reductionism, Marxism has tended to fall prey to an antiethical form of reductionism, which in arguing the dominance of social over natural factors literally spirits the biological out of existence altogether."[5] Soper then calls for a "social biology." We can equally call for a "social chemistry," "social hydrology," and so on, that is, a "social ecology," which for socialists means "socialist ecology."

The Greens are forcing the Reds to pay close attention to the material interchanges between people and nature and to the general issue of biological exploitation, including the biological exploitation of labor, and also to adopt an ecological sensibility. Some Reds have been trying to teach the Greens to pay closer attention to capitalist production relationships, competition, the world market, etc.—to sensitize the Greens to the exploitation of labor and teaching both Greens and Reds to pay attention to the sphere of reproduction and women's labor.

What does a green socialism mean politically? Green consciousness would have us put "Earth first," which can mean anything you want it to mean politically. As mentioned earlier, what most Greens mean in practice most of the time is the politics of localism. By contrast, pure red theory and practice historically has privileged the "central."

To sublate socialism and ecology does not mean in the first instance defining a new category which contains elements of both socialism and ecology but which is in fact neither. What needs to be sublated politically is localism (or decentralism) and centralism, that is, self-determination and the overall planning, coordination, and control of production. To circle back to the main theme, localism per se won't work politically and centralism has self-destructed. To abolish the state will not work; to rely on the liberal democratic state in which "democracy" has merely a procedural or formal meaning will not work, either. The only political form that might work, that might be eminently suited to both ecological problems of site specificity and global issues, is a democratic state—a state in which the administration of the division of social labor is democratically organized.[6]

Finally, the only *ecological* form that might work is a sublation of two kinds of ecology, the "social biology" of the coastal plain, the plateau, the local hydrological cycle, etc., and the energy economics, the regional and international "social climatology," etc., of the globe—that is, in general, the sublation of nature's economy defined in local, regional, and international terms. To put the conclusion somewhat differently, we need "socialism" *at least* to make the social relations of production transparent, to end the rule of the market and commodity fetishism, to end the exploitation of human beings by other human beings; we need "ecology" *at least* to make the social productive forces transparent, to end the degradation and destruction of the earth.

NOTES

1. This is a crude simplification of green thought and politics, which vary from country to country, and which are also undergoing internal changes. In the United States, for example, where Marxism historically has been relatively hostile to ecology, "Left green" is associated with anarchism or libertarian socialism.

2. This slogan was coined by a conservative cofounder of the German Greens and was popularized in the United States by antisocialist New Age Greens Fritjof Capra and Charlene Spretnak. Needless to say, it was never accepted by Left Greens of any variety.

3. "Mainstream environmentalists" is used to identify those who are trying to save capitalism from its ecologically self-destructive tendencies. Many individuals who call themselves "environmentalists" are alienated by, and hostile to, global capitalism, and also do not necessarily identify with the "local" (see below).

4. Martin O'Connor writes, "One of the striking ambivalencies of many writers on 'environmental' issues is their tendency to make recourse to authoritarian solutions, e.g., based on ethical elitism. An example is the uneasy posturings found in the collection by Herman Daly in 1973 on *Steady-State Economics.*"

5. Quoted by Ken Post, "In Defense of Materialistic History," *Socialism in the World* 74/75 (1989): 67.

6. I realize that the idea of a "democratic state" seems to be a contradiction in terms, or at least immediately raises difficult questions about the desirability of the separation of powers; the problem of scale inherent in any coherent description of substantive democracy; and also the question of how to organize, much less plan, a nationally and internationally regulated division of social labor without a universal equivalent for measuring costs and productivity (however "costs" and "productivity" are defined) (courtesy of John Ely).

PART IV

ECOFEMINISM

16

THE TIME FOR ECOFEMINISM

FRANÇOISE D'EAUBONNE
Translated by Ruth Hottell

NEW PERSPECTIVES

In September 1973 a movement was born in France—a movement closer to the Belgian Unified Feminist party than to the French MLF [Movement de Libération de Femmes].[1] The new group, the Feminist Front, was formed by several women from an MLF group, others from the subgroup Evolution (founded in 1970 after "The Women's Estates General" and in reaction to it), and above all, by independent women from all parties and all movements. The group's statutes adhered to the 1901 law for associations, and accordingly, the new movement marks a tendency which is much more legalistic and even reformist.

In contrast with the Italian movement, which remains somewhat divided, there is no antagonism between this new front and the MLF, which it supports in some of its activities and invites to participate in the front's work. Its dream (rather utopian still, it must be admitted, given the average age, the settled nature, and the bourgeois status of its current members, notwithstanding the continuing possibility of an infusion of new blood) is to serve as a link between all the women's movements and associations, leading toward a massive "sorority" (the Americans say sisterhood) which would further women's causes and their liberation in a determinant fashion. The front believes in means as moderate as those espoused by Betty Friedan: parliamentary representation, obligatory home-economics courses for boys as well as girls, professional promotion, and equal salaries. A woman's right to choose abortion is supported without qualifications, the right to divorce is

From: Françoise d'Eaubonne, "Le Temps de L'Ecoféminisme," *Le Féminisme ou la Mort* (Paris: Pierre Horay, 1974), 215–52, excerpts.

upheld against certain reactionary projects, but the sexual revolution still causes a great discomfort for these prudent ones—or for these prudes.

Another group has arrived on the scene even more recently, formed from a group of the Revolutionary Feminists and also adhering the statutes of the law of 1901: "The League of Women's Rights." This new group is much more radical, although just as resolutely legalistic. Its goal is to fight on legal grounds, enclosing the male society in its own contradictions, with the help of a collective of women lawyers and judges, in order to have women proclaimed a group susceptible to racist treatment and, thus, able to protest against discrimination on constitutional grounds.

We do not yet know the future contributions of these groups which have so suddenly and so quickly emerged after three years of MLF existence. But thanks to the first group, an attempt at synthesis is possible between two struggles previously thought to be separated, feminism and ecology.

Even though Shulamith Firestone had already alluded to the ecological content inherent in feminism in *The Dialectic of Sex: The Case for Feminist Revolution*, this idea had remained at the embryo stage until 1973. It was picked up again by certain members of the Feminist Front, which first included it in its manifesto, then removed it. The authors then separated themselves from such a timorous new movement and founded an information center: the Ecology-Feminism Center, destined to become later, as a part of their project of melding an analysis of and the launching of a new action: *ecofeminism*.

In regards to this question, I would add the idea: "from revolution to mutation," which inspired the title of a portion of this work. Nor would it be superfluous to recall one of the questions posed earlier by the Revolutionary Feminists, a group within the MLF: "We need to know nonetheless if the movement will be a *mass movement to which all women will potentially belong as a specifically exploited group, or if it will be just another subgroup?*"

It is in keeping with this spirit that militants, as much from the Feminist Front as from the League of Women's Rights, concentrate their strength on mobilizing and sensitizing as many of their "sisters" as possible about the relatively restricted and immediate objectives, working toward reasonable goals which can seem "reassuring" (outside the acronym MLF, which has already become "frightening" for many), all the while not forgetting (at least for the "league," younger and more dynamic than the FF) the eventual targets: the disappearance of the salaried class (beyond equal salaries), the disappearance of competitive hierarchies (beyond access to promotion), the disappearance of the family (beyond the control of procreation), and, above and beyond all of these, a *new humanism* born with the irreversible end of the male society, and which by definition must work through the ecological problem (or rather the extreme ecological peril).

For the moment, certainly, the mobilization of women around "specifically feminine" issues can take, even on a legal level, a tone of exigency which widely surpasses the ancient demand for "rights."

We want, say these followers of the most recent feminism, to depart from what certain subversive German groups call the "anti-authoritarian quagmire," without meanwhile sinking into bureaucracy or elitism; we also want to reach the workers and to throw out the bases in the provinces; but as immediate and concrete as these projects are, we know that, above all, we urgently need to remake the planet around a totally new model. This is not an ambition, this is a necessity; the planet is in danger of dying, and we along with it.

Besides the authoritarian socialists or leftists of various degrees whose drone I have situated here as "principal struggle and secondary struggle," there are analysts and agitators, possible companions in struggle much more enlightened than these neo-Stalinists; they never cease calling for the "totalization" of the combat and protesting that everything which is "fragmentary" compromises the final goal—to destroy the Carthage of the System. These activists do not place themselves in the minefield of the "class struggle"; rather, they emphasize the necessity of a global awareness. It is not a question of forwarding one's particular demands, but of introducing new areas of consciousness:

> It is logical that individuals begin with the real experience of their own alienation in order to define the movement of their revolt; but once that is defined, nothing other than integration into the cultural firmament of the system is possible for them. . . . Those of you who are normal, stop limiting yourselves to your normality; homosexuals, stop limiting yourselves to your ghettos; women, stop limiting yourselves to your femininity or to your counter-femininity. Invade the world, exhaust your dreams.[2]

There is a lot of truth in this exhortation, for it was by limiting themselves to their supposed "counter-femininity," that is, the demands for such-and-such a right, that yesterday's feminists imagined themselves to be the magic transformers of their femininitude, that they buried themselves, having gained, for all practical purposes, nothing but air. I completely agree that women's struggle should be global, totalitarian—even when it presents itself from a modestly reformist aspect—or not be at all. But the authors of *Grand Soir* are committing the same error as the society they are fighting by conflating the category "women" with "homosexuals," "lunatics,"[3] and other minorities in revolt against their alienation. In so doing, they forget that the question for women is not that of some minority but that of a majority reduced to the status of a minority, and the only one to be so treated; and, what is more, the only one of the two sexes with the ability of accepting, refusing, slowing down, or accelerating the reproduction of humanity, with whom rests at the moment, even if they are not totally aware of it, the sentence of death or the salvation of humanity in its entirety.

The sole totalitarian combat capable of overturning the System, instead of simply changing it once again for another, and of shifting finally from the spent "revolution" to the mutation that our world calls for, is the women's combat, that of all women; and not only because they were placed in the situation described in the preceding pages, because the iniquity and absurdity outrage their very souls and demand the overthrow of unbearable excess; that is legitimate, but it remains sentimental. The point, quite simply, is that it is no longer a matter of well-being, but of necessity; not of a better life, but of escaping death; and not of a "fairer life," but of the sole possibility for the entire species *to still have a future.*

U-Thant [former UN secretary-general], the French scholars from the Museum of Natural History, the European Council, the *Unesco Courier* have said it again and again to all the highest international authorities; Conrad Lorenz sums it up: "For the first time in human history, no society can pick up the reins." (Obviously he is thinking, like everyone else, of a male society; that is, a society composed of representation, competition, and industrialization, in short, aggression and sexual hierarchy.) The Futurologists Conference, which took place several months ago in Rome and received a great deal of press—more than the war in the Middle East—has repeated it, even if France is keeping silent.[4]

What were these Cassandras shouting from the rooftops? Quite frankly that the point of no return had practically been reached, that one cannot stop a vehicle careening at a hundred miles an hour toward a brick wall when one is only sixty feet away from it, and that everything could end with a very virile "Prepare to meet your maker!" or "Keep clear of industrialized zones!"

Why this flight? To what extent does this colossal declaration of failure coincide with the feminine wish to snatch the car's steering wheel from the hands of male society, with the intention not of driving in its place but of jumping from the car?

These new perspectives of feminism do not detach themselves from the MLF, but they stand apart within it, due less to the more classical and not underground[5] language or the acceptance of a fledgling organization and the concerns of the feminine "masses" than to the global objective, an answer to the critique of parcellization[6] referred to earlier, and which would even be a new humanism: *ecofeminism.*

The reasoning is simple. Practically everyone knows that the two most immediate threats of death today are overpopulation and the destruction of natural resources; fewer are aware of the entire responsibility of the male System—the System as male (and not capitalist or socialist)—in creating these two perilous situations; but very few have yet discovered that each one of the *two* menaces is the logical outcome of one of the *two* parallel discoveries which gave power to men fifty centuries ago: reproduction, and their capability of sowing the earth as they do women.

Until that time, women alone possessed a monopoly on agriculture, and the men believed women were impregnated by the gods. Upon discovering the two possibilities at once—agricultural and procreational—man launched what Wolfgang Lederer has called "the great upheaval" for his own benefit. Having seized control of the soil, thus of *fertility* (later, industry), and of woman's womb (thus of fecundity), it was logical that the overexploitation of the one and the other would result in this double peril, menacing and parallel: overpopulation (a glut of births) and destruction of the environment (a glut of products).

The only mutation which can save the world, therefore, is the "great upheaval" of male power which brought about, first, agricultural overexploitation, then lethal industrial expansion. Not a "matriarchy," of course, or "power to women," but the destruction of power by women, and finally, the way out of the tunnel: egalitarian administration of a world being reborn (and no longer "protected" as the soft, first-wave ecologists believed).

ECOLOGY AND FEMINISM

I had already defended this point of view, before the birth of the Feminist Front and before reading Shulamith Firestone, in 1972, in a letter to *Nouvel-Observateur*, after an ecologist conference where the absence of women was particularly blatant. This declaration brought me numerous, unexpected letters. Some advocated a demand for "women's power"[7] and accused me of reverse sexism. Pierre Samuel, in his otherwise just and honest book, *Ecologie: Cycle infernal ou détente*,[8] referred to my work as "dangerous exaggeration," and, in the end, aligned himself with the concept which was to become the one behind the Ecology-Feminism Center founded by the dissidents from the Feminist Front.

But, is worldwide catastrophe really at our doorstep? Stopping short of the noble remarks of somebody like Lartéguy, who, laced with the figures I was citing on television, shrugged his shoulders and said, "Oh, specialists are always mistaken in their predictions,"[9] nevertheless a lot of people say, "Someone will invent something." All this talk allows them to put the Horse of the Apocalypse in a bottle, like a goldfish.

Consider, however, this remark: "Ecology is going to replace Vietnam among the essential preoccupations for students." Who said this and when? The *New York Times* . . . in 1970. That year, the son of the woman physician Weill-Hall, who founded "family planning," rallied Americans in an effort to struggle against collective suicide, the correlation of which he saw clearly in the problem his mother attacked.

Ecology, the "science that studies the relations of living beings among themselves and the physical environment in which they are evolving," includes, by definition, the relations between the sexes and the ensuing birthrate. Because of the horrors which menace us, the most intense interest is oriented toward the exhaustion of resources and the destruction of the environment which is why it is time to recall that other element, the one which so closely ties together the question of women and of their combat.

Shall I give some technical details?

In America, the alarm was sounded in 1970 by the antiestablishment youth of Earth Day, who moved immediately to ecoterrorism—they bury cars and physically fight with loggers. Celestine Ward, in her feminist book *Women Power,* says that "they want to breathe in accordance with the cycle of the cosmos." No—they want to breathe, period. Judge for yourselves:

Each year, Irène Chédeaux[10] tells us, America must eliminate 142 tons of smoke, 7 million old cars, 30 million tons of paper, 28,000 million bottles, 48,000 million tin cans. "Each one of the 8 million New Yorkers breathes as much noxious material as if he had smoked 37 cigarettes, and he knows now that there are more rats in the dumps than humans in the city." . . .

A totally new sort of protester has just been born: the ecoguerrilla. At night, students saw down billboards that disfigure the countryside. The newspaper *Actuel* reminds us that in Berkeley in 1967 young engineers founded Ecology Action. They called themselves "apolitical"; we can imagine what political interest groups might react to their goals:

To suppress absurd practices: new model cars every year. . . .

To control necessary but potentially dangerous practices: the treatment and distribution of food, the construction and organization of housing, the production and distribution of energy.

To eliminate destructive practices: social regimentation, covering the earth with concrete, asphalt and buildings, the dissemination of waste, *wars.* It can quite simply be called a revolution, comments those signing the article. But *which one?* . . .

The global population at the end of the Roman Empire is estimated to have been 250 to 300 million, less than the population of modern-day China. Outside China and India, which always had the second-highest birthrate, the rest of the human species was distributed mainly throughout the Mediterranean basin. Epidemics, famine, and natural disasters varied the demographic levels, and global growth was insignificant. It took sixteen centuries before the world population reached 500 to 550 million in the seventeenth century. Therefore, we can evaluate, *grosso modo*, the level of growth in the following way:

- 250 million in sixteen centuries (twice the original number). By 1750, we had reached 700 million. Thus:
- 200 million in one century only.
- Next, growth increases. The discoveries of civilization, work to decrease mortality rates, greater longevity gave us, in 1950, 2,500 million inhabitants. New figures:
- 1,800 million inhabitants in two centuries, that is, 900 million a century instead of the 200 million of preceding centuries.

"The situation is so grave that it demands predictions," says Elizabeth Draper in *Conscience et contrôle des naissances*.[11] But the expansion has only intensified; we can no longer count by centuries, but by decades; in 1960 we already numbered 3 billion (twenty years before the expected date, since experts predicted 3,000 to 3,600 million by 1980!) The rate of growth is 500 million in ten years: in ten years, the world population increased, between 1950 and 1960, by the number of the total population in the seventeenth century! . . .

. . . Let's look at the American example; Robin Clarke, UNESCO expert, gives the following:

During the course of a lifetime an American baby will consume:

100 million liters of water,
28 tons of iron and steel,
25 tons of paper,
50 tons of food,
10,000 bottles, 17,000 cans, 27,000 capsules, (maybe) 2 or 3 cars, 35 tires,
will burn 1,200 barrels of petroleum,
will discard 126 tons of garbage,
and will *produce* 10 tons of radiant particles.

This means, the expert concludes, that *the birth of an American baby carries twenty-five times more significance for the ecology than that of a Hindu baby*.[12] It remains for us to evaluate, which Robin Clarke did not think of doing, the chances for survival of such a distressing world economy where, according to the region, one individual will consume during the course of sixty or seventy years twenty-five times more than another, for it is possible that the rebellion of the third world, having enough of being the Büchenwald of the planet, will change the statisticians' figures. Even if there is no third world rebellion, where will this American baby-turned-adult find to consume, burn, and throw away all that Robin Clarke predicts for him, since it has already been predicted that the earth will become a desert in twenty to thirty years at the rate we are going now? ("In 1985, announces Professor René Dubos, the entire planet will be nothing more than a barren desert.") Will massive death from malnutrition put an end to the exponential rate of demographic growth? And of urban concentration? . . .

. . . The world is beginning to accept the idea of abortion for other reasons which make women violently demand their right to exercise control over their own bodies, over their future, over their procreation; it is thanks to worry over the exponential rate described and analyzed earlier that the male society is experiencing some tendency to question itself and to accept demands which are brought about by totally different motives. But what is remarkable about this situation is that this interest, if it is met, will lead to the realization of a situation more favorable for an oppressed caste—women—than for the caste of the oppressors; and the latter group knows it quite well. That is why they hesitate to accord something they themselves must desire: a stop to an insane growth in the birthrate which, along with the destruction of the environment, signs the death warrant for everyone. At the same time, obtaining this fortunate halt by giving women freedom of access to contraception and abortion is, for men, the conviction that *women won't stop there* and will begin to control their own lives; this is, recalling Fourier, a scandal of such violence that it is capable of undermining the bases of society. Hence the hesitations, the contradictions, the reforms and obstacles, the steps forward, the backlash; this comic grimace on the face of power conveys the extreme interior opposition which is tipping the male society apart, at all levels, in all countries.

"Making money, getting rich, exploiting man and nature in order to climb to the most expensive places on the social ladder. . . . As long as a society organizes its production for the goal of converting the resources of man and of nature into profits, no equitable and planned system of ecological balance can exist.[13]

It is evidence itself. At the base of the ecological problem are found the structures of *a certain power*. Like that of overpopulation, it is a problem of *men*; not only because it is men who hold world power and because for a century now they could have instituted radical contraception; but because power is, at the lower level, allocated in such a way to be exercised by men over women. In the domain of ecology as well as that of overpopulation, we see conflicts contradictory to capitalism harshly confronting one another, even though these problems far surpass the scope of capitalism and the socialist camp suffers from just as many, for the good reason that in both sexism still reigns. Under these conditions, where the devil can we find—even in the case of prolonged pacific coexistence of both economic camps—a possibility of implementing an "equitable and planned" system on a planetary scale, whether it be ecological or demographic? Lorenz is right: no (male) society can pick up the reins.

The first problem: stopping new births. The male society has begun to be afraid, and, in turn, Poland, Romania, and Hungary have adopted the solution of "liberalizing" abortion; in the capitalist camp, England followed on 24 October 1967; Japan penalizes a surplus of births. Four American states, New York among them, "have followed in the path of the frank and massive yes" (*Guérir*,[14] August 1971). On the other hand, in the Latin Catholic countries, we find a sole timid overture in France; in Spain, even therapeutic abortion is forbidden: Let the mother die, so long as the fetus lives, even if it has only a minute chance of not following its mother in death. These examples represent some of the principal contradictions of a civilization panicked at the necessity of consenting to what will prolong humanity but also toll the first bell for its old putrid patriarchal forms.

If these problems are men's problems, it is because their origin is masculine: it is the male society, built by males and *for* males—we must repeat this tirelessly.

It would be derisory to play at the little game of historical "ifs"; What would have happened if women had not lost the war of the sexes at the distant time when phallocracy was born—thanks to the passing of agriculture to the male sex? Without what August Bebel calls "the defeat of the feminine sex," and the modern Wolfgang Lederer, author of *Gynophobia*,[15] calls "the great upheaval," what would have happened? We would do well to keep ourselves from entering further into this fantasy (or else we will be in the domain of a charming science-fiction novel): "Ah! If the sky fell, a lot of sparrows would be caught," my grandmother used to say. A lone negative response emerges: perhaps humanity would have vegetated at an infantile stage, perhaps we would have never known either the jukebox or a spaceship landing on the moon, but the environment would have never known the current massacre, and even the word "ecology" would have remained in the little cervical box of *Homo sapiens*, like the word "kidney" or "liver," which never cross the lips or the pen if no suffering or pain is felt from these organs.

"Pollution," "destruction of the environment," "runaway demography" are all men's words, corresponding to men's problems: those of a male culture. These words would have no place in a female culture, directly linked to the ancient ancestry of the great mothers. That culture may have been simply a miserable chaos, much like the ones of the Orient which, as phallocratic as it may be, brings out much more anima than animus. It seems that neither of the two cultures would have been satisfactory, insofar as it would have been sexist also; but the ultimate negativity of a culture of women would have never been this, this extermination of nature, this systematic destruction—with maximum profit in mind—of all the nourishing resources.

It is interesting, returning to the famous stress of the white rat, to note that in the animal kingdom as in the human kingdom it is the female who tends to refuse procreation and not the male, although the instinct of preservation should be present throughout the species.

If we consider the behavior of the males in power in our society, what do we see? Conscious of the peril of overpopulation, they strive to make us believe that it is a "third world problem" and concentrate their efforts on the most disadvantaged spot of the planet, consequently the one which consumes the least. (Pierre Samuel, in *Ecologie: Cycle infernal ou détente*,[16] notes that the United States alone, with 6 percent of the world population, consumes 45 percent of the global resources!). . . .

Before capitalism, the last arrival, aging and resistant, before feudalism, before phallocracy, feminine power, which never reached the dimension or the stature of matriarchy, was founded on the possession of agriculture; but it was an autonomous possession, probably accompanied by a sexual segregation; and that is why there was never a true matriarchy. Men controlled the pastorate and the hunt, women agriculture; each of these two armed groups confronted the other; such is the origin of the supposed legend of the Amazons.[17] When the family arrived on the scene, woman could still treat power from a position of power, as long as agricultural functions continued to make her sacred; the discovery of the process of fertilization—of the womb and of the soil—tolled the last bell. Thus began the iron age of the second sex. It has certainly not ended today. But the earth, symbol and former preserve of the womb of the great mothers, has had a harder life and has resisted longer; today, her conqueror has reduced her to agony. This is the price of phallocracy.

In a world, or simply a country, in which women (and not, as could be the case, *a* woman) found themselves truly in power, the first act would be to limit and space out births. For a long time, *well before overpopulation*, that is what they have been trying to do. The proof is the existence of anticonceptual folklore, where the most frighteningly dangerous procedures border on pure superstition. (Note: not only to avoid having a child, but *so that the husband would stay away from the bed*, oh Freud and the "feminine tendency to evade sexuality"!) These conjuration rites, obviously, were never cited by male scholars, whereas lists of opposite rites—those of fertility—exist everywhere. It is only on the planetary scale (not even national), that man deigns to notice the overpopulation; woman notices the rabbit effect at the family level. What methods did she have at her disposal to make it known? . . .

To summarize—if the class struggle, demography, ecology are all the problems and affairs of men, it is due to "the great defeat of the feminine sex" which took place throughout the planet in 3000 BC. After the demise of the Amazons and agriculture, the guarantor of power, shared for a certain time between the sexes in the Hittite, Cretan, and Egyptian civilizations, little by little the wealth of the earth became masculine at the time when woman, tied to the family, no longer had recourse to vanished Amazonian ways. Patriarchal and masculine power peaked in the Bronze Age, with the discovery of what would later become industry. Women were then put under strict surveillance by the victorious sex, which still suffered from fear and distrust of them; they were exiled from all sectors other than the family ghetto; not only from power and from work outside the home, but even from areas in which man seemed to have no fear of competition: physical sports (ancient Greece, except Lacedaemonia), theater (feudal England, Japan), art and culture, higher education (practically the world over, apart from a few oft-cited exceptions). . . .

How, then, do we approach the problem of maximal profit which sacrifices the collective interest to private interest, or the race for power which takes the place of collective interest in revolutionary instances; how can this problem be resolved as long as mental structures remain as they are: that is to say, informed by fifty centuries of masculine planetary civilization, overexploitative and destructive of the resources?

The proof that no revolution directed and accomplished by men can achieve the necessary mutation is that none has ever gone further than replacing one regime by another, one system by another in accordance with the existing structures, and that none has ever even envisaged *the possibility of going further*, of departing from the infernal cycle of production-consumption which is the alibi

for this enormous mass of work—useless, alienating, mystified, and mystifying
—the very base of the male society wherever it exists. . . .

The case is clear—we are in no way pleading for an illusory superiority of
women over men, or even for the "values" of the Feminine, which exist only on
a cultural level and not at all on a metaphysical one. We are saying: Do you want
to live or die? If you refuse planetary death, you will have to accept the revenge
of women, for their personal interests join those of the human community,
whereas the males' interests, on an individual basis, are separate from those of
the general community, and this holds true even at the level of the current male
System. For proof, we have merely to consider the contradiction between the
supreme instances of its power, pushing women into production (and they have
announced: "1975: The Year of Women"), and the private interests of the males
living under this same power, who furiously resist the idea of depriving themselves of their personal maids. We have merely to consider the contradiction
between the effort of this same power to diffuse and aid contraception in the goal
of using, for its production, the feminine time taken away from the nourishing
function, and the same indignant resistance from individual males against the
fact that their females could control their procreation!

This brings me back to the beginning of this work: awareness of femininitude, of the misfortune of being a woman, is taking place today in a contradiction and an ambiguity which announce the end of the same misfortune. Starting
at real-life events, at radical subjectivity, her experience as a species treated as a
minority, separated, reified, *looked at*, a woman of my generation discovers that
her "little problem," her "secondary question," this so-minute detail of the subversive front, indeed, her "fragmentary struggle," is no longer content to link
with but identifies directly with the number one question, with the original
problem; the basis, even, of the indispensable need to change the world, not just
to improve it, but *so that there can still be a world.*

What revenge for the sole human majority to be treated as a minority! Until
now it was difficult for women to comprehend the source of the misfortune of
femininitude they limited themselves to demands for "fragments" of world management before getting to the roots, free sexual disposition, which suddenly
revealed a sense of totality. She who until now had been not man's "companion"
but at once the alchemical crucible of his reproduction, his beast of burden, his
scapegoat, his spittoon; with whom he sometimes amused himself by covering her
in stones and by proclaiming her to be his Holy Grail; the one who always brought
about in him, by virtue of a constant threat of victory, a hostile distrust which has
gone as far in certain cultures as the hatred that engenders ritual mutilation
(Africa) and death (sexocide of witches in the Middle Ages, of the "debauched"
of the Mediterranean basin or the Orient) and that has evolved into a true "gynophobia," as Lederer puts it, she has now become in law and in possibility, if not
in fact, what she was in prephallocratic times—the sole controller of procreation.
As soon as this right can be freely lived through massive contraception and unin-

hibited access to abortion, the disappearance of half the human nightmare will depend on her as she implements the "stress of the white female rat."

This immense power which she is going to acquire, which is already coming within sight, has nothing in common with the one that organizes, decides, represents, and oppresses and which still remains in male's hands; it is exactly in this area that she can most efficiently bring about its defeat and sound the death knell of ancient oppression. In short, according to a slogan of the Ecology-Feminism Center, we have to tear the planet away from the male today in order to restore it for humanity of tomorrow. That is the only alternative, for if the male society persists, there will be no tomorrow for humanity.

Her very life threatened, as well as the one she passes on (that she *chooses* whether or not to pass on), she who is the keeper of this life source, in whom the forces of the future are realized and through whom they pass, is thus doubly concerned in finding the fastest solution to the ecological problem. And, what is more, she represents, in the purest Marxist fashion in the world, this producing class frustrated from its production by male distribution, since this source of collective wealth (procreation) is possessed by a minority, males, the feminine species being the human majority.

The specialists themselves recognize it; along with Edgar Morin in the ecological conference of the *Nouvel Observateur* (where no women appeared), they admit that "we are starting to comprehend that the abolition of capitalism and the liquidation of the bourgeoisie merely give way to a new oppressive structure." Reimut Reiche had already brought out and explained this fact in *Sexuality and Class Struggle* that the "core" resists all revisions of the order. That core, as I have shown here, is the phallocracy. It is at the base of an order which can only assassinate nature in the name of profit if it is capitalist, and in the name of progress if it is socialist. The problem for women is, first, demography, then nature, and thus the world; her urgent problem, the one in common with youth, is autonomy and control of her destiny. If humanity is to survive, she must resign herself to this fact.

Consequently, with a society finally in the Feminine, which will mean nonpower (and not power to women), it will be proven that no other human category could have accomplished the ecological revolution, for no other was as directly concerned at all levels. And the two sources of wealth diverted for male profit will once again become an expression of life and no longer an elaboration of death; and the human being will finally be treated first as a person, and not above all else as a male or female.

And the planet placed in the feminine will flourish for all.

1971–74

NOTES

1. Translator's note: The initials indicate the French expression *"Mouvement de Libération des Femmes."* For background information and updated details concerning the feminist groups mentioned here, see Claire Duchen, *Feminism in France* (Boston: Routledge and Kegan Paul, 1986); Claire Duchen, *French Connections: Voices from the Women's Movement in France* (Amherst: University of Massachusetts Press, 1987); Elaine Marks and Isabelle de Courtivron, eds., *New French Feminisms* (New York: Schocken Books, 1981); and Nicole Ward Jouve, *White Woman Speaks with Forked Tongue* (New York: Routledge, 1991). These authors discuss the period between the 1970s and the 1990s, discussing the beginnings and evolution of the various groups. Marks and de Courtivron succinctly describe the origins of the MLF: "The 'Mouvement de Liberation des Femmes' . . . commonly referred to as the MLF, is not an organization. It is the name invented by the French press during the summer of 1970 to identify the diverse radical women's groups that had been visible in Paris, Lyon, and Toulouse since the fall of 1968" (p. 30). Although the MLF roughly corresponds, therefore, to the women's liberation movement in the United States, the initials MLF will be used throughout this translation due to the specificity to French feminist activity.

2. *Le Grand Soir,* a prosituationist paper, copied by *Le Fléau Social,* May 1973. (Translator's note: The work has not appeared in English. Its title translates as *The Social Plague.*)

3. The antipsychiatric movement has engendered an "International Association of Looney Lunatics" which contests "normality" even more radically than do the movements for the liberation of homosexuals. (Translator's note: The original text reads: "Internationale des Fous Furieux." I have chosen to translate "fous furieux" as "looney lunatics" to approximate the strength of the alliteration. In doing so, I may have obscured the fury implicit in "furieux"—a concept which is particularly important in this context.)

4. See *Halte à la croissance* (Paris: Fayard, 1972). (Translator's note: The work appeared first in English: Donella H. Meadows, Dennis L. Meadows, Jorgen Randers, and William W. Behrens III, *The Limits to Growth: A Report for the Club of Rome's Project on the Predicament of Mankind* (London: Earth Island, 1972). *Halte à la croissance* is a translation into French of a part of it written by Janine Delaunay.

5. Translator's note: "Underground" is in English and italicized in the original text.

6. Translator's note: On p. 10 of the French text, in a section entitled "Femininitude," d'Eaubonne discusses the criticism espoused by "certain extremists of the left" who call women "a category among others of oppressed people," and claim that a struggle targeting women only is "parcellary."

7. Translator's note: The original text gives the expression in English.

8. Translator's note: Published in Paris at Union Générale de l'édition, 1973; the book has not appeared in English. The title translates as *Ecology: Vicious Circle or Alleviation.*

9. Actually, the figures claimed to be exaggerations have since been greatly surpassed.

10. In *Anticlichés sur l'Amerique* (Paris: Robert Laffont, 1971). The general proportion of these figures has risen 2 to 6 percent in noxious substances. (Translator's note: The title of this work translates as *Anti-Clichés about America.* It has not appeared in English.)

11. Elizabeth Draper, *Conscience et contrôle des naissances* (Paris: Robert Laffont, 1971). She predicted six billion inhabitants by the year 2000. . . . Translator's note: The book was published first in English as *Birth Control in the Modern World: The Role of the Individual in Population Control* (London: Allen and Unwin, 1965).

12. The Right's answer to the classical argument: overpopulation, third world problem.

13. *Actuel*, October 1971.

14. Translator's note: The title translates as "to heal."

15. [See August Bebel, *Women in the Past, Present, and Future* (San Francisco: G. B. Benham, 1897).] Wolfgang Lederer, *Gynophobia ou la peur des femmes* (Paris: Payot, 1971). Translator's note: The work appeared a few years earlier in English: Wolfgang Lederer, *The Fear of Women* (New York: Harcourt Brace Jovanovich, 1968).

16. Translator's note: See note 8.

17. Pierre Gordon, *Initiation Sexuelle et Morale Religieuse* (Paris: Presse universitaire de France, 1945), and my work *Le Féminisme*. Translator's note: Gordon's work has not appeared in English. The title translates as *Sexual Initiation and Religious Morality*.

17

THE ECOFEMINIST CONNECTION

ARIEL KAY SALLEH

Author's note: This much-cited article, first printed in *Environmental Ethics* (1984), was written a year before as a playfully mocking activist speech for a pre-feminist environmentalist audience. The genre should be clear to anyone with a feeling for pace and color in language. But once in publication, the politically loaded irony of taunts like "women already flow with the system of nature" was lost on North American academics. The piece was made to do duty as a philosophical essay and I myself was painted as "pre-feminist," holding to naive essentialist or unexamined views of womanhood! Over the ensuing decade, *Environmental Ethics* ran responses from Wittbecker (1986), Zimmerman (1987), Fox (1989), and my rejoinders "The Ecofeminism/Deep Ecology Debate: A Reply to Patriarchal Reason" (1992) and "Class, Race, and Gender Discourse in the Ecofeminism/Deep Ecology Debate" (1993). —A. K. S.

I offer a feminist critique of deep ecology as presented in the seminal papers of Naess and Devall. I outline the fundamental premises involved and analyze their internal coherence. Not only are there problems on logical grounds, but the tacit methodological approach of the two papers are inconsistent with the deep ecologists' own substantive comments. I discuss these shortcomings in terms of a broader feminist critique of patriarchal culture and point out some practical and theoretical contributions which eco-feminism can make to a genuinely deep ecology problematic.

. . . beyond that perception of otherness lies the perception of psyche, policy and cosmos, as metaphors of one another . . .

John Rodman[1]

From: "Deeper than Deep Ecology: The Ecofeminist Connection," *Environmental Ethics* 6, no. 4 (Winter 1984): 339–45.

In what sense is ecofeminism "deeper than deep ecology"? Or is this a facile and arrogant claim? To try to answer this question is to engage in a critique of a critique, for deep ecology itself is already an attempt to transcend the short-sighted instrumental pragmatism of the resource-management approach to the environmental crisis. It argues for a new metaphysics and an ethic based on the recognition of the intrinsic worth of the nonhuman world. It abandons the hard-headed scientific approach to reality in favor of a more spiritual consciousness. It asks for voluntary simplicity in living and a nonexploitive, steady-state economy. The appropriateness of these attitudes as expressed in Naess's and Devall's seminal papers on the deep ecology movement is indisputable.[2] But what is the organic basis of this paradigm shift? Where are Naess and Devall "coming from," as they say? Is deep ecology a sociologically coherent position?

The first feature of the deep ecology paradigm introduced by Naess is replacement of the Man/Nature dualism with a *relational total-field image*, where man is not simply "in" his environment, but essentially "of" it.[3] The deep ecologists do not appear to recognize the primal source of this destructive dualism, however, or the deeply ingrained motivational complexes which grow out of it. Their formulation uses the generic term *Man* in a case where use of a general term is not applicable. Women's monthly fertility cycle, the tiring sym-biosis of pregnancy, the wrench of childbirth and the pleasure of suckling an infant, these things already ground women's consciousness in the knowledge of being coterminous with Nature. However tacit or unconscious this identity may be for many women, bruised by derogatory patriarchal attitudes to motherhood, including modern male-identified feminist ones, it is nevertheless "a fact of life." The deep ecology movement, by using the generic term Man, simultaneously presupposes the difference between the sexes in an uncritical way, and yet over-looks the significance of this difference. It overlooks the point that if women's lived experiences were recognized as meaningful and were given legitimation in our culture, it could provide an immediate "living" social basis for the alterna-tive consciousness, which the deep ecologist is trying to formulate and introduce as an abstract ethical construct. Women already, to borrow Devall's turn of phrase, "flow with the system of nature."

The second deep ecology premise, according to Naess, is a move away from anthropocentrism, a move toward *biological egalitarianism* among all living species. This assumption, however, is already canceled in part by the implicit contradiction contained in Naess's first premise. The master-slave role, which marks man's relation with nature, is replicated in man's relation with woman. A self-consistent biological egalitarianism cannot be arrived at unless men become open to both facets of this same urge to dominate and use. As Naess rightly, though still somewhat anthropocentrically, points out, the denial of dependence on Mother/Nature and the compensatory drive to mastery which stems from it, have only served to alienate man from his true self. Yet the means by which

Naess would realize this goal of species equality is through artificial limitation of the human population. Now putting the merits of Naess's "ends" aside for the moment, as a "means" this kind of intervention in life processes is supremely rationalist and technicist, and quite at odds with the restoration of life-affirming values that is so fundamental to the ethic of deep ecology. It is also a solution that interestingly enough cuts right back into the nub of male dependence on women as mothers and creators of life—another grab at women's special potency, inadvertent though it may be.

The third domain assumption of deep ecology is the *principle of diversity and symbiosis*: an attitude of live and let live, a beneficial mutual coexistence among living forms. For humans the principle favors cultural pluralism, an appreciation of the rich traditions emerging from Africa, China, the Australian Aboriginal way, and so on. These departures from anthropocentrism, and from ethnocentrism, are only partial, however, if the ecologist continues to ignore the cultural inventiveness of that other half of the human race, women; or if the ecologist unwittingly concurs in those practices which impede women's full participation in his own culture. The annihilation of seals and whales, the military and commercial genocide of tribal peoples, are unforgivable human acts, but the annihilation of women's identity and creativity by patriarchal culture continues as a fact of daily existence. The embrace of progressive attitudes toward nature does little in itself to change this.

Deep ecology is an *anticlass posture*; it rejects the exploitation of some by others, of nature by man, and of man by man, this being destructive to the realization of human potentials. However, sexual oppression and the social differentiation that this produces is not mentioned by Naess. Women again appear to be subsumed by the general category. Obviously the feminist ecological analysis is not "in principle" incompatible with the anticlass posture of deep ecology. Its reservation is that in bypassing the parallel between the original exploitation of nature as object-and-commodity resource and of nurturant woman as object-and-commodity resource, the ecologist's anticlass stance remains only superficially descriptive, politically and historically static. It loses its genuinely deep structural critical edge. On the question of political praxis though, there is certainly no quarrel between the two positions. Devall's advocacy of loose activist networks, his tactics of nonviolent contestation, are cases in point.[4] Deep ecology and feminism see change as gradual and piecemeal; the violence of revolution imposed by those who claim "to know" upon those who "do not know" is an anathema to both.

The fight against *pollution and resource depletion* is, of course, a fundamental environmental concern. And it behooves the careful activist to see that measures taken to protect resources do not have hidden or long-term environmental costs which outweigh their usefulness. As Naess observes, such costs may increase class inequalities. In this context he also comments on the "after hours" environmentalist syndrome frequently exhibited by middle-class professionals.

Devall, too, criticizes what he calls "the bourgeois liberal reformist elements" in the movement—Odum, Brower, and Lovins, who are the butt of this remark. A further comment that might be made in this context, however, is that women, as keepers of *oikos*, are in a good position to put around-the-clock ecological consciousness into practice. Excluded as many still are from full participation in the social-occupational structure, they are less often compromised by the material and status rewards, which may silence the activist professional. True, the forces of capitalism have targeted women at home as consumer par excellence, but this potential can just as well be turned against the systematic waste of industrialism. The historical significance of the domestic labor force in moves to recycle, boycott, and so on, has been grossly underestimated by ecologists.

At another level of analysis entirely, but again on the issue of pollution, the objectivist attitude of most ecological writing and the tacit mind-body dualism which shapes this, means that its comprehension of "pollution" is framed exclusively in external material terms. The feminist consciousness, however, is equally concerned to eradicate ideological pollution, which centuries of patriarchal conditioning have subjected us all to, women and men. Men, who may derive rather more ego gratification from the patriarchal status quo than women, are on the whole less motivated to change this system than women are. But radical women's consciousness-raising groups are continually engaging in an intensely reflexive political process; one that works on the psychological contamination produced by the culture of domination and helps women to build new and confident selves. As a foundation for social and political change, this work of women is a very thorough preparation indeed.

The sixth premise of Naess's deep ecology is the *complexity, not complication principle*. It favors the preservation of complex interrelations which exist between parts of the environment, and inevitably, it involves a systems-theoretical orientation. Naess's ideal is a complex economy supported by division, but not fragmentation of labor; worker alienation to be overcome by opportunities for involvement in mental and manual, specialized and nonspecialized tasks. There are serious problems of implementation attached to this vaguely sketched scenario, but beyond this, the supporting arguments are also weak, not to say very uncritical in terms of the stated aims of the deep ecology movement. The references to "soft future research," "implementation of policies," "exponential growth of technical skill and intervention," are highly instrumental statements which collapse back into the shallow ecology paradigm and its human chauvinist ontology. What appears to be happening here is this: the masculine sense of self-worth in our culture has become so entrenched in scientist habits of thought, that it is very hard for men to argue persuasively without recourse to terms like these for validation. Women, on the other hand, socialized as they are for a multiplicity of contingent tasks and practical labor functions in the home and out, do not experience the inhibiting constraints of status validation to the same extent. The traditional feminine role runs counter to the exploitive technical rationality, which is currently the requisite masculine norm. In place of the disdain that the feminine role receives from all quarters, "the separate reality" of this role could well be taken seriously by ecologists and reexamined as a legitimate source of alternative values. As Snyder suggests, men should try out roles which are not highly valued in society and one might add, particularly this one, for herein lies the basis of a genuinely grounded and nurturant environmentalism. As one ecofeminist has put it:

> If someone has laid the foundations of a house, it would seem sensible to build on those foundations, rather than import a prefabricated structure with no foundations to put beside it.[5]

A final assumption of deep ecology described by Naess is the importance of *local autonomy* and *decentralization*. He points out that the more dependent a region is on resources from outside its locality, the more vulnerable it is ecologically and socially: for self-sufficiency to work, there must be political decentralization. The drive to ever-larger power blocs and hierarchical political structures is an invariant historical feature of patriarchal societies, the expression of an impulse to compete and dominate the Other. But unless men can come to grips honestly with this impulse within themselves, its dynamic will impose itself over and over again on the anatomy of revolution. Women, if left to their own devices, do not like to organize themselves in this way. Rather, they choose to work in small, intimate collectivities, where the spontaneous flow of communication "structures" the situation. There are important political lessons for men to learn from observing and participating in this kind of process. And until this learning takes place, notions like autonomy and decentralization are likely to remain hollow, fetishistic concepts.

Somewhat apologetically, Naess talks about his ecological principles as "intuitive formulations" needing to be made more "precise." They are a "condensed codification" whose tenets are clearly "normative"; they are "ecophilosophical," containing not only norms but "rules," "postulates," "hypotheses," and "policy" formulations. The deep ecology paradigm takes the form of "subsets" of "derivable premises," including at their most general level "logical and mathematical deductions." In other words, Naess's overview of ecosophy is a highly academic and positivized one, dressed up in the jargon of current science-dominated standards of acceptability. Given the role of this same cultural scientism in industry and policy formulation, its agency in the very production of the ecocrisis itself, Naess's stance here is not a rationally consistent one. It is a solution trapped in the given paradigm. The very term *norm* implies the positivist split between fact and value, the very term *policy* implies a class separation of rulers and ruled. Devall, likewise, seems to present purely linear solutions—"an objective approach," "a new psychology"; the language of cost-benefit analysis, "optimal human carrying capacity," and the language of science, "data on hunter gatherers," both creep back in. Again, birth "control programs" are recommended; "zoning" and "programming," the language of technocratic managerialism. "Principles" are introduced and the imperative *should* rides roughshod through the text. The call for a new epistemology is somehow dissociated in this writing from the old metaphysical presuppositions, which prop up the argument itself.

In arguing for an ecophenomenology, Devall certainly attempts to bypass this ideological noose— "Let us think like a mountain," he says—but again, the analysis here rests on what is called "a gestalt of person-in-nature": a conceptual effort, a grim intellectual determination "to care"; "to show reverence" for Earth's household and "to let" nature follow "its separate" evolutionary path. The residue of specular instrumentalism is overpowering: yet the conviction remains that a radical transformation of social organization and values is imminent: a challenge to the fundamental premises of the dominant social paradigm. There is a concerted effort to rethink Western metaphysics, epistemology, and ethics here, but this "rethink" remains an idealism closed in on itself because it fails to face up to the uncomfortable psychosexual origins of our culture and its crisis. Devall points by turn to White's thesis that the environmental crisis derives from the Judeo-Christian tradition, to Weisberg's argument that capitalism is the root cause, to Mumford's case against scientism and technics. But for the ecofeminist, these apparently disparate strands are merely facets of the same motive to control, which runs a continuous thread through the history of patriarchy. So, it has been left to the women of our generation to do the theoretical housework here— to lift the mat and sweep under it exposing the deeply entrenched epistemological complexes which shape not only current attitudes to the natural world, but attitudes to social and sexual relations as well.[6] The accidental convergence of feminism and ecology at this point in time is no accident.

Sadly, from the ecofeminist point of view, deep ecology is simply another self-congratulatory reformist move; the transvaluation of values it claims for itself is quite peripheral. Even the Eastern spiritual traditions, whose authority deep ecology so often has recourse to—since these dissolve the repressive hierarchy of Man/Nature/God—even these philosophies pay no attention to the inherent Man/Woman hierarchy contained within this metaphysic of the Whole. The suppression of the *feminine* is truly an all-pervasive human universal. It is not just a suppression of real, live, empirical women, but equally the suppression of the feminine aspects of men's own constitution which is the issue here. Watts, Snyder, Devall—all want education for the spiritual development of "personhood." This is the self-estranged male reaching for the original androgynous natural unity within himself.

The deep ecology movement is very much a spiritual search for people in a barren secular age; but how much of this quest for self-realization is driven by ego and will? If, on the one hand, the search seems to be stuck at an abstract cognitive level, on the other, it may be led full circle and sabotaged by the ancient compulsion to fabricate perfectability. Men's ungrounded restless search for the alienated Other part of themselves has led to a society where not life itself, but "change," bigger and better, whiter than white, has become the consumptive end. The dynamic to overcome this alienation takes many forms in the postcapitalist culture of narcissism—material and psychological consumption like karma-cola, clown workshops, sensitivity training, bioenergetics, gay lib, and surfside six. But the deep ecology movement will not truly happen until men are brave enough to rediscover and to love the woman inside themselves. And we women, too, have to be allowed to love what we are, if we are to make a better world.

NOTES

1. John Rodman, "The Liberation of Nature?" *Inquiry* 20 (1977): 83–145; quoted by Bill Devall in "The Deep Ecology Movement," *Natural Resources Journal* 20 (1980): 317.

2. Arne Naess, "The Shallow and the Deep, Long-Range Ecology Movement," *Inquiry* 16 (1973): 95–100; Bill Devall, "The Deep Ecology Movement," pp. 299–322.

3. See Ariel Kay Salleh, "Of *Portnoy's Complaint* and Feminist Problematics," *Australian and New Zealand Journal of Sociology* 17 (1981): 4–13; "Ecology and Ideology," *Chain Reaction* 31 (1983): 20–21; "From Feminism to Ecology," *Social Alternatives* (1984).

4. And on this connection, see Ariel Kay Salleh, "The Growth of Ecofeminism," *Chain Reaction* 36 (1984): 22–28. See also comments in "Whither the Green Machine?" *Australian Society* 3 (1984): 15–17.

5. Ann Pettitt, "Women Only at Greenham," *Undercurrents* 57 (1982): 20–21.

6. Some of this feminist writing is discussed in Ariel Kay Salleh, "Contributions to the Critique of Political Epistemology," *Thesis Eleven* 8 (1984): 23–43.

18

ECOSOCIAL FEMINISM
AS A GENERAL THEORY
OF OPPRESSION

VAL PLUMWOOD

The uneasy present relationship between the various social change movements has been reflected in the vigorous and often bitter debate which has been taking place in the area of green theory and political ecology. Three of the main articulated positions involved in this dialogue, which has taken place in a major way in the United States, are each linked to critiques of domination associated with the respective movements. The positions—social ecology, Deep Ecology, and ecological feminism—are usually treated as presenting alternative and competitive analyses of the destruction of the biosphere. Thus Deep Ecology is perhaps the best-known branch of what has been called "deep green theory," a set of positions or critiques treating anthropocentrism[1] and the human domination of nature as one of the major roots of environmental problems. Social ecology, whose best-known exponent is Murray Bookchin, draws on radical tradition and focuses on an analysis of ecological problems in terms of human social hierarchy, while ecofeminism sees androcentrism or the domination of women as a model for other kinds of domination and as linked especially to the domination of nature.

The debate involves real issues of genuine importance, but is also motivated by competitive reductionism and has an unnecessarily dismissive character. An example will illustrate what is meant by "reductionism" here. Back in the days when Marxism was king of radical discourses, other discourses and critiques, such as those of the women's movement and the environment movement, were reduced to subject status to be subsumed, incorporated into the kingdom of the

From: Ronnie Harding, ed., *Ecopolitics V Proceedings,* Centre for Liberal and General Studies, University of New South Wales, Kensington, New South Wales, Australia, 1992, 63–72.

sovereign. Their insights and problems were recognized and accorded legitimacy and attention just to the extent that they could be so absorbed (for example, those aspects of the feminist critique which could be reduced to questions of "class").[2] The reducing position claims to have the "fundamental" or "master" critique into which the reduced others can be absorbed as colonies. What cannot be reduced is dismissed.

Thus Bookchin flatly denies the legitimacy and relevance of the new rival critique of anthropocentrism, the human domination of nature (which of course is the hardest thing for the older radical tradition to try to take account of). The domination of nature, he assures us, came after the domination of human by human and is entirely secondary to it. His practice coheres with his theoretical view of the inferior status of this critique, since in *Remaking Society* he rarely mentions nonhuman nature without attaching the word "mere" to it. (Thus deep ecologists are said to want to "equate the human with mere animality," to "dissolve humanity into a mere species within a biospheric democracy," and reduce humanity "to merely one life form among many").[3] The egalitarian approach of Deep Ecology is roundly condemned as debasing to humans and involving a denial of their special qualities of rationality. Many ecological writers (especially Elizabeth Dodson Gray) have made the point that we construe difference in terms of hierarchy, and that it is not a question of denying human difference but of ceasing to treat it as the basis of superiority and domination.[4] Despite this, Bookchin persistently interprets the denial of human hierarchy over nature as the denial of human distinctness. His concept of humans as "second nature" (nature rendered self-conscious) makes it difficult to conceptualize conflicts between the interests of humans and the interests of "first" nature. Thus Bookchin writes of second nature in an ecological society as "first nature rendered self-reflexive, a thinking nature that knows itself and can guide its own evolution."[5]

The critique of human domination of nature seems to me to be a new and inestimably important contribution to our understanding of domination. However, the alternative position in the debate presented by leading exponents of Deep Ecology is hardly less unsatisfactory or incomplete than Bookchin's account of social ecology and is equally intent on masculine strategies of colonization or on denying connection. Leading deep ecologist Warwick Fox makes repeated counterclaims to "most fundamental" status for his own critique of the domination of nature, arguing that it accounts for forms of human domination also.[6] At the same time (and inconsistently) Fox attacks critiques of other forms of domination as irrelevant to environmental concern, claiming, for example, that feminism has nothing to add to the conception of environmental ethics.[7] This exemplifies a common tendency among deep ecologists to dismiss human hierarchy as simply irrelevant to the problem of the destruction of nature and to distance themselves from other radical critiques. The mistreatment of nature is seen as a problem created by the blanket category of humans, among whom no relevant distinction need be made in explanation of environmental problems.

Each of the three major critiques would appear to share an approach via the concept of *domination* which could provide ground for a new synthesis and a common political orientation. However, as it stands, this potential for meeting remains undeveloped mainly because Deep Ecology has chosen to develop in ways which suppress the potential for a truly *political* understanding of its theme of human/nature domination, that is, as involving the operations of power. Thus Deep Ecology under the guidance of Warwick Fox has chosen for its core concept the notion of *identification*, understood as an individual psychic act rather than a political practice, yielding a theory which is both individualist (in the sense that it fails to look beyond the individual) and psychological-reductionist (in the sense that it discounts factors beyond psychology).[8] A similarly apolitical understanding is given to its core concept of ecological selfhood. Part of the motivation for this way of interpreting the central concepts seems to be desire to distance from other radical traditions and movements which are seen as less respectable and dangerously political. Such an account of Deep Ecology lends itself to being served up as a religious or spiritual garnish for a main course that turns out to be an uncritical approach to change via liberal political theory.[9]

AN ALTERNATIVE COOPERATIVE APPROACH

That fact that the debate is necessary and important should not obscure the fact then that much of its divisiveness is unnecessary. For example, the recognition of nature and animals as oppressed is *in no way incompatible* with the recognition of that domination as linked to, and continuous with, human domination. In fact it greatly enlarges and extends our understanding of each to see them as so linked. The point that it is possible to have essentially linked forms of oppression, and correspondingly relations of cooperation between movements which can fruitfully explore connection without attempting reduction, has been implicit in the approach of many ecofeminists. The concern to avoid absorption or colonization, not to lose what is new and distinctive in the environmental critique, is a legitimate one, but does not demand either isolationist or competitive reductionist stances in relation to other movements. The choice between reducing other critiques or being reduced by them is a false one. The barriers to a theoretical synthesis are political, not theoretical.

It is possible to respect the differences which different struggles and forms of oppression give rise to, to recognize their irreducibility, without paying the price of isolationism. As with individuals, so with movements, connection is risky but nevertheless essential for many reasons. First, some of us understand ourselves as oppressed in several different but connected ways, and work for a world in which sexism, the destruction of nature, racism, and militarism, for example, do not exist. For us movement connection is not an intellectual matter alone; its necessity is felt in the bones, in the attempt to understand and change our lives, in which various forms of oppression may be exemplified. Incompatibilities between critiques in such contexts appear as direct and personal discomforts and conflicts which impel us to change them. But to seek connection is not to require uniformity, or the absorption of each critique and movement into a single oceanic theory or movement, as I argue below.

From such a third perspective many of the criticisms Deep Ecology and social ecology have made of each other seem valid, but can be avoided by such a third position. Thus Deep Ecology has, on this view, been right in criticizing social ecology's human chauvinism and continued subscription to traditional doctrines of human supremacy and difference.[10] But social ecology is similarly correct in its criticism of the insensitivity of Deep Ecology to differences within the category of humans and disregard for the role of human hierarchy in creating environmental problems.[11]

The strategies of isolation and colonization followed by both critiques are both bad methodology and bad politics. They are bad methodology because they involve a false choice, and bad politics because they pass up important opportunities for connection and strengthening. They are bad politics also because it is essential for critiques which purport to treat hierarchy to be prepared to meet others on a basis of equality, not with an agenda of inferiorizing or absorbing them. Maximizing chances for change must involve broadening the base of those who desire change, who can see how change is relevant to their lives, and this involves maximizing connection with a wide variety of issues and social change movements.

ECOFEMINISM AND CONNECTION

Unless social and Deep Ecology rework their positions, it seems that if we are to obtain such a theoretical base adequate to encompass and link the concerns of the environment movement, then it is to ecological feminism that we will have to turn. Ecofeminism is a very diverse position, and there are ecofeminists who are closely associated with Deep Ecology and other ecofeminists who are close to social ecology, as well as others close to radical and other forms of feminism. Some forms of feminism and ecofeminism, principally those emerging from radical feminism, have a reductionist slant of their own, taking patriarchy to be the basis of all hierarchy, the basic form of domination to which other forms (including not only the domination of nature but capitalism and other forms of human social hierarchy) can be reduced. But many other feminists do not find this approach convincing, and see women's oppression as one (although perhaps a key one and certainly an irreducible one), among a number of forms of oppression.[12] Their approach has centered on enriching and networking various different critiques and exploring connections with the critique of the domination of women. Thus unlike the other two positions, ecofeminism as a general position is not committed to reducing or dismissing either the critique of anthropocentrism and the domination of nature with which Deep Ecology is concerned or that of the human hierarchy with which social ecology is concerned.[13] Many try to combine both perspectives. Thus Karen Warren writes that "transformative feminism would expand upon the traditional conception of feminism as a movement to end women's oppression by recognizing and making explicit the interconnections between all systems of oppression. . . . Feminism, properly understood, is a movement to end all forms of oppression. . . . A transformative feminism would build on these insights [of socialist and black feminism] to develop a more expansive and complete feminism, one which ties the liberation of women to the liberation of all systems of oppression."[14]

Major forms of ecofeminism have been concerned with cooperative rather than competitive movement strategies. The domination of women is of course central to the ecofeminist understanding of domination, but is also an illuminating model for many other kinds of domination, since the oppressed are often both feminized and naturalized. The ecofeminism of writers such as Rosemary Ruether has always stressed the links between the domination of women, of human groups such as blacks, and of nature. "An ecological ethic" she writes, "must always be an ethic of ecojustice that recognizes the interconnection of social domination and the domination of nature."[15] Thus the work of many ecofeminists foreshadows the development of ecosocial feminism as a general theory of oppression.

THE NETWORK OF OPPRESSION

Ecofeminism has particularly stressed that the treatment of nature and of women as inferior has supported and "naturalized" not only the hierarchy of male to female but the inferiorization of many other groups of humans seen as more closely identified with nature. It has been used to justify, for example, the supposed inferiority of black races or indigenes (conceived as more animal), the supposed inferiority of "uncivilized" or "primitive" cultures, and the supposed superiority of master to slave, boss to employee, mental to manual worker. For Western society, which has particularly employed a strongly genderized concept of nature as a way of imposing a hierarchical order on the world, feminization and naturalization have been crucial and connected strands supporting pervasive human relations of inequality and domination both within Western society and between Western society and non-Western societies. The interwoven dualisms of Western culture, of human/nature, mind/body, male/female, reason (civilization)/nature, have been involved here to create a logic of interwoven oppression consisting of many strands coming together.[16]

This interweaving occurs not only at the level of ideology, of the dualistic conceptions of the world, but also at the level of material practices, of production and reproduction. Domination must be seen as material and cultural, not as happening just at the level of ideas. Maria Mies has come closest to giving a general account of how this interweaving happens at the material level, and has traced a plausible historical path from the monopoly of means of coercion in the hands of male hunting bands, to the development of protomilitary forms of organization and weaponry.[17] This enables a form of parasitism directed first against women and other tribal groups, who are exploited as agricultural laborers or slaves, and later against wider groups as a process of accumulation. In this form of accumulation (a growth-driven process envisaged as much wider than capitalism itself and including most known forms of socialism), domination has been directed in more recent times against a variety of human groups, especially women and the colonized, and also against nature itself. Mies's framework integrates without reduction, anarchist, antimilitarist, class, race, anticolonialist, feminist, and environmental concerns.

The example of the sealing industry, the first form of accumulation witnessed on this continent [Australia], will serve to illustrate some of these processes, and especially how the inferiorization of the sphere of nature and the feminine combines with the definition of oppressed groups as part of this sphere to yield both an ideology and a material practice of linked oppression. The history of the convicts, of the Aborigines and of the seals and whales whose deaths fueled these processes is interwoven at all levels. At the level of *production*, the convict system helped maintain the savagely repressive internal order of the class and property structure of Britain, the product of a long-term previous accumulation process. The slaughter of seals and whales provided fuel, oil, and a commercial basis for the convict transportation industry.

It took Australia's first export industry a mere eight years from the first sealing expedition in 1798 to reduce the numbers of seals in Bass Strait and Tasmania to a point below the levels capable of commercial exploitation. When the seals were finished, the sealers moved on to other states and to New Zealand. Sealers typically killed all sizes and ages, clubbing or stabbing seals as they came ashore. From 1806, when William Collins set up a bay whaling station in the Derwent (where whales were reported to be so thick in season that collisions with boats were a problem), the seal story was immediately repeated with whales. Female southern right whales (or bay whales), now some of the world's rarest whales, were killed along with their young when they entered the bays to give birth. Within a few decades the industry could boast the virtual local extinction of bay whales. Again the industry moved on to repeat the same story in New Zealand.[18]

The ships which had delivered their human cargoes of convicts to the disciplinary system went sealing or whaling and returned profitably to the "civilized" world with holds filled with oil or skins. In turn, the runaway convicts and soldiers provided suitably hardened and desperate workers for these industries and helped clear the country of the despised natives, described by the *Hobart Town Gazette* of 1824 as "the most peaceful creatures in the world."[19] The industry involved the abduction and enslavement of large numbers of Aboriginal women, who were subject to cruelty and to rape, and the killing of other Aborigines. Settlement along these lines led to the virtual annihilation of Aboriginal Tasmanians, who survive today as a distinct grouping with mixed ancestry, claiming Aboriginal identity but almost entirely dispossessed of their culture and lands.[20]

At the level of *culture*, the ideology which linked these common oppressions stressed the inferiority of the order of "nature," which was construed as barbaric, alien, and animal, and also as passive and female. It was contrasted with the truly human realm, marked by patriarchal, Eurocentric, and body-hating concepts of reason and "civilization," maximally distanced from "nature." Aborigines were seen as part of this inferior order, supposedly being "in a state of nature," and without culture. Aborigines "lived like the beasts of the forest," writes Cook at Adventure Bay in 1770; they were "strangers to every principle of social order."[21] Early journal reports consistently stressed their nudity and propensity to leave open to public view "those parts which modesty directs us to conceal." Where clothes are construed as the mark of civilization and culture, nudity confirms an animal and cultureless state, a reduction to body. The ideology of nature, reason, and civilization made it possible to deny kinship and see the natural world as an inferior realm open to merciless exploitation. The unclothed bodies of Aborigines and the technological economy of Aboriginal life meant that they could be seen in terms of such an ideology as not fully human, but as part of the realm of nature, to be treated in much the same way as the seals. Such was the philosophical basis not only for the annihilation of Tasmanian Aborigines but for the annexation of Australia under the rubric of *Terra nullius*, a land without occupiers, under which Australians still live.[22]

Such a history bears witness to a particular way of treating the sphere of *reproduction*, the sphere our culture has associated with women and the feminine principle. (I use reproduction here, following Carolyn Merchant, to include the reproduction of all nature, as well as human reproduction.)[23] The systematic wiping out of breeding colonies and killing of animals, whales, and seals in the act of giving birth seems to involve contempt for the very processes of life. It implies ignorance or denial of human dependence on these processes, and a view of humans as apart, outside of nature, which is treated as limitless provider. Such an economy is only possible in a culture which has systematically backgrounded, "disappeared," the sphere of reproduction. The inferiorization and denial of this sphere of reproduction involves the inferiorization and denial of women's labor, including reproductive labor.[24] The real heart of the problem of sustainability lies in this kind of consciousness with respect to the reproductive economy, a consciousness which is the product of an alienated and masculinized account of human identity.

These connections are not just historical curiosities. Aborigines are still killed, not now through open mass annihilation, with guns and poisoned flour, but through police violence. *Terra nullius* is still the fiction under which the Australian commonwealth is constituted. Seals too are now legally "protected," but the slaughter of seals, in large enough numbers to keep the population in danger, continues unabated into the present.[25] The sphere of reproduction remains unacknowledged. The network of oppression stretches from the past into the present.

A COOPERATIVE MOVEMENT STRATEGY: METHODOLOGY AND POLITICS

The conception of oppression as a network of multiple, interlocking forms of domination raises a number of new methodological dilemmas and requires a number of adjustments for liberation movements. The associated critiques cannot, for example, simply be added together, for there are too many discrepancies between them. Should we say, for example, that opposition struggle involves one movement or many? Each answer to the one/many dilemma has its problems.

One way to deal with the multiplicity of oppressions is to say that each involves all, for example, that feminism should be thought of as a movement to end all forms of oppression.[26] But this seems to imply an oceanic view of the movements as submerged in a single great movement, for example, that there should not and cannot be an autonomous women's movement concerned primarily with women's oppression (or indeed any other autonomous movement). But this would deny the specificity of, for example, women's oppression and the need for accounts to relate to lived experience, as well as the possibility of difference of direction and conflicts of interest between movements (for example, ethnic, race, and sexual oppression). And even if struggles have a common origin point, enemy, or conceptual structure, it does not follow that they then become the same struggle. The women's movement especially has had good historical reason to distrust the submergence of women's struggle in the struggles of other movements, and has wanted to insist on the importance of movement autonomy and separate identity. And if a struggle which is too narrow and aimed at only a small part of an interlocking system will fail, so too will one which is too broad and lacking in a clear focus and a basis in personal experience. On the other hand, treating the women's movement as isolated from other struggles is equally problematic, because there is no neutral, apolitical concept of the human or of society in which women can struggle for equality, and no pure, unqualified form of domination which is simply male and nothing else which oppresses them. And since most women are oppressed in multiple ways, as particular kinds of women,[27] women's struggle is inevitably interlinked with other struggles.

The dilemma is created by setting up a choice between viewing liberation struggles as a shifting multiplicity only fortuitously connected (as in poststructuralism), versus viewing them as a monolithic, undifferentiated, and unified system. But if there are reasons for seeing it both as a multiple structure and as a unified structure, any model which does not recognize both these aspects is distorting. It is possible to bypass this one/many dilemma if these forms of oppression are seen as very closely, perhaps essentially, related, and working together to form a single system without losing a degree of distinctness and differentiation. A good working model which is easily visualized and which enables such an escape from the one/many dilemma is that of oppressions as forming a net or web. In a web there are both one and many, both distinct foci and strands with room for some independent movement of the parts, but a unified overall mode of operation, forming a single system. The objections which some feminists have raised to what has been called "dual systems theory" in the case of capitalism and patriarchy[28] focus on the links and unified operation of the web rather than the differentiated aspects of the structure. The interconnectedness of forms of oppression provides another reason for viewing these oppressions as forming a *single mutually supporting system*. The sorts of considerations which tell against the oceanic view provide a reason for viewing them as forming a *differentiated system*, with distinct parts which can and must be focused upon separately as well as together, as in a web.[29] Bell hooks's conception of feminism as retaining its own identity but as necessarily overlapping with and participating in a wider struggle captures the politics implied by the weblike nature of oppression and enables a balance between the requirements of identity politics and the requirements of connected opposition which arises from the connected nature of oppressions.[30]

If oppressions form a web,[31] it is a web which now encircles the whole globe and begins to stretch out to the stars, and whose strands grow ever tighter and more inimical to life as more and more of the world becomes integrated into the system of the global market and subject to the influence of its global culture. In the methodology and strategy for dealing with such a web, it is essential to take account of both its connectedness and the capacity for independent movement among the parts. Rarely can it be said, "Once we have cut this section, solved this problem, all the rest will follow, other forms of oppression will wither away." A web can continue to function and repair itself despite damage to localized parts of its structure. The parts can even be in conflict and perhaps move for a limited time in opposite directions.

The strategies for dealing with such a web require cooperation. A cooperative movement strategy suggests a methodological principle for both theory and action, that whenever there is a choice of strategies or of possibilities for theoretical development, then other things being equal those strategies and theoretical developments which take account of or promote this wider, connected set of objectives are to be preferred to ones which do not. This should be regarded as a minimum principle of cooperative strategy. But it is one which the major green positions of social and Deep Ecology currently fail.

Thus Deep Ecology has chosen the company of American nature mysticism and of religious Eastern traditions such as Buddhism over that of various radical movements, including feminism, it might have kept better company with. Elsewhere I have argued[32] that Deep Ecology gives various accounts of the ecological self, as indistinguishable (holistic), as expanded and as transpersonal, and that all of these are problematic both from the perspective of ecological philosophy, and from that of other movements. Deep Ecology could do virtually everything it needs to do with a different account of the self as relational which does form a relevant connecting base for other movements. Social ecologists have rightly pointed to some of the political implications of this choice, that they lead away from connections with the radical movements and traditions, and lead toward these being seen as only accidentally connected to environmental concerns. That is, Deep Ecology has chosen a theoretical base which leads to a weakening of connection with other radical movements that are then seen as only accidentally connected, which from a different perspective appear essentially connected. Social ecology, however, also follows a noncooperative strategy in choosing to discount the critique of anthropocentrism, since, as we have seen, there is no incompatibility between rejecting both human hierarchy and also the human domination of nature.

If there is some reason for hope in our current situation, I believe it mainly lies in this: that we now have the possibility of obtaining a much more complete and connected understanding of the web of domination than we have ever had before, and hence a much more comprehensive and connected oppositional practice. What may be especially significant about this point in history is not only that the current global power of the web places both human and biological survival itself for the first time in the balance, but that several critically important parts of its fabric have recently become for the first time the subject of widespread conscious, self-reflective opposition. Domination must always be conceived of in terms of an open and not a closed set. We can never afford to become complacent, to feel that ours is the complete tally, the final form of understanding. Nevertheless the picture now seems much more comprehensive. Some forms of socialist feminism have already moved to integrate colonial/racial domination as a third parameter for analysis, but have still to take explicit account of nature.[33] A further merger with ecofeminism to form a social ecofeminism would provide a broader and deeper basis of oppositional theory and practice, and fill out some crucial connections.

Previous major oppositional theories such as Marxism now seem so defective because they had so many blind spots, such an incomplete, reductionistic, and fragmentary understanding of the web.[34] And that very incompleteness, the failure to address a wide enough range of concerns led to their failure in practice as well as in theory, leaving domination ever ready to renew and consolidate itself in a different but related form, as state and bureaucratic tyranny, as sexism, as militarism, as power over nature. But this does not mean that we should abandon the entire set of radical traditions, as Porritt suggests (and certainly not that we should abandon them because they are presently unfashionable), but rather that we must come to an understanding of them as limited and partial. The problems of human inequality and hierarchy which the radical traditions addressed over the centuries have not gone away and are taking new and even more sinister "environmental" forms. Their visions of human equality and the immense creative energies they harnessed over long periods of history have helped to form the green vision of a world where all species matter, and are still highly relevant to our understanding of the web. We must somehow balance recognition of the power and strength of past radical traditions with recognition of the need for major revision and reworking, and so come to build better.

NOTES

1. The critique of anthropocentrism or "deep green theory" has developed over a period of twenty years and has included Deep Ecology as a proper subset. See, for example, Arne Naess, "The Shallow and the Deep, Long-Range Ecology Movement: A Summary," *Inquiry* 16 (1973): 95–100; Arne Naess, *Ecology, Community, and Lifestyle* (New York: Cambridge University Press, 1989); Val Plumwood, "Critical Notice: John Passmore's *Man's Responsibility for Nature*," *Australasian Journal of Philosophy* 53 (1975): 171–85; Richard Routley and Val Plumwood, "Against the Inevitability of Human Chauvinism," in *Ethics and Problems of the 21st Century*, ed. K. E. Goodpaster and K. M. Sayre (Notre Dame, IN: University of Notre Dame Press, 1979); Richard Routley and Val Plumwood, "Human Chauvinism and Environmental Ethics," in *Environmental Philosophy*, ed. Don Mannison, Michael McRobbie, and Richard Routley, Monograph Series 2, Philosophy RSSS, 96–189 (Canberra: Australian National University, 1980); Bill Devall and George Sessions, *Deep Ecology: Living as if Nature Mattered* (Salt Lake City: Peregrine Smith Books, 1985).

2. This tendency is discussed by various contributors, especially Sandra Harding, "What Is the Real Material Base of Patriarchy and Capital," in *Women and Revolution*, ed. Lydia Sargent (Boston: South End Press, 1981).

3. Murray Bookchin, *Remaking Society* (Montreal: Black Rose Books, 1989), p. 44, see also pp. 39, 42. See also Murray Bookchin, *Kick It Over*, Special Supplement (1988), 6A; Murray Bookchin, "Social Ecology versus Deep Ecology," *Green Perspectives* 4/5 (Summer 1987).

4. Elizabeth Dodson Gray, *Green Paradise Lost: Re-mything Genesis* (Wellesley, MA: Roundtable Press, 1979), p. 19.

5. Murray Bookchin, *The Philosophy of Social Ecology* (Montreal: Black Rose Books, 1990), p. 182.

6. Warwick Fox, "The Deep Ecology–Ecofeminism Debate and Its Parallels," *Environmental Ethics* 11, no. 1 (1989): 5–25.

7. On the irrelevance of feminism, see Fox, "The Deep Ecology–Ecofeminism Debate," p. 14. For a discussion of this point, see Karen Warren, "The Power and Promise of Ecofeminism," *Environmental Ethics* 12 (1988): 123–46, see esp. 144–45.

8. Fox, "The Deep Ecology–Ecofeminism Debate"; Warwick Fox, *Towards a Transpersonal Ecology: Developing New Foundations for Environmentalism* (Boston: Shambala, 1990).

9. This treatment was suggested by the remarks of Jonathon Porritt's "Green Politics," address to the Ecopolitics V conference, University of New South Wales, Sydney, Australia, April 1991.

10. For a general critique of Bookchin's position, see Robyn Eckersley, "Divining Evolution: The Ecological Ethics of Murray Bookchin," *Environmental Ethics* 11 (1989): 99–116.

11. George Bradford, "Return of the Son of Deep Ecology," *Fifth Estate* 24, no. 1 (1989): 5–35; Brian Tokar, "Exploring the New Ecologies: Social Ecology, Deep Ecology and the Future of Green Political Thought," *Fifth Estate* 24, no. 1 (1989): 18–19.

12. See the work of ecofeminists such as Rosemary Radford Ruether, *New Woman, New Earth* (Minneapolis: Seabury Press, 1975); Karen Warren, "Feminism and Ecology: Making Connections," *Environmental Ethics* 9 (1987): 3–20; Warren, "Power and Promise of Ecofeminism"; and Ynestra King, "The Ecology of Feminism and the Feminism of Ecology," in *Healing the Wounds*, ed. Judith Plant (Philadelphia: New Society Publishers, 1989); Ynestra King, "Healing the Wounds: Feminism, Ecology, and the Nature/Culture Dualism," in *Reweaving the World*, ed. Irene Diamond and Gloria Feman Orenstein (San Francisco: Sierra Club Books, 1990).

13. This orientation to connection is not grasped by those Deep Ecology critics of ecofeminism who take their own imperialist urges to be universal and understand ecofeminism as the claim that "there is no need to worry about any form of human domination other than that of androcentrism." See Fox, "The Deep Ecology–Ecofeminism Debate," p. 18, and Michael Zimmerman, "Feminism, Deep Ecology, and Environmental Ethics," *Environmental Ethics* 9 (1987): 21–44, see esp. p. 37.

14. Warren, "Feminism and Ecology," p. 18.

15. Rosemary Radford Ruether, "Toward an Ecological-Feminist Theology of Nature," in *Healing the Wounds*, ed. Judith Plant (Philadelphia: New Society Publishers, 1989), pp. 145–50, quotation on p. 149.

16. See also Joan L. Griscom, "On Healing the Nature/History Split in Feminist Thought," *Heresies* 13, no. 4 (1981): 4–9. Of course, other node points on the net might equally well be starting points for generalization. The model does not imply that all strands are equally important.

17. Maria Mies, *Patriarchy and Accumulation on a World Scale* (London: Zed Books, 1986).

18. James Bonwick, *The Last of the Tasmanians* (London, 1870).

19. Ibid., quotation on p. 45.

20. Val Plumwood, "SealsKin," *Meanjin* 51, no. 1 (1992): 45–58.

21. Bonwick, *Last of the Tasmanians*, quotation on p. 56.

22. A similar ideology has covered other slaughters in other new worlds. The Spanish priest Las Casas, historian of Columbus's conquests, noted that the Christians despised the natives and held them as fit objects for enslavement "because they are in doubt as to whether they are animals or beings with souls" (Frederick Turner, *Beyond Geography: The Western Spirit against the Wilderness* [New Brunswick, NJ: Rutgers University Press, 1986], p. 142).

23. Carolyn Merchant, *Ecological Revolutions* (Chapel Hill: University of North Carolina Press, 1989), pp. 6–7.

24. Hazel Henderson, *Creating Alternative Futures: The End of Economics* (New York: Berkeley Publishing, 1978); Marilyn Waring, *Counting for Nothing* (London: Allen and Unwin/Port Nicholson Press, 1988).

25. See Andrew Darby, "Seal Kill: The Slaughter in Our Southern Seas," *Good*

Weekend, January 5, 1991, pp. 12–15. As many as three thousand fur seals are killed by the Tasmanian fishing industry each year, many in macho shootouts that wipe out whole colonies. Others are killed even more cruelly as their playfulness leads them to become entangled in the plastic fish-bait packaging and discarded nets tossed overboard from boats. The industry dumps plastic garbage of all kinds, which is a killer of marine life and is found on the most remote beaches.

26. Warren, "Power and Promise of Ecofeminism," p. 133.

27. Elizabeth V. Spelman, *Inessential Woman: Problems of Exclusion in Feminist Thought* (Boston: Beacon Press, 1988).

28. Mies, *Patriarchy and Accumulation*; Iris Young, "Beyond the Unhappy Marriage: A Critique of Dual Systems Theory," in *Women and Revolution*, ed. Lydia Sargent (Boston: South End Press, 1981), pp. 43–70.

29. The net or web analogy is an alternative to the pillar analogy of some ecofeminists, as in "racism, sexism, class exploitation and ecological destruction form interlocking pillars upon which the structure of patriarchy rests" (Sheila Collins quoted in Warren, "Feminism and Ecology," p. 7). This seems unnecessarily limiting and quantified.

30. bell hooks, *Talking Back* (Boston: South End Press, 1989), p. 22.

31. The web model was of course suggested by Foucault (Michel Foucault, "Disciplinary Power and Subjection" in *Power/Knowledge: Selected Interviews and Other Writings of Michel Foucault, 1972–1977*, ed. Cohn Gordon (New York: Pantheon, 1980), p. 234. His version has recently been criticized by Hartsock (Nancy Hartsock, "Foucault on Power: A Theory for Women?" in *Feminism/Postmodernism* [New York: Routledge, 1990]). As Hartsock notes, taking individuals to be the agents of such power relations can result in a view of power as so pervasive and reciprocal that domination disappears and power is "everywhere and therefore nowhere" (p. 170). But we do not have to understand the strands of the net, as Hartsock argues Foucault does, as an all-pervading and utterly diffuse set of power relations operating equally between all individuals, with no central focus or organizing principle. Rather, as I argue (in Val Plumwood, *Feminism and the Mastery of Nature* [New York: Routledge, 1993]), we can understand them as the large-scale multiple but linked social structures of oppression.

32. Val Plumwood, "Nature, Self, and Gender: Feminism, Environmental Philosophy, and Critique of Rationalism," *Hypatia* 6, no. 1 (1991): 3–27.

33. Donna Haraway, "A Manifesto for Cyborgs," in *Feminism/Postmodernism*, ed. Linda J. Nicholson, 190–233 (New York: Routledge, 1990); Donna Haraway, *Simians, Cyborgs and Women* (London: Free Association Press, 1991); Hartsock, "Foucault on Power," p. 164.

34. Val Plumwood, "On Karl Marx as an Environmental Hero," *Environmental Ethics* 3 (1981): 237–44.

19

ECOFEMINIST MOVEMENTS

NOËL STURGEON

Ynestra King, one of the founders of US ecofeminism, has called it the "third wave of the women's movement," indicating her sense, at one time, that this most recent manifestation of feminist activity was large and vital enough to parallel the first-wave nineteenth-century women's movement and the second-wave women's liberation movement of the 1960s and 1970s.[1] . . . I want to attempt some descriptions and definitions of ecofeminism as a movement[2] and as a set of theories.

Most simply put, ecofeminism is a movement that makes connections between environmentalisms and feminisms; more precisely, it articulates the theory that the ideologies that authorize injustices based on gender, race, and class are related to the ideologies that sanction the exploitation and degradation of the environment.[3] In one version of its origins, the one I will privilege . . . ecofeminism in the United States arises from the antimilitarist direct action movement of the late seventies and eighties, and develops its multivalent politics from that movement's analysis of the connections between militarism, racism, classism, sexism, speciesism, and environmental destruction. But, as I will also show, ecofeminism has multiple origins and is reproduced in different inflections and deployed in many different contexts. In particular . . . I will argue that ecofeminism has roots in both feminism and environmentalism.

Given both its attempt to bridge different radical political positions and its historical location as at least one of many third-wave women's movements, US ecofeminism aims to be a multi-issue, globally oriented movement with a more

---

From: *Ecofeminist Natures: Race, Gender, Feminist Theory and Political Action* (New York: Routledge, 1997), pp. 23–30, excerpts.

diverse constituency than either of its environmentalist or feminist predecessors. Ecofeminism is thus a movement with large ambitions and with a significant, if at the moment largely unorganized, constituency. Many people are interested in the scope of ecofeminism, its drawing together of environmentalism and feminism. Environmentalism is one of the most popular and significant locations for radical politics today; it attracts people because of the seemingly apocalyptic nature of our ecological crises and the many ways in which environmental problems affect people's daily lives, as well as the sense of its global relevance. As a feminist movement, ecofeminism reworks a long-standing feminist critique of the naturalization of an inferior social and political status for women so as to include the effects on the environment of feminizing nature. Coupled with environmentalism, this version of feminism gains a political cachet not easily matched by other radical political locations, particularly for young US feminists who already think of themselves as environmentalists, having been more or less socialized as such. Ecofeminism is a significant and complex political phenomenon, a contemporary political movement that has far-reaching goals, a popular following, and a poor reputation among many academic feminists, mainstream environmentalists, and some environmental activists of color. Part of what I want to do . . . is to understand the sources of that poor reputation and to explore the reasons for the failure of ecofeminism to live up to its potential.

ECOFEMINIST GENEALOGIES

A name that can usefully if partially describe the work of Donna Haraway and Mary Daly, Alice Walker and Rachel Carson, Starhawk and Vandana Shiva,[4] ecofeminism is a shifting theoretical and political location that can be defined to serve various intentions. The present chaotic context of the relatively new and diverse political positionings that go under the name of "ecofeminism" allows me to construct . . . a series of definitions and historical trajectories of the movement, ones I recognize as always interested and certainly contestable.[5] . . . I will piece together stories about ecofeminist beginnings and evolution by tracing the use of the word "ecofeminism" as it appears in political actions, organizations, conferences, publications, and university courses. Not a history so much as a genealogy, embedded in this tracing is an effort to tease out the label's shifting meanings and political investments in order to delineate the construction of ecofeminism as an object of knowledge, as a political identity, and as a set of political strategies within the convergence of local and global environmentalisms, academic and activist feminisms, and anticolonialist and antiracist movements.[6] . . .

Both an activist and an academic movement, ecofeminism has grown rapidly since the early eighties and continues to do so in the nineties. As activists, ecofeminists have been involved in environmental and feminist lobbying efforts, in demonstrations and direct actions, in forming a political platform for a US Green party, and in building various kinds of ecofeminist cultural projects (such as ecofeminist art, literature, and spirituality). They have taken up a wide variety of issues, such as toxic waste, deforestation, military and nuclear weapons policies, reproductive rights and technologies, animal liberation, and domestic and international agricultural development. In academic arenas, scholars who are either identified with or interested in ecofeminism have been active in creating and critiquing ecofeminist theories. A wave of publications in the area, including several special issues of journals, indicates research activity on ecofeminism in religious studies, philosophy, political science, art, geography, women's studies, and many other disciplines.[7]

... Ecofeminism can be seen primarily as a feminist rebellion within male-dominated radical environmentalisms, where I have found it popping up in almost every arena, often without communication between these slightly or greatly different versions of ecofeminism. Thus, one can find ecofeminists appearing within the antinuclear movement, social ecology, bioregionalism, Earth First!, the US Greens, animal liberation, sustainable development, and, to a lesser extent, the environmental justice movement. . . .

The origins of this varied activity called "ecofeminism" have been described in different ways.[8] Certainly, an ecological critique was an important part of women's movements worldwide from the mid-1970s, particularly those concerned with nuclear technology, neocolonialist development practices, and women's health and reproductive rights. In my reading of these developments, ecofeminism in the United States arose in close connection with the nonviolent direct action movement against nuclear power and nuclear weapons. Until the Women's Pentagon Actions in 1980, however, there were numerous events and groups connected with ecofeminism that were concerned with a number of issues, militarism being only one of many.

The earliest event I've seen described as making the connection between women and the environment was in 1974, at the Women and the Environment conference at UC Berkeley organized by Sandra Maburg and Lisa Watson. An ecofeminist newsletter, *W. E. B.: Wimmin of the Earth Bonding*, published four issues from 1981 to 1983, concerned with feminist and lesbian back-to-the-land communities, health, appropriate technology, and political action.[9]

Most influentially, however, US ecofeminism's initiating event was the Women and Life on Earth: Ecofeminism in the 1980s conference at Amherst in 1980, organized by Ynestra King (then of the Institute for Social Ecology), Anna Gyorgy (an organizer in the antinuclear Clamshell Alliance), Grace Paley (a feminist writer and pacifist activist), and other women from the antinuclear, environmental, and lesbian-feminist movements.[10]

The Women and Life on Earth conference organized panels and workshops on the alternative technology movement (staffed by the group Women In Solar Energy, or WISE), organizing, feminist theory, art, health, militarism, racism, urban ecology, theater, as well as other topics: eighty workshops in all. Over 650 women attended, far beyond the expected hundred or so.[11] Speakers included Patricia Hynes of WISE; Lois Gibbs, then of the Love Canal Homeowners Association and later of the Citizen's Clearinghouse for Hazardous Waste (CCHW);[12] and Amy Swerdlow, feminist activist and historian.[13] The conference generated an ongoing Women and Life On Earth (WLOE) group in Northampton, Massachusetts, which published a newsletter entitled *Tidings*, as well as several other WLOE groups in New York, Cape Cod, and other areas in the northeastern United States.[14]

Several other ecofeminism conferences and organizations were either inspired by Women and Life On Earth or assisted by WLOE organizers. A conference already in the planning stages in 1980, Women and the Environment: The First West Coast Ecofeminist Conference drew five hundred women, who listened to talks by Angela Davis, Anna Gyorgy, China Galland, and Peggy Taylor. Workshops were offered on "alternative energy, global view, planning, health, organizing media, no nukes, and peace."[15] In London, a Women For Life On Earth (WFLOE) group formed, inspired by the Amherst conference, and organized a conference in 1981. Energy from that conference spawned numerous WFLOE groups, twenty-six in the United Kingdom and nine in other countries, including Australia, Canada, France, Japan, and West Germany.[16] WFLOE put out a newsletter at least until winter 1984, organized a number of gatherings, and supported the Greenham Common peace camp. Organizers of WFLOE, Stephanie Leland and Leonie Caldecott, edited the first ecofeminist anthology, *Reclaim the Earth: Women Speak Out for Life on Earth,* in 1983.

From the Women and Life on Earth conference at Amherst also grew the organizing efforts for the Women's Pentagon Actions (WPA) of 1980 and 1981, in which large numbers of women demonstrated and engaged in civil disobedience. As defined by the Unity Statement of the WPA,[17] the politics behind these early ecofeminist actions were based on making connections between militarism, sexism, racism, classism, and environmental destruction (however unevenly the action may have addressed these issues).[18] Influenced by the writings of Susan Griffin,[19] Charlene Spretnak,[20] Ynestra King,[21] and Starhawk, a set of political positions that began to be called ecofeminism developed among women sympathetic to the politics of the WPA and other antimilitarist and environmental actions. Many women involved in later antimilitarist direct actions thus began to call themselves ecofeminists in the middle eighties as a way of describing their interlocking political concerns.[22] In fact, an article in the 1981 issue of *Tidings*, the newsletter of WLOE and the WPA, states that organizers decided not to get involved with a Mother's Day Coalition for Disarmament March in Washington, DC, because "The Mother's Day action is a single issue action and not explicitly feminist." Furthermore, the march was not organized using a "participatory feminist process."[23] Thus, even after the WPA, "ecofeminism" referred not to antimilitarism alone but to a particular kind of feminism, radically democratic antimilitarism that made connections to other political issues. Rather than arising from "the peace movement," ecofeminists deeply influenced the nature of feminist peace politics in the 1980s.

As the label became more common among feminist antimilitarist activists, a concomitant interest in ecofeminism was emerging in the academy. The two arenas were intertwined at the Ecofeminist Perspectives: Culture, Nature, Theory conference in March 1987 at the University of Southern California (USC), organized by Irene Diamond and Gloria Orenstein. This well-attended conference was the beginning of a rapid flowering of ecofeminist art, political action, and theory that continues today.[24] This conference also marked the point where the word "ecofeminism" began to be used outside the antimilitarist movement to describe a politics that attempted to combine feminism, environmentalism, antiracism, animal liberation, anticolonialism, antimilitarism, and nontraditional spiritualities.

During the years following the USC conference, US ecofeminists became active in the international arena, intervening in the process of the globalization of environmentalism. In 1991, a World Women's Conference for a Healthy Planet in Miami, Florida, was organized by the Women's Environmental Development Organization, or WEDO. For political reasons . . . WEDO did not explicitly identify as "ecofeminist," but its rhetoric and vision were clearly in the ecofeminist tradition. This conference brought together women from all over the world to discuss environmental issues in the context of women's knowledge, women's needs, and women's activism. It served as a springboard for an ecofeminist presence at the UN Conference on Environment and Development at Rio de Janeiro in 1992, which had some influence on the international deliberations about solutions to worldwide environmental problems. Besides this activity in an international arena, there have been other important ecofeminist conferences, such as the Eco-visions: Women, Animals, the Earth, and the Future conference in Alexandria, Virginia, in March 1994 (which emphasized connections between feminism, environmentalism, and animal liberation), and the Ecofeminist Perspectives conference at University of Dayton, Ohio, in March 1994 (which emphasized ecofeminist interventions into environmental philosophy). In all these events, organizers stressed ecofeminism's ability to make connections between various radical politics. Which part of this multivalent politics is emphasized or even included varies widely and remains deeply contested among those that identify as ecofeminists. In particular, until the late eighties, antispeciesist theories were underdeveloped portions of the ecofeminist tool kit. Theories of the connections between heterosexism and naturism remain underdeveloped within ecofeminism as of this writing.[25]

WOMEN AND NATURE, FEMINISM AND ENVIRONMENTALISM

Within this multivoiced and vibrant set of political positions were very different theorizations of the connections between the unequal status of women and the life-threatening destruction of the environment. A constant and ongoing focus of ecofeminist theorizing, as well as critiques of ecofeminism, has been how to conceptualize the "special connection" between women and nature often presumed by the designation *ecofeminism*. Very briefly and generally, I will outline five ways this relationship is described. Though I isolate these analyses as positions, in operation they are often combined and intertwined.

One position involves an argument that patriarchy equates women and nature, so that a feminist analysis is required to understand fully the genesis of environmental problems. In other words, where women are degraded, nature will be degraded, and where women are thought to be eternally giving and nurturing, nature will be thought of as endlessly fertile and exploitable.

Another position, which is really the other side of the position just described, argues that an effective understanding of women's subordination in Western cultures requires an environmentalist analysis. In a culture that is in many ways anti-nature, which constructs meanings using a hierarchical binarism dependent on assumptions of culture's superiority to nature, understanding women as more "natural" or closer to nature dooms them to an inferior position. Furthermore, in a political economy dependent on the freedom to exploit the environment, a moral and ethical relation to nature is suspect. If women are equated with nature, their struggle for freedom represents a challenge to the idea of a passive, disembodied, and objectified nature.

A third position argues for a special relationship between women and nature using a historical, cross-cultural, and materialist analysis of women's work. By looking at women's predominant role in agricultural production and the managing of household economies worldwide (cooking, cleaning, food production, and purchasing of household goods, healthcare, and childcare), this position maintains that environmental problems are more quickly noticed by women and impact women's work more seriously.[26]

A fourth position argues that women are biologically close to nature, in that their reproductive characteristics (menstrual cycles, lactation, birth) keep them in touch with natural rhythms, seasonal and cyclical, life- and death-giving. Ecofeminists who are comfortable with this position feel that women potentially have greater access than men do to sympathy with nature, and will benefit themselves and the environment by identifying with nature.

A fifth position is taken by feminists who are interested in constructing resources for a feminist spirituality and who have found these resources in nature-based religions: paganism, witchcraft, goddess worship, and Native American spiritual traditions. Because such nature-based religions historically contain strong images of female power and place female deities as at least equal to male deities, many persons who are searching for a feminist spirituality have felt comfortable with the appellation of "ecofeminist."

Before proceeding, I want to point to just one of the most obvious contradictions within ecofeminism: the serious lack of agreement between positions one and two and position four. The first two positions see the equation of women and nature as patriarchal; the fourth position sees this equation as empowering to women and as providing resources for a feminist environmentalism. Some variations of position five, concerned with feminist spirituality, also see the equation of women and nature as empowering. This contradiction is obscured by reductive depictions of ecofeminism as "essentialist" without noting the existence of strong constructionist positions within ecofeminism. That this contradiction—between the critique of the connection between women and nature and the desire for a positive version of that connection—is so deeply embedded illuminates the consistent recurrence to essentialist notions of women and nature that ecofeminism encounters in its attempt to construct a collective subject within a social movement. It is also what prevents me from assigning one or the other of the positions described above to one or another ecofeminist author; in most cases, these different analyses of the connections between women and nature are operating at the same time. One of my contentions . . . is that white ecofeminist discourses about "indigenous" women function to obscure this particular division within ecofeminism. Thus, particular ecofeminist discourses of racial difference sidestep the contradictions between particular theorizations of the connection between women and nature. . . . But . . . there has been a greater effort within ecofeminist theory to make connections between women and nature rather than between feminism and environmentalism as political movements, even though . . . such movement connections are often at stake in the production of these theories. The subtext of movement contexts influences theoretical constructions in which essentialist connections between women and nature are more frequent than they otherwise might be.

To construct these and other variations of the theoretical connections between women and nature, or between environmentalism and feminism, ecofeminists have drawn on a number of feminist theories that, while not necessarily aimed at answering questions about the relationship between feminist and environmental politics, provided crucial analytical tools. Feminist philosophical critiques of forms of abstract rationality that reify divisions between culture and nature, mind and emotion, objectivity and subjectivity; psychoanalytic theories of the ways in which masculinist anxiety about women's reproductive capacities structures male-dominated political and economic institutions; feminist rethinkings of Christian theology; critiques of the patriarchal nature of militarism; feminist anthropological research; feminist critiques of science; feminist analyses of the sexual objectification of women and feminist poststructuralist theories of constructed subjectivities and critiques of essentialism: these are only a few of the vital feminist resources for ecofeminist theories.[27] Despite its reliance on central feminist theories, most strongly reflected in position two above, ecofeminist theory remains in a tenuous relation to feminist theory. . . .

Feminist antiracist theory was also an important resource for ecofeminists, providing a foundation from which to analyze the ways in which hierarchies were created and maintained as well as a guide to constructing a movement that attempts to be inclusive and antiracist. Antiracism was thus a political position apparent in the very beginnings of ecofeminism as theory and as practice, even though it has been a movement that is predominantly white. At the same time, there are many women of color who are either prominent in the movement or who serve as role models for white ecofeminists. To further complicate the picture, many environmental activists are women of color who do not identify as ecofeminists, given that the genealogy of the label arises from the white feminist antimilitarist movement and that US ecofeminism has continued to be a movement largely of white, middle-class women.[28]

NOTES

1. Ynestra King, personal communication, May 1990, repeated in several public speeches. The concept of ecofeminism as a "third wave" is echoed by Val Plumwood, who usefully qualifies the claim by stating: "It is not a tsunami, or freak tidal wave which has appeared out of nowhere sweeping all before it. Rather, it is prefigured in and builds on work not only in ecofeminism but in radical feminism, cultural feminism, and socialist feminism over the last decade and a half." *Feminism and the Mastery of Nature* (London and New York: Routledge, 1993), p. 39.

2. As I have mentioned in the introduction, I do not see ecofeminism as a "social movement" in most traditional senses, i.e., a particular mobilization around a specific grievance that acquires organizational form. Neither do I see it purely as an "intellectual movement," the other way the term is often used—that is, a set of ideas elaborated by a school of thinkers and writers.

3. This definition paraphrases Greta Gaard, "Living Interconnections with Animals and Nature," in *Ecofeminism: Women, Animals, Nature*, 1–12, esp. p. 1 (Philadelphia: Temple University Press, 1993).

4. Donna Haraway, a white socialist feminist deeply influenced by poststructuralism, explicitly aligns herself with ecofeminism in "Situated Knowledges: The Science Question in Feminism and the Privilege of Partial Perspective," in *Simians, Cyborgs, and Women: The Reinvention of Nature*, 201 (New York: Routledge, 1991), and in Haraway's interview with Marcy Darnovsky entitled, "Overhauling the Meaning Machines," *Socialist Review* 21, no. 2 (1991): 65–84, esp. pp. 69–70, 78. Mary Daly's radical feminist classic, *Gyn/Ecology* (Boston: Beacon Press, 1978; 1990) is now considered by many to be one foundation for ecofeminist theory. Alice Walker, a prominent best-selling African American writer, has contributed explicitly to ecofeminist antimilitarist and animal liberationist concerns, most clearly through her pieces, "Only Justice Can Stop a Curse," in *In Search of Our Mother's Gardens*, 338–42 (San Diego: Harcourt Brace Jovanovich, 1983), and "Am I Blue?" in *Living By The Word*, 3–8 (San Diego: Harcourt Brace Jovanovich, 1988). Rachel Carson, a natural scientist who was not an explicit feminist, is claimed as an ecofeminist foremother because of her book *Silent Spring*, which arguably intitiated the first nonconservationist environmental movement in America (see Grace Paley's dedication to Rachel Carson in *Reweaving the World: The Emergence of Ecofeminism*, ed. Irene

Diamond and Gloria Feman Orenstein, ii [San Francisco: Sierra Club Books, 1990]). Starhawk, a pagan, witch, activist in the nonviolent antimilitarist direct-action movement, writer, and theorist, has been an important influence on ecofeminism; see her *Dreaming the Dark* (Boston: Beacon Press, 1982), *The Spiral Dance: A Rebirth of the Ancient Religion of the Great Goddess* (San Francisco: Harper and Row, 1988), and *Truth or Dare* (San Francisco: Harper and Row, 1985). Vandana Shiva is a theoretical physicist who is also the director of an environmental research institute in Dehra Dun, India; her book *Staying Alive: Women, Ecology and Development in India* (London: Zed Press, 1988) is an important ecofeminist text.

5. However, my description is not simply an arbitrary construction. Both my own participation in the ecofeminist movement as an activist and theorist since 1984, and my experience as the editor of the *Ecofeminist Newsletter* (published annually from 1990–1996), give me a broad and immediate sense of the movement and ongoing personal contact with a wide variety of people who call themselves "ecofeminists." In the following section of this chapter, I deliberately avoid the typologizing of ecofeminisms as radical, cultural, Marxist, socialist, and poststructuralist. . . . Here, I will just say that such typologies would work against the genealogical method I employ in this chapter.

6. The term "genealogy" in its current usage is derived from Nietzsche via Michel Foucault in the latter's essay "Nietzsche, Genealogy, History," in *Language, Counter-Memory, Practice*, ed. and trans D. F. Bouchard, 139–64 (Ithaca, NY: Cornell University Press, 1977).

7. The last few years have seen a rapid increase in the literature on ecofeminism, in the context of a growing body of environmental literature. An analysis of the publication history of ecofeminist literature indicates a trend from more marginal "movement-oriented" publications to more scholarly journals and university presses. Journals that have devoted special issues to the topic are: *Heresies* 13 (1981); *New Catalyst* 10 (Winter 1987–1988); *Woman of Power* (Spring 1988); *Studies in the Humanities* 15, no. 2 (1988); *Hypatia: Journal of Women and Philosophy* 6, no. 1 (1991); *American Philosophical Association Newsletter on Feminism and Philosophy* 2 (Fall 1991); and *Society and Nature* 2, no. 1 (1993). Besides those listed above, journals that have published numerous articles on ecofeminism include *Capitalism, Nature, Socialism*; *Environmental Ethics*; *Environmental Review*; *The Trumpeter*; *Women and Environments*; *Women's International Network News*; and *Women's Studies International Forum*. A partial, chronological listing of books on ecofeminism would include Rosemary Radford Ruether, *New Woman/New Earth: Sexist Ideologies and Human Liberation* (New York: Seabury Press, 1975); Susan Griffin, *Woman and Nature: The Roaring inside Her* (San Francisco: Harper and Row, 1978); Elizabeth Dodson Gray, *Green Paradise Lost* (Wellesley, MA: Roundtable Press, 1979); Carolyn Merchant, *The Death of Nature: Women, Ecology and the Scientific Revolution* (San Francisco: Harper and Row, 1980); Brian Easlea, *Science and Sexual Oppression: Patriarchy's Confrontation with Women and Nature* (London: Weidenfeld and Nicholson, 1981); Leonie Caldecott and Stephanie Leland, eds., *Reclaim the Earth: Women Speak Out for Life on Earth* (London: The Women's Press, 1983); Andreé Collard with Joyce Contrucci, *Rape of the Wild: Man's Violence against Animals and the Earth* (Bloomington: Indiana University Press, 1988); Vandana Shiva, *Staying Alive: Women, Ecology and Development in India* (London: Zed Books, 1988); Irene Dankelman and Joan Davidson, *Women and Environment in the Third World* (London: Earthscan Publica-

tions, 1988); Judith Plant, ed., *Healing the Wounds: The Promise of Ecofeminism* (Philadelphia: New Society Publishers, 1989); Carolyn Merchant, *Ecological Revolutions: Nature, Gender and Science in New England* (Chapel Hill: University of North Carolina Press, 1989); Irene Diamond and Gloria Feman Orenstein, eds., *Reweaving the World: The Emergence of Ecofeminism* (San Francisco: Sierra Club Books, 1990); Janet Biehl, *Finding Our Way: Rethinking Ecofeminist Politics* (Boston: South End Press, 1991); Carol Adams, *The Sexual Politics of Meat: A Feminist-Vegetarian Critical Theory* (New York: Continuum Press, 1991); Rosemary Radford Ruether, *Gaia and God: An Ecofeminist Theology of Earth Healing* (San Francisco: Harper and Row, 1992); Mary Mellor, *Breaking the Boundaries: Toward a Feminist Green Socialism* (London: Virago Press, 1992); Greta Gaard, ed., *Ecofeminism: Women, Animals, Nature* (Philadelphia: Temple University Press, 1993); Carol Adams, ed., *Ecofeminism and the Sacred* (New York: Continuum Press, 1993); Val Plumwood, *Feminism and the Mastery of Nature* (New York: Routledge, 1993); Maria Mies and Vandana Shiva, *Ecofeminism* (London and Atlantic Highlands, NJ: Zed Press, 1993); Vera Norwood, *Made from This Earth: American Women and Nature* (Chapel Hill: University of North Carolina Press, 1993); Karen Warren, ed., *Ecological Feminism* (New York: Routledge, 1994); Irene Diamond, *Fertile Ground: Women, Earth, and the Limits of Control* (Boston: Beacon Press, 1994); Vandana Shiva, ed., *Close to Home: Women Reconnect Ecology, Health and Development Worldwide* (Philadelphia: New Society Publishers, 1994); Rosi Braidotti, Ewa Charkiewicz, Sabine Häusler, Saskia Wieringa, *Women, the Environment and Sustainable Development* (London: Zed Books, 1994); Carol Adams, *Neither Man nor Beast: Feminism and the Defense of Animals* (New York: Continuum Press, 1994); Vandana Shiva and Inguna Moser, eds., *Biopolitics: A Feminist and Ecological Reader on Biotechnology* (London: Zed Books, 1995); Carol Adams and Josephine Donovan, eds., *Animals and Women: Feminist Theoretical Explorations* (Durham, NC, and London: Duke University Press, 1995); Carolyn Merchant, *Earthcare: Women and the Environment* (London and New York: Routledge, 1996); Karen Warren, ed., *Ecological Feminist Philosophies* (Indianapolis: Indiana University Press/Hypatia, 1996), and Karen Warren, ed., *Ecofeminism: Women, Culture, Nature* (Bloomington: Indiana University Press, 1997). A number of books on ecofeminism are forthcoming at this writing, including Ynestra King, *Ecofeminism: The Reenchantment of Nature* (Boston: Beacon); Chaia Heller, *The Revolution That Dances: From a Politics of Desire to a Desirable Politics* (Littleton, CO: Aigis Publications). Manuscripts in process that I know of are those by Greta Gaard, *Ecological Politics: Ecofeminists and the Greens* (Philadelphia: Temple University Press, forthcoming); and Christine Cuomo (on ecofeminist ethics). For a sampling of the periodical literature on ecofeminism, see Carol Adams and Karen Warren, "Feminism and the Environment: A Selected Bibliography," *American Philosophical Association Newsletter on Feminism and Philosophy* 90, no. 3 (Fall 1991): 148–57. A popular interest in ecofeminism is indicated by special issues of *Utne Reader* 36 (November/December 1989) and *Ms.* 2, no. 2 (1991); the sporadic, uneven column on ecofeminism in *Ms.*, as well as the growing interest in ecofeminism evinced by trade publishers (Beacon, Harper and Row, Vintage, etc.). The word "ecofeminism" became a Library of Congress subject heading around 1992.

　　8. The most thorough historian of ecofeminism to date is Carolyn Merchant. See her section entitled "Ecofeminism," in *Radical Ecology: The Search for a Livable World*, 183–210 (New York: Routledge, 1992), and *Earthcare: Women and the Environment* (New

York: Routledge, 1995), esp. chs. "Earthcare: Women and the American Environmental Movement" (pp. 139–66) and "Conclusion: Partnership Ethics: Earthcare for a New Millennium" (pp. 209–24). Other accounts of ecofeminism's beginnings and development can be found in Ynestra King, "The Eco-Feminist Imperative," in *Reclaim the Earth*, ed. Caldecott and Leland, pp. 12–16, and "Ecological Feminism," *Z Magazine* 1, nos. 7/8 (1988): 124–27; Charlene Spretnak, "Ecofeminism: Our Roots and Flowering," in *Reweaving the World*, ed. Diamond and Orenstein, pp. 3–14; Braidotti et al., "Ecofeminism: Challenges and Contradictions," in *Women, the Environment and Sustainable Development*, pp. 161–68; and Greta Gaard and Lori Gruen, "Ecofeminism: Toward Global Justice and Planetary Health," *Society and Nature* 2, no. 1 (1993): 1–35. Many of these accounts (except for King's) start with the coining of the word "ecofeminism" in 1974 by Françoise d'Eaubonne and cite her *Le Féminisme ou la Mort* (Paris: Pierre Horay, 1974), though Braidotti cites "Feminism or Death?" in *New French Feminisms: An Anthology*, ed. Elaine Marks and Isabelle de Courtivron, 64–67 (Amherst: University of Massachusetts Press, 1980). Aside from the 1980 essay cited above, which does not explicitly mention ecofeminism, d'Eaubonne's work was not available in English translation until 1994, in an essay translated by Ruth Hottel as "The Time for Ecofeminism," in *Key Concepts in Critical Theory: Ecology*, ed. Carolyn Merchant, 174–97 (Atlantic Highlands, NJ: Humanities Press, 1994). Though undoubtedly d'Eaubonne's 1974 formulation was an early use of the term, since her work was not available in English translation until 1994, the notion of her authorship of the term appears to have been introduced by Karen Warren in "Toward an Ecofeminist Ethic," *Studies in the Humanities* 15 (1988): 140–56; after that, d'Eaubonne appears as the coiner of the word in most accounts. Since d'Eaubonne's formulation enters histories of US ecofeminism well after the word comes to signify a set of interlocking concerns about the status of women and degradation of the environment articulated by feminist antimilitarist activists in 1980, I am inclined to give Ynestra King the credit for the invention of the word in its US context. Ariel Salleh comments that the delay in translating d'Eaubonne to English signifies the US imperialist context of the production of feminist knowledge, while centering d'Eaubonne as the founder of ecofeminism in turn closes off possible non-Western origins for the word. She states that "the term 'ecofeminism' (was) spontaneously appearing across several continents in the 1970s" but for "politico-economic reasons . . . ecofeminists working from more visible niches in the dominant English-speaking culture have tended to get their views broadcast first." See Salleh's book review of Vandana Shiva's *Staying Alive*, *Hypatia* 6, no. 1 (1991): 206.

9. My thanks to Ann Megisikwe (Ann Filemyr) (who, along with Marjaree Chimera, edited *W. E. B.*) for telling me about the newsletter and providing me with copies. Another important ecofeminist newsletter was *E. V. E.* (Ecofeminist Visions Emerging) published in New York City by Cathleen and Colleen McGuire from 1991–1993. The newsletter I edit, the *Ecofeminist Newsletter*, was a similar effort. Because of the widespread, grassroots, and decentralized nature of the early period of ecofeminism's development, it is extremely difficult to track down materials documenting the movement. It is very likely that there were many more groups and publications than I name in this section.

10. See Spretnak, "Our Roots and Flowering," for a fuller list of organizers; also see Barbara Epstein, *Political Protest and Cultural Revolution: Nonviolent Direct Action in the 1970s and 1980s* (Berkeley: University of California Press, 1991), p. 161.

11. Ynestra King, "Where the Spiritual and Political Come Together," *Women for Life on Earth* (Winter 1984): 4. King has given different figures in "What Is Ecofeminism?" *Nation* (December 12, 1987): 730, claiming eight hundred attendees and two hundred workshops. I am inclined to stick to the description dated closer to the conference itself.

12. CCHW is presently an important group in the environmental justice movement.

13. Other speakers were Ynestra King and Catherine Carlotti. I am citing speakers whose speeches I have copies of, but there were many more. I thank Riley Dunlap for lending me his archive on Women and Life on Earth.

14. *Tidings* (May 1981): 1–16.

15. Anna Gyorgy, "Evaluating Eco-Feminism West Coast," *Tidings* (May 1981): 14.

16. *Women for Life on Earth* (Winter 1984): 58–59.

17. The Unity Statement, including its original illustrations depicting women of all races and ages, has been reprinted in Lynne Jones, ed., *Keeping the Peace* (London: Women's Press, 1983), pp. 42–43. For descriptions of the action, see Ynestra King, "All Is Connectedness," in *Keeping the Peace*, pp. 40–63, and Rhoda Linton and Michele Whitham, "With Mourning, Rage, Empowerment and Defiance: The 1981 Women's Pentagon Action," *Socialist Review* 12, nos. 3/4 (1982): 11–36.

18. For a discussion of the complex political agenda of the WPA, see T. V. Reed, "Dramatic Ecofeminism: The Women's Pentagon Action as Theater and Theory," in *Fifteen Jugglers, Five Believers: Literary Politics and the Poetics of American Social Movements*, 120–41 (Berkeley: University of California Press, 1992). In particular, the WPA actions were criticized for the "essentialism" of their rhetoric connecting women and nature. See Ellen Willis's columns in the *Village Voice* 25, no. 25 (June 18–24, 1980): 28, and 25, no. 29 (July 16–22, 1980): 34. Additionally, and more relevant to my argument in ch. 3, many feminist activists of color identified the feminist antimilitarist movement as a white-dominated movement.

19. Particularly Griffin's *Women and Nature: The Roaring inside Her* (New York: Harper and Row, 1978).

20. Especially as the editor of *The Politics of Women's Spirituality: Essays on the Rise of Spiritual Power within the Feminist Movement* (New York: Anchor Books, 1982).

21. An important ecofeminist theorist, King has usefully collected many of her classic essays in *What Is Ecofeminism?* (New York: Ecofeminist Resources, 1990).

22. See Judith McDaniel, ed., *Reweaving the Web of Life: Feminism and Nonviolence* (Philadelphia: New Society Publishers, 1982), for several early formulations of the connections between feminism and environmentalism stemming from feminist antimilitarism. Note the reworking of this title in Diamond and Orenstein's explicitly ecofeminist anthology, *Reweaving the World*.

23. Anonymous, *Tidings* (May 1981): 14.

24. See Irene Diamond and Gloria Feman Orenstein's description of the conference and its importance; "Ecofeminism: Weaving the Worlds Together," *Feminist Studies* 14 (Summer 1988): 368–70. There have been a number of important ecofeminist conferences since.

25. Greta Gaard's essay, "Toward a Queer Ecofeminism," thus promises to break new and exciting ground when it is published (*Hypatia*, forthcoming). In this essay, she notes that "the May 1994 special issue of the Canadian journal *UnderCurrents* is the first to address the topic of "Queer Nature." Gaard goes on (in n. 1) to note that though several of the essays in this special issue initiate an exploration of a "queer ecofeminism,"

none of them specifically develop connections between queer theory and ecofeminism, which is the purpose of her essay.

26. This position is especially common in the ecofeminist analyses that operate within the political and academic arena called "Women, Environment, and Development."

27. For a detailed description of the different theories useful to ecofeminism, see Greta Gaard and Lori Gruen, "Global Justice and Planet Health," *Society and Nature* 2, no. 1 (1993): 1–35.

28. Gwyn Kirk, "Blood, Bones, and Connective Tissue: Grassroots Women Resist Ecological Destruction," paper presented at the National Women Studies Association, Austin, June 1992; Giovanna Di Chiro, "Defining Environmental Justice: Women's Voices and Grassroots Politics," *Socialist Review* 22, no. 4 (October–December 1992): 93–130.

20

TOWARDS A FEMINIST GREEN SOCIALISM

MARY MELLOR

What men value has brought us to the brink of death. What women find worthy may bring us back to life.

Marilyn Waring[1]

The world of nurturance and close human relationships is the sphere where the basic human needs are anchored and where models for *humane* alternatives can be found. This world, which has been carried forward mainly by women, is an existing alternative culture, a source of ideas and values for shaping an alternative path of development for nations and all humanity.

Hilkka Pietilä[2]

Socialists have long sought a world in which people offer mutual support to each other without demanding cash payment or profit. Greens look for a sustainable, decentralized world of face-to-face interaction. Both decry the institutions of modern society for being destructive, exploitative, and alienating. Programs and blueprints for a new society are constructed to take us "forward" or bring us "back" to a green or socialist world. It is odd that they do not see that this world exists already. Most of women's lives are spent in a decentralized world beyond the market where basic needs prevail. Women, particularly in subsistence economies, are rooted to the earth and the reproduction of life in their daily work.

The world that socialists and Greens are seeking to escape from is a male world. It is mainly male green and socialist writers who seek a new autonomy for

From *Breaking the Boundaries: Towards a Feminist Green Socialism* (London: Virago Press, 1992), pp. 249–51, 276–81.

the individual (self-actualization, self-realization) in the comfort of small-scale, mutual community life: Gorz, Bahro, Schumacher, Robertson, Trainer, and even Marx. There may well be women who have written in a similar vein, but they are either "hidden from history" or much less numerous. This latter is the more likely, for women are right to be suspicious of visions of Greens and socialists that do not embrace an explicit feminism. The attack on modernity that has been launched with such passion from many quarters has rarely acknowledged that it is a modernity created by men from dominant cultures. It is a modernity that women have barely begun to share. Charles Fourier once said that women's liberation was the yardstick by which a civilization should be judged; this was echoed by Lenin but subsequently sidelined by later Marxists and socialists. Equally, while Greens applaud the feminine principle, they pay much less attention to the mechanisms by which men dominate women.

The task for a feminist green socialism is to move from a world built upon the interests and experience of men, a male-experience world, a ME-world, to a world built upon the interests and experience of women, a women's-experience world, a WE-world. A WE-world based upon women's experience would need to be both decentralized and safe, ecologically, physically and socially. Domestic life and paid work would have to be integrated, as women cannot move far from their homes because of their domestic and caring responsibilities. The boundaries between paid and unpaid work are in any event anachronistic from a woman's perspective. As all production would need to take place near the home, dangerous, polluting work could not be carried out, as it would affect the health of the local community. People would also not want their local environment disfigured by destructive forms of production. Shops and other social facilities would need to be within easy reach of the local community and not stuck on the edge of town, available only to the car driver. People, not traffic, would have priority on the roads. Public transport would be universally available.

A world based on women's experience would ensure people's physical and social safety. Personal violence would never be treated as a "private" matter. Streets and public spaces would be patrolled if there were any danger to children, minority groups, or women. A WE-world would see its primary role as producing well-rounded and creative human beings and a well-integrated community life. People would not be asked to uproot themselves every few years to chase promotion or to meet the needs of a transnational corporation. If someone were needed in a new location, a local person would have to be appointed and trained. People would not be forced to migrate hundreds or thousands of miles to look for work, never seeing their family or community for years on end. Emotional needs would be given equal priority with physical needs; people would listen to each other, sympathize, and empathize. . . .

RECOVERING THE WE-WORLD

> Seeking to promote a new vision of democratic communities, with real resources for sharing the caring . . . feminists dreamt of a world where relationships beyond the couple would be invested with . . . meaning, commitment and passion.
>
> Lynne Segal[3]

A feminist green socialism must be underpinned by the values that have hitherto been imposed upon women: altruism, selfless caring, the desire to help other people realize their potential. This will not be achieved unless women's sexual liberation is at the center of a feminist green socialist politics. As Ann Ferguson has argued, the heterosexual family has played a central role in securing the "sex/affective" labor of women.[4] If the boundaries of the present destructive and distorted ideas of masculinity and femininity are to be broken, women must have control of their bodies both sexually and in terms of reproduction. However, sexual liberation is meaningless without economic liberation; to achieve their sexual and reproductive liberation, women must be economically independent of men. If men are to learn to live in the WE-world, they will have to abandon the benefits and the constraints of both patriarchy and masculinity.[5]

Ann Ferguson sees the elimination of what she calls the "sex-gender" divisions in society through "gynandry." She prefers this word to androgyny. Both mean transcending the boundaries of the sexes, but androgyny implies the priority of male values and experience, whereas gynandry stresses the importance of women's values and experience. If we are to move towards gynandry, men must acquire women's skills of "sex/affective labor," that is "nurturance, emotional sensitivity, receptivity to the desires of others, skills in the healing arts."[6] Ferguson argues that gynandrous people could develop only under a society run on the basis of "council socialism," where decision making was decentralized to the lowest possible level. Types of work would not be segregated on any basis, and everyone would receive an equal income or a basic income if they were too old or young to work. All workers would work two fifteen-hour "blocks" each week: one in what Ferguson calls "standard sector work" (construction, factories, doctoring, teaching, skilled clerical, and white-collar work) and a second in service work (garbage collection; recycling; cleaning; physical care of young, sick, and elderly; and educational support work). Service work would be rotated. Education would be a free and lifelong process, and everyone would be required to share in caring for the next generation.

Children would not be segregated from the adult world and would have rights to income and care. These would be provided by the adults who share the child's home, no matter what form the household may take. With a high level of communal support for childcare, this should not be an onerous burden. Ferguson suggests that there should be the equivalent of the Cuban Domestic Family Code, whereby both men and women are held responsible for housework and providing an income for children. Ann Ferguson's ideal community reflects many of the attributes of bell hooks's description of a black neighborhood:

> Many people raised in black communities experienced this type of community-based child care. Black women who had to leave the home and work to help provide for families could not afford to send children to day care centers and such centers did not always exist. They relied on people in their communities to help. She did not need to go with her children every time they walked to the playground to watch them because they would be watched by a number of people living near the playground. People who did not have children often took responsibility for sharing in childrearing. In my own family, there were seven children and when we were growing up it was not possible for our parents to watch us all the time or even give that extra special individual attention children sometimes desire. Those needs were often met by neighbors and people in the community.[7]

Most working-class communities would provide a similar support system, although almost always provided by women. Bell hooks claims that in her community caring was not taken for granted, it was valued and appreciated. When she went to university, hooks was surprised by the way in which white students rejected the care they had received as they prepared themselves for a competitive, affluent, individualized society. Emotional and physical support was pushed below the surface, out of sight. People pretended they had "got there on their own."

A feminist green socialism would recognize the essential values of care and nurturing in human development, and our mutual responsibility for them. However, we do not just need to build ourselves as individuals, we need to build a supportive social environment and a sustainable relationship with the natural world.

CARING FOR THE WORLD

> As we face slow environmental poisoning . . . we can hope that the prospect of the extinction of life on the planet will provide a universal impetus to social change.
>
> Ynestra King[8]

We do not have much time if we are to avoid increased human suffering and ecological destruction on a global scale. The present world system is driven by the competitive and individualistic values of patriarchal capitalism, based on private ownership, profit and self-interest, sustained by militarized nation-states. It is a system that has produced tremendous developments as well as tremendous destruction. It is not a system of organization or values appropriate for a sustainable future. Only a feminist green socialism is equal to this task. Feminist, because it acknowledges the centrality of women's life-producing and life-sustaining work and focuses upon the predominance of men in destructive institutions. Green, because it argues that we should act and think globally to regain a balance between the needs of humanity and the ability of the planet to sustain them. Socialist, because it recognizes the rights of all peoples of the world to live in a socially just and equitable community.

We need to deconstruct from above and reconstruct below. We need to free ourselves from the structures and institutions that divide us, while building our connectedness with each other and the natural world. For some people that will mean difficult and dangerous struggles with powerful institutions. Workers' struggles with multinational companies, forest-dwellers' and forestry workers' struggles against landowners who are willing to use murder as a weapon of repression. Poor women's daily struggle for livelihood and in defense of their land. The violence women face at the hands of men. Environmental and peace campaigners who risk their lives and liberty. Some people may be able to break away completely from modern industrial society to become green "Benedictines" in small-scale communities, as Rudolf Bahro has suggested. This will not be a solution for the mass of the population; most of us will be able to change things only in little ways, using small spaces, possibly only within ourselves and our own lives. Even now, in the shantytowns of Brazil, Colombia, and Mexico, women are pooling their resources into a "common pot"; in Kenya, women are drawing on traditional mutual aid networks to obtain supplies of fuel and water. Bell hooks records a similar experience in poor black communities where a cooperative value system emerged

> that challenged notions of individualism and private property so important to the maintenance of white supremacist, capitalist patriarchy. Black folk created in marginal spaces a world of community and collectivity where resources were shared.[9]

Within the richer societies there are possibilities in consumer action, through boycotts or selective consumption. People can begin to free themselves from total dependence upon the wider economy through more flexible approaches to work and income, such as a basic income scheme or local production and trading schemes; by joining alternative organizations such as producer or consumer cooperatives or forming local mutual aid networks. Local authorities could be empowered to act more creatively to stimulate local autonomy, through using local savings and labor. As well as finding marginal spaces locally to create a world of community and collectivity, we must reach out to those who are struggling for their livelihood across the globe and work with them to break down the boundaries that are destroying us all. We must build multicultural and international alliances that will be "bridges of power" and solidarity.[10] If we can do nothing else, we must constantly push and question: push at boundaries and question authorities. What is being done, in whose interest, to whose benefit? What will be the cost in ecological, social, and personal terms? By our pushing and questioning we must reclaim time. The world is being hurtled to oblivion at the pace of the fastest man, the man who is least personally encumbered, who is most nurtured and who takes account of the fewest restraining factors. There is a destructive conflict between the need of the individual to achieve as much as possible before (he) dies and the need of the planet to move as slowly as possible towards its inevitable death. For most women, sustaining the future generation is more important than personal achievements. Women, moving at the pace of the slowest child or the most needy adult, are closer to the sustainable pace of the planet.

At the same time, we must not reject modernity out of hand. If we stop development now, we are freezing it at the present levels of advantage and disadvantage. The many benefits of science, technology, and large-scale organization have not yet passed down to the majority of the people of this planet; that is the cruelty of the patriarchal, racist, capitalist system. The majority of people in the world have never had the chance to determine the way in which the benefits of human knowledge and creativity are used. It is not for the well-heeled white male middle class of the North to cry halt. It is fine for a well-educated person with the whole of Western (or Eastern) culture in their heads to go and whittle wood on a Welsh hillside, but communing with nature will not have the same meaning for those who have never had those advantages.

A sustainable egalitarian society will not be achieved if we abandon the ideals of progress and rational control of human history for an irrational and atavistic attachment to a mythical or mystical past or future. What is needed is a change of priorities: from men to women, from rich to poor, from North to South, from nature exploitation to nature stewardship. However, none of these will be achieved on its own; that is the reality of our interconnectedness. And that is why we need a feminist green socialism.

NOTES

1. Marilyn Waring, *If Women Counted* (London: Macmillan, 1989), p. 11.

2. Hilkka Pietilä, "Alternative Development with Women in the North," paper given to the Third International Interdisciplinary Congress of Women, Dublin, July 6–10, 1987, p. 26; previously published in Johan Galtung and Mars Friberg, eds., *Alternativen* (Stockholm: Akademilitteratur, 1986).

3. Lynne Segal, *Slow Motion* (London: Virago, 1990), p. 279.

4. Ann Ferguson, *Blood at the Root* (London: Pandora Press, 1989).

5. Segal, *Slow Motion*; John Stoltenberg, *Refusing to Be a Man* (London: Fontana, 1990).

6. Ferguson, *Blood at the Root*, p. 231.

7. bell hooks, *Talking Back* (London: Sheba, 1989), p. 144.

8. Ynestra King, "Toward an Ecological Feminism and a Feminist Ecology," in *Machina ex Dea*, ed. Joan Rothschild, 125 (Oxford: Pergamon, 1983).

9. hooks, *Talking Back*, p. 76.

10. Lisa Albrecht and Rose M. Brewer, *Bridges of Power: Women's Multi-cultural Alliances* (Philadelphia: New Society Publishers, 1990).

PART V

ENVIRONMENTAL JUSTICE

21

THE IMPORTANCE OF ENVIRONMENTAL JUSTICE

PETER WENZ

People play multiple roles in environmental systems. We are all constituents, as well as observers of the environment. So when we discuss the environment, we are always to some extent discussing ourselves. *Environmental Justice* is primarily about theories of distributive justice. There are three reasons for concentrating on contexts of environmental concern. . . . [First], we are involved in the environment more than we sometimes realize, and greater self-awareness can facilitate prudent behavior. Second, theories of distributive justice have not been related as often to the environmental area as to some other areas of concern. Third, and most important, environmental concerns involve relationships not only among people who live in the same society at the same time; but also among people who live in different societies at the same time, between people of the present and those of the future, between human and nonhuman animals, and between people and the biosphere in general. Because environmental concerns are uniquely global . . . theories of distributive justice are tested most thoroughly for their comprehensiveness when they are applied to environmental matters. . . .

Justice usually becomes an issue in contexts in which people's wants or needs exceed the means of their satisfaction. . . . The difference that an ample supply can make is illustrated well in the case of water. People need water wherever they live; but water is scarce in some places, ample in others. Where it is scarce, societies have devised elaborate methods of apportioning the water among those who need and desire it. Infringing upon the water rights of another is considered a serious injustice. In societies with advanced legal systems, such as in the United States, the apportionment of water in areas of scarcity is gov-

From: *Environmental Justice* (Albany: State University of New York Press, 1988), pp. xi–xii, 621, excerpts.

erned by intricate legal rules. But where water is plentiful, the situation is entirely different. In some parts of England, residents are charged a quarterly fee for the maintenance of the water works and sewage installation. The fee does not vary according to the amount of water used. In fact, there is no water meter to determine the amount used. Because there is enough water to serve everyone's wants and needs, people do not care how much they or their neighbors consume.

In sum, questions about justice arise concerning those things that are, or are perceived to be, in short supply relative to the demand for them. In these situations, people are concerned about getting their fair share, and arrangements are made, or institutions are generated, to allocate the scarce things among those who want or need them.

These generalizations are subject to two qualifications. First, the people sharing the scarce good must care enough about what they receive to desire their fair share. . . . In a drought, I am likely to want my house to be allocated its fair share of water by those at our local utility. Water is very important to me. I am not personally acquainted with most people in my town. I am not so benevolent as to want those among them who are no worse off than I to be given more water than I am given. So limited benevolence is, along with scarcity, an element of situations in which issues of justice arise.

The second qualification is this: Arrangements or institutions designed to allocate scarce things make sense only for those things that people are able to distribute. . . . The people at City Water, Light, and Power are able to distribute water from the reservoir. But the ability people have to distribute rain, good fortune, and perfect pitch is very limited. These things may be scarce, and people may want their fair shares, but there are no arrangements or institutions designed to allocate them, because people lack the power of distribution. . . .

THE NEED FOR COORDINATED ENVIRONMENTAL RESTRAINT

Allowing people to make, take, or receive whatever they can get—so long as they do not directly attack, brutalize, or steal from others—will sometimes make scarcity worse, hurting everyone in the long run. Garrett Hardin calls this "The Tragedy of the Commons." Imagine a pasture that can be used in common by many herdsmen. It is a limited resource, but one capable of supplying enough food for the animals that the herdsmen depend upon for their livelihood. Suppose that one of the herdsmen wants to increase his income. He can do this by doubling his herd. He will have to work harder to care for the larger herd, but he feels that the increased income is worth the effort. He is not directly attacking, brutalizing, or stealing from anyone else, so his extra income is earned within the rules that prohibit these things. The pasture is not appreciably damaged by the grazing of these extra animals because, though they double the individual's herd, they do not significantly increase the total number of animals grazing on the common pasture.

Since it is a common pasture, other herdsmen who wish to increase their income are also free to double or triple their flocks. However, as more and more animals graze in the common pasture, the flora is ruined owing to overgrazing. The result is the destruction of the common resource, the pasture, on which all had depended for a livelihood. No one committed barbarous acts against anyone else; yet, while literally minding their own business, they ruined the basis of that business. Their efforts to increase their incomes were self-defeating. In situations like this one, allowing people to take whatever they want of a scarce resource results in its destruction. The appropriation made by each person must be coordinated with those of others to ensure that the collective appropriation is not excessive and ruinous. So in these situations it is practical to determine each person's fair share of the collective good in order to avoid the tragedy of the commons. Such a determination can be made only by reference to an agreed standard of justice. Philosophical investigations into the nature and principles of justice are required.

Many environmental resources are like the pasture in Hardin's story. The oceans, the air, and the ozone layer are as important to our lives as was the pasture to the herdsmen. Yet no one owns them. If, in the pursuit of her own pleasure, gain, or preferred lifestyle, each person is free to use or despoil these natural resources in any way that does not directly brutalize other human beings, everyone will suffer in the long run. The ozone layer, for example, protects us all from solar radiation that can cause cancer. Suppose that the use of aerosol sprays diminishes the ozone layer. No single individual's use of aerosol sprays has an appreciable effect on the layer, so anyone can use such sprays without harming anyone else. But if millions use these sprays over long periods of time, the protection provided by the layer against harmful solar radiation could be significantly diminished. This could harm everyone, including those who make and use aerosol sprays. So if millions of people would like to use these sprays, some restraint must be exercised. One way of effecting such restraint would be to devise and enforce a system that permits people to use only limited amounts of aerosol spray. But what should the limit be? As in the case of the common pasture, it makes sense for each to be given her fair share, whatever that may be. Practicality again suggests that mutually agreeable principles of justice be discovered and employed in order to determine everyone's fair share.

This reasoning applies to pollution, generally. Whether it is a banana peel tossed from a car window on the highway or a gram of carbon monoxide emitted from the car's exhaust, every litter bit does *not* hurt. It is the concentration of litter bits that hurts, and this concentration usually hurts those who do the polluting as well as others. Our waste products are integral to our life processes. In limited concentrations, they aid the life process as a whole. But our preferred lifestyles and activities tend to generate kinds and levels of waste that pollute our environment and can eventually make it uninhabitable. So restraint is necessary in a great many matters, ranging from our use of the national parks to our consumption of fossil fuels. And again, when restraint is necessary to preserve the environment, it seems that everyone should receive a fair share, and be restrained to a fair degree, in accordance with reasonable principles of justice. This is environmental justice. . . .

THE VULNERABILITY OF MODERN SOCIETIES

It is an essential condition of a nation-state, especially of one that seeks to provide some liberty for its people, that the vast majority of people perceive the divisions of benefits and burdens to be reasonably just. The importance of this condition, like the importance of many essentials, is most clearly evident when the condition is not met. Consider, for example, the situation in Northern Ireland. In the mid-1960s, the condition of Catholics in Northern Ireland resembled that of blacks in the South of the United States. They went to separate schools, received poorer educational opportunities, had lower incomes, were discriminated against in employment, and held fewer government jobs. A movement of peaceful protest against these conditions was begun in the late 1960s. It was modeled on the protest of Martin Luther King Jr. in the American South. Unlike the government of the United States, which compromised with the civil rights movement, the British government made few concessions. . . .

If a modern, vulnerable social order is perceived by a significant percentage of its members to be irremediably and grossly unjust, it can be crippled or destroyed by relatively few people, with the active and passive support of others. This has happened in Northern Ireland. The application of force is inadequate to prevent this. The modern social order requires the voluntary cooperation of the vast majority, especially in a relatively free society. In order to receive this cooperation, the social order must be perceived as tolerably just. Discussions of the nature and principles of justice are therefore a practical necessity. We live in a society that many believe to be characterized by significant injustices. The society is vulnerable to radical disruption from within. But beliefs concerning justice are sufficient to mollify the vast majority of people so as to dissuade them from engaging in or supporting terrorism. . . . The perception that society's institutions and policies are reasonably just is necessary for such (relatively) willing submission to restraint. . . .

THE NEED FOR ENVIRONMENTAL JUSTICE

The conditions of justice recur frequently with respect to the environment. Arrangements must often be made to allocate access to activities and commodities so as to ensure that the uses people make of the environment are compatible with one another, and with the environment's continued habitability.

For example, it is now widely believed that burning large amounts of coal that contain significant quantities of sulfur results in acid rain. This rain is blamed for the defoliation of forests in the northeastern United States, southeastern Canada, and southern Germany. Many people use these forests as sources of recreation. They are also used by the lumber industry, and serve to retard soil erosion. So there is pressure to decrease significantly the amount of high sulfur coal that is burned, or to require that smokestacks of furnaces using this coal be fitted with scrubbers that prevent a very high percentage of the sulfur from escaping into the atmosphere. The coal is burned hundreds of miles away from the affected forests. It is used to power factories and to provide people with electricity. Significant reductions in the use of this coal would adversely affect the owners of the mines from which the coal is extracted, as well as the workers in those mines. The factories that currently use this coal would have either to use alternate, usually more expensive fuels, or else install expensive scrubbers. Either course could make their products more expensive. Products that become too expensive will no longer be marketable, and the factories that produce them will have to shut down, putting many people out of work. Electricity that is produced for household consumption would have to become more expensive for the same reasons.

Because the areas where the mines are located and where the coal is used are so far from the areas where the forests are damaged, the people who benefit from the current use of high sulfur coal, and who would be adversely affected by proposed changes, are for the most part different from the direct beneficiaries of forest preservation. Their interests are opposed. The more that one group gets of what it wants, the less the other group can get.

Decisions of public policy are required concerning the use of high sulfur coal. People will not feel well served by their government if this policy, which could put them out of work, or result in a mudslide covering their houses, is not clearly defensible. People will want to know why they should have to make the sacrifices that are required of them, and how these sacrifices compare to those that are required of others. The government will have to employ defensible principles of justice in fashioning its environmental policy if those affected by it are to believe that the sacrifices required of them are justified.

The same is true of most other environmental policies. They require people to make important sacrifices. Many policies are designed to deal with situations resembling the tragedy of the commons. As already noted, the air we breathe is like the commons. Uncontrolled automobile emissions of carbon monoxide would jeopardize the health of many people, especially children, the aged, and those with emphysema. Emissions could be reduced by improving mass transit systems at public expense, on the assumption that the use of mass transit would then replace some automobile use. The improvement in air quality would in this case be paid for by the taxpaying public. This may seem equitable, since clean air is a public good. But members of the public would not benefit equally from this policy. Those living in urban areas would benefit most for two reasons: The air quality would be improved most in those areas, and urban residents would have available inexpensive public transportation. Should all taxpayers have to pay for benefits that accrue disproportionately to urban residents?

Alternatively, automobile emissions could be reduced by placing heavy taxes on gasoline, thereby discouraging its use. The improvement in air quality would be more widespread because automobile use would be reduced generally, not just in urban areas. But do rural areas require such improvement? And are the burdens allocated equitably? On this plan, the user pays. This would hit hardest lower-income people and those in rural areas who must use their cars to get to work. Rich people would be virtually unaffected.

Many other plans could be devised to reduce air pollution caused by automobile emissions, and various plans can be combined with one another. Although all are aimed at reducing air pollution, each plan, and each combination of plans, benefits different people and/or places different burdens on different groups. Because these benefits and burdens can be significant, it will be necessary to assure people that they are receiving their fair share of benefits and are not being required unfairly to shoulder great burdens. The social fabric will not be destroyed by any one environmental policy that is perceived to be unjust. But the number and extent of environmental policies has increased and will continue to increase considerably. The perception that these policies are consistently biased in favor of some groups and against others could undermine the voluntary cooperation that is necessary for the maintenance of social order. Voluntary cooperation is especially necessary if the social order is to be maintained in a relatively open society where authoritarian measures are the exception rather than the rule. Thus, because social solidarity and the maintenance of order in a relatively free society require that people consider their sacrifices to be justified in relation to the sacrifices of others, environmental public policies will have to embody principles of environmental justice that the vast majority of people consider reasonable.

22.

CONFRONTING
ENVIRONMENTAL RACISM

ROBERT BULLARD

Communities are not all created equal. In the United States, for example, some communities are routinely poisoned while the government looks the other way. Environmental regulations have not uniformly benefited all segments of society. People of color (African Americans, Latinos, Asians, Pacific Islanders, and Native Americans) are disproportionately harmed by industrial toxins on their jobs and in their neighborhoods. These groups must contend with dirty air and drinking water—the by-products of municipal landfills, incinerators, polluting industries, and hazardous waste treatment, storage, and disposal facilities.

Why do some communities get "dumped on" while others escape? Why are environmental regulations vigorously enforced in some communities and not in others? Why are some workers protected from environmental threats to their health while others (such as migrant farmworkers) are still being poisoned? How can environmental justice be incorporated into the campaign for environmental protection? What institutional changes would enable the United States to become a just and sustainable society? What community organizing strategies are effective against environmental racism? . . . The pervasive reality of racism is placed at the very center of the analysis.

INTERNAL COLONIALISM AND WHITE RACISM

The history of the United States has long been grounded in white racism. The nation was founded on the principles of "free land" (stolen from Native Americans and

From: Robert Bullard, ed., *Confronting Environmental Racism: Voices from the Grassroots* (Boston: South End Press, 1993), pp. 15–24, 38–39.

Mexicans), "free labor" (cruelly extracted from African slaves), and "free men" (white men with property). From the outset, institutional racism shaped the economic, political, and ecological landscape, and buttressed the exploitation of both land and people. Indeed, it has allowed communities of color to exist as internal colonies characterized by dependent (and unequal) relationships with the dominant white society or "mother country." In their 1967 book, *Black Power*, Stokely Carmichael and Charles Hamilton were among the first to explore the "internal" colonial model as a way to explain the racial inequality, political exploitation, and social isolation of African Americans. As Carmichael and Hamilton write:

> The economic relationship of America's black communities [to white society] . . . reflects their colonial status. The political power exercised over those communities goes hand in glove with the economic deprivation experienced by the black citizens.
>
> Historically, colonies have existed for the sole purpose of enriching, in one form or another, the "colonizer"; the consequence is to maintain the economic dependency of the "colonized."[1]

Generally, people of color in the United States—like their counterparts in formerly colonized lands of Africa, Asia, and Latin America—have not had the same opportunities as whites. The social forces that have organized oppressed colonies internationally still operate in the "heart of the colonizer's mother country."[2] For Robert Blauner, people of color are subjected to five principal colonizing processes: they enter the "host" society and economy involuntarily; their native culture is destroyed; white-dominated bureaucracies impose restrictions from which whites are exempt; the dominant group uses institutionalized racism to justify its actions; and a dual or "split labor market" emerges based on ethnicity and race. Such domination is also buttressed by state institutions. Social scientists Michael Omi and Howard Winant go so far as to insist that "every state institution is a racial institution."[3] Clearly, whites receive benefits from racism, while people of color bear most of the cost.

ENVIRONMENTAL RACISM

Racism plays a key factor in environmental planning and decision making. Indeed, environmental racism is reinforced by government, legal, economic, political, and military institutions. It is a fact of life in the United States that the mainstream environmental movement is only beginning to wake up to. Yet, without a doubt, racism influences the likelihood of exposure to environmental and health risks and the accessibility to healthcare. Racism provides whites of all class levels with an "edge" in gaining access to a healthy physical environment. This has been documented again and again.

Whether by conscious design or institutional neglect, communities of color in urban ghettos, in rural "poverty pockets," or on economically impoverished Native American reservations face some of the worst environmental devastation in the nation. Clearly, racial discrimination was not legislated out of existence in the 1960s. While some significant progress was made during this decade, people of color continue to struggle for equal treatment in many areas, including environmental justice. Agencies at all levels of government, including the federal EPA, have done a poor job protecting people of color from the ravages of pollution and industrial encroachment. It has thus been an uphill battle convincing white judges, juries, government officials, and policy makers that racism exists in environmental protection, enforcement, and policy formulation.

The most polluted urban communities are those with crumbling infrastructure, ongoing economic disinvestment, deteriorating housing, inadequate schools, chronic unemployment, a high poverty rate, and an overloaded healthcare system. Riot-torn south-central Los Angeles typifies this urban neglect. It is not surprising that the "dirtiest" zip code in California belongs to the mostly African American and Latino neighborhood in that part of the city.[4] In the Los Angeles basin, over 71 percent of the African Americans and 50 percent of the Latinos live in areas with the most polluted air, while only 34 percent of the white population does.[5] This pattern exists nationally as well. As researchers D. R. Wernette and L. A. Nieves note:

> In 1990, 437 of the 3,109 counties and independent cities failed to meet at least one of the EPA ambient air quality standards . . . 57 percent of whites, 65 percent of African Americans, and 80 percent of Hispanics live in 437 counties with substandard air quality. Out of the whole population, a total of 33 percent of whites, 50 percent of African Americans, and 60 percent of Hispanics live in the 136 counties in which two or more air pollutants exceed standards. The percentage living in the 29 counties designated as nonattainment areas for three or more pollutants are 12 percent of whites, 20 percent of African Americans, and 31 percent of Hispanics.[6]

Income alone does not account for these above-average percentages. Housing segregation and development patterns play a key role in determining where people live. Moreover, urban development and the "spatial configuration" of communities flow from the forces and relationships of industrial production which, in turn, are influenced and subsidized by government policy.[7] There is widespread agreement that vestiges of race-based decision making still influence housing, education, employment, and criminal justice. The same is true for municipal services such as garbage pickup and disposal, neighborhood sanitation, fire and police protection, and library services. Institutional racism influences decisions on local land use, enforcement of environmental regulations, industrial facility siting, management of economic vulnerability, and the paths of freeways and highways.

People skeptical of the assertion that poor people and people of color are targeted for waste disposal sites should consider the report the Cerrell Associates provided the California Waste Management Board. In their 1984 report, *Political Difficulties Facing Waste-to-Energy Conversion Plant Siting*, they offered a detailed profile of those neighborhoods most likely to organize effective resistance against incinerators. The policy conclusion based on this analysis is clear. As the report states: "All socioeconomic groupings tend to resent the nearby siting of major facilities, but middle and upper socioeconomic strata possess better resources to effectuate their opposition. Middle and higher socioeconomic strata neighborhoods should not fall within the one-mile and five-mile radius of the proposed site."[8]

Where then will incinerators or other polluting facilities be sited? For Cerrell Associates, the answer is low-income, disempowered neighborhoods with a high concentration of nonvoters. The ideal site, according to their report, has nothing to do with environmental soundness but everything to do with lack of social power. Communities of color in California are far more likely to fit this profile than are their white counterparts.

Those still skeptical of the existence of environmental racism should also consider the fact that zoning boards and planning commissions are typically stacked with white developers. Generally, the decisions of these bodies reflect the special interests of the individuals who sit on these boards. People of color have been systematically excluded from these decision-making boards, commissions, and governmental agencies (or allowed only token representation). Grassroots leaders are now demanding a shared role in all the decisions that shape their communities. They are challenging the intended or unintended racist assumptions underlying environmental and industrial policies.

TOXIC COLONIALISM ABROAD

To understand the global ecological crisis, it is important to understand that the poisoning of African Americans in south-central Los Angeles and of Mexicans in border *maquiladoras* have their roots in the same system of economic exploitation, racial oppression, and devaluation of human life. The quest for solutions to environmental problems and for ways to achieve sustainable development in the United States has considerable implications for the global environmental movement.

Today, more than nineteen hundred *maquiladoras*, assembly plants operated by American, Japanese, and other foreign countries, are located along the two thousand-mile US-Mexican border.[9] These plants use cheap Mexican labor to assemble products from imported components and raw materials, and then ship them back to the United States.[10] Nearly half a million Mexicans work in the *maquiladoras*. They earn an average of $3.75 a day. While these plants bring jobs, albeit low-paying ones, they exacerbate local pollution by overcrowding the border towns, straining sewage and water systems, and reducing air quality. All this compromises the health of workers and nearby community residents. The Mexican environmental regulatory agency is understaffed and ill-equipped to adequately enforce the country's laws.[11]

The practice of targeting poor communities of color in the third world for waste disposal and the introduction of risky technologies from industrialized countries are forms of "toxic colonialism," what some activists have dubbed the "subjugation of people to an ecologically-destructive economic order by entities over which the people have no control."[12] The industrialized world's controversial third world dumping policy was made public by the release of an internal, December 12, 1991, memorandum authored by Lawrence Summers, chief economist of the World Bank. It shocked the world and touched off a global scandal. Here are the highlights:

"Dirty" Industries: Just between you and me, shouldn't the World Bank be encouraging MORE migration of the dirty industries to the LDCs [Less Developed Countries]? I can think of three reasons:

1) The measurement of the costs of health impairing pollution depends on the foregone earnings from increased morbidity and mortality. From this point of view a given amount of health impairing pollution should be done in the country with the lowest cost, which will be the country with the lowest wages. I think the economic logic behind dumping a load of toxic waste in the lowest wage country is impeccable and we should face up to that.

2) The costs of pollution are likely to be non-linear as the initial increments of pollution probably have very low cost. I've always thought that under-polluted areas in Africa are vastly UNDER-polluted; their air quality is probably vastly inefficiently low compared to Los Angeles or Mexico City. Only the lamentable facts that so much pollution is generated by non-tradable industries (transport, electrical generation) and that the unit transport costs of solid waste are so high prevent world welfare-enhancing trade in air pollution and waste.

3) The demand for a clean environment for aesthetic and health reasons is likely to have very high income elasticity. The concern over an agent that causes a one in a million change in the odds of prostate cancer is obviously going to be much higher in a country where people survive to get prostate cancer than in a country where under 5 [year-old] mortality is 200 per thousand. Also, much of the concern over industrial atmosphere discharge is about visibility impairing particulates. These discharges may have very little direct health impact. Clearly trade in goods that embody aesthetic pollution concerns could be welfare enhancing. While production is mobile the consumption of pretty air is a non-tradable.

The problem with the arguments against all of these proposals for more pollution in LDCs (intrinsic rights to certain goods, moral reasons, social concerns, lack of adequate markets, etc.) could be turned around and used more or less effectively against every Bank proposal.

BEYOND THE RACE-VERSUS-CLASS TRAP

Whether at home or abroad, the question of who *pays* and who *benefits* from current industrial and development policies is central to any analysis of environmental racism. In the United States, race interacts with class to create special environmental and health vulnerabilities. People of color, however, face elevated toxic exposure levels even when social class variables (income, education, and occupational status) are held constant.[13] Race has been found to be an independent factor, not reducible to class, in predicting the distribution of (1) air pollution in our society,[14] (2) contaminated fish consumption,[15] (3) the location of municipal landfills and incinerators,[16] (4) the location of abandoned toxic waste dumps,[17] and (5) lead poisoning in children.[18]

Lead poisoning is a classic case in which race, not just class, determines exposure. It affects between three and four million children in the United States—most of whom are African Americans and Latinos living in urban areas. Among children five years old and younger, the percentage of African Americans who have excessive levels of lead in their blood far exceeds the percentage of whites at all income levels.[19]

The federal Agency for Toxic Substances and Disease Registry found that for families earning less than six thousand dollars annually, an estimated 68 percent of African American children had lead poisoning, compared with 36 percent for white children. For families with incomes exceeding fifteen thousand dollars, more than 38 percent of African American children have been poisoned, compared with 12 percent of white children. African American children are two to three times more likely than their white counterparts to suffer from lead poisoning independent of class factors.

One reason for this is that African Americans and whites do not have the same opportunities to "vote with their feet" by leaving unhealthy physical environments. The ability of an individual to escape a health-threatening environment is usually correlated with income. However, racial barriers make it even harder for millions of African Americans, Latinos, Asians, Pacific Islanders, and Native Americans to relocate. Housing discrimination, redlining, and other market forces make it difficult for millions of households to buy their way out of polluted environments. For example, an affluent African American family (with an income of fifty thousand dollars or more) is as segregated as an African American family with an annual income of five thousand dollars.[20] Thus, lead poisoning of African American children is not just a "poverty thing."

White racism helped create our current separate and unequal communities. It defines the boundaries of the urban ghetto, barrio, and reservation, and influences the provision of environmental protection and other public services. Apartheid-type housing and development policies reduce neighborhood options, limit mobility, diminish job opportunities, and decrease environmental choices for millions of Americans. It is unlikely that this nation will ever achieve lasting solutions to its environmental problems unless it also addresses the system of racial injustice that helps sustain the existence of powerless communities forced to bear disproportionate environmental costs.

THE LIMITS OF MAINSTREAM ENVIRONMENTALISM

Historically, the mainstream environmental movement in the United States has developed agendas that focus on such goals as wilderness and wildlife preservation, wise resource management, pollution abatement, and population control. It has been primarily supported by middle- and upper-middle-class whites. Although concern for the environment cuts across class and racial lines, ecology activists have traditionally been individuals with above-average education, greater access to economic resources, and a greater sense of personal power.[21]

Not surprisingly, mainstream groups were slow in broadening their base to include poor and working-class whites, let alone African Americans and other people of color. Moreover, they were ill equipped to deal with the environmental, economic, and social concerns of these communities. During the 1960s and 1970s, while the "Big Ten" environmental groups focused on wilderness preservation and conservation through litigation, political lobbying, and technical evaluation, activists of color were engaged in mass direct action mobilizations for basic civil rights in the areas of employment, housing, education, and healthcare. Thus, two parallel and sometimes conflicting movements emerged, and it has taken nearly two decades for any significant convergence to occur between these two efforts. In fact, conflicts still remain over how the two groups should balance economic development, social justice, and environmental protection.

In their desperate attempt to improve the economic conditions of their constituents, many African American civil rights and political leaders have directed their energies toward bringing jobs to their communities. In many instances, this has been achieved at great risk to the health of workers and the surrounding communities. The promise of jobs (even low-paying and hazardous ones) and of a broadened tax base has enticed several economically impoverished, politically powerless communities of color both in the United States and around the world.[22] Environmental job blackmail is a fact of life. You can get a job, but only if you are willing to do work that will harm you, your families, and your neighbors.

Workers of color are especially vulnerable to job blackmail because of the greater threat of unemployment they face compared to whites and because of their concentration in low-paying, unskilled, nonunionized occupations. For example, they make up a large share of the nonunion contract workers in the oil, chemical, and nuclear industries. Similarly, over 95 percent of migrant farmworkers in the United States are Latino, African American, Afro-Caribbean, or Asian, and African Americans are overrepresented in high-risk, blue-collar, and service occupations for which a large pool of replacement labor exists. Thus, they are twice as likely to be unemployed as their white counterparts. Fear of unemployment acts as a potent incentive for many African American workers to accept and keep jobs they know are health threatening. Workers will tell you that "unemployment and poverty are also hazardous to one's health." An inherent conflict exists between the interests of capital and that of labor. Employers have the power to move jobs (and industrial hazards) from the Northeast and Midwest to the South and the Sunbelt, or they may move the jobs offshore to third world countries where labor is even cheaper and where there are even fewer health and safety regulations. Yet, unless an environmental movement emerges that is capable of addressing these economic concerns, people of color and poor white workers are likely to end up siding with corporate managers in key conflicts concerning the environment.

Indeed, many labor unions already moderate their demands for improved work-safety and pollution control whenever the economy is depressed. They are afraid of layoffs, plant closings, and the relocation of industries. These fears and anxieties of labor are usually built on the false but understandable assumption that environmental regulations inevitably lead to job loss.[23]

The crux of the problem is that the mainstream environmental movement has not sufficiently addressed the fact that social inequality and imbalances of social power are at the heart of environmental degradation, resource depletion, pollution, and even overpopulation. The environmental crisis can simply not be solved effectively without social justice. As one academic human ecologist notes, "Whenever [an] in-group directly and exclusively benefits from its own overuse of a shared resource but the costs of that overuse are 'shared' by out-groups, then in-group motivation toward a policy of resource conservation (or sustained yields of harvesting) is undermined."[24]

THE MOVEMENT FOR ENVIRONMENTAL JUSTICE

Activists of color have begun to challenge both the industrial polluters and the often indifferent mainstream environmental movement by actively fighting environmental threats in their communities and raising the call for environmental justice. This groundswell of environmental activism in African American, Latino, Asian, Pacific Islander, and Native American communities is emerging all across the country. While rarely listed in the standard environmental and conservation directories, grassroots environmental justice groups have sprung up from Maine to Louisiana and Alaska.

These grassroots groups have organized themselves around waste facility siting, lead contamination, pesticides, water and air pollution, native self-government, nuclear testing, and workplace safety.[25] People of color have invented and, in other cases, adapted existing organizations to meet the disproportionate environmental challenges they face. A growing number of grassroots groups and their leaders have adopted confrontational direct action strategies similar to those used in earlier civil rights conflicts. Moreover, the increasing documentation of environmental racism has strengthened the demand for a safe and healthy environment as a basic right of all individuals and communities.[26]

Drawing together the insights of both the civil rights and the environmental movements, these grassroots groups are fighting hard to improve the quality of life for their residents. As a result of their efforts, the environmental justice movement is increasingly influencing and winning support from more conventional environmental and civil rights organizations. For example, the National Urban League's *1992 State of Black America* included—for the first time in the seventeen years the report has been published—a chapter on the environmental threats to the African American community.[27] In addition, the NAACP, ACLU, and NRDC led the fight to have poor children tested for lead poisoning under Medicaid provisions in California. The class-action lawsuit *Matthews v. Coye*, settled in 1991, called for the state of California to screen an estimated 500,000 poor children for lead poisoning at a cost of $15 to $20 million.[28] The screening represents a big step forward in efforts to identify children suffering from what federal authorities admit is the number one environmental health problem of children in the United States. For their part, mainstream environmental organizations are also beginning to understand the need for environmental justice and are increasingly supporting grassroots groups in the form of technical advice, expert testimony, direct financial assistance, fund-raising, research, and legal assistance. Even the Los Angeles chapter of the wilderness-focused Earth First! movement worked with community groups to help block the incinerator project in south-central Los Angeles. . . .

CONCLUSION

The mainstream environmental movement has proven that it can help enhance the quality of life in this country. The national membership organizations that make up the mainstream movement have clearly played an important role in shaping the nation's environmental policy. Yet, few of these groups have actively involved themselves in environmental conflicts involving communities of color. Because of this, it's unlikely that we will see a mass influx of people of color into the national environmental groups any time soon. A continuing growth in their own grassroots organizations is more likely. Indeed, the fastest-growing segment of the environmental movement is made up by the grassroots groups in communities of color which are increasingly linking up with one another and with other community-based groups. As long as US society remains divided into separate and unequal communities, such groups will continue to serve a positive function.

It is not surprising that indigenous leaders are organizing the most effective resistance within communities of color. They have the advantage of being close to the population immediately affected by the disputes they are attempting to resolve. They are also completely wedded to social and economic justice agendas and familiar with the tactics of the civil rights movement. This makes effective community organizing possible. People of color have a long track record in challenging government and corporations that discriminate. Groups that emphasize civil rights and social justice can be found in almost every major city in the country.

Cooperation between the two major wings of the environmental movement is both possible and beneficial, however. Many environmental activists of color are now getting support from mainstream organizations in the form of technical advice, expert testimony, direct financial assistance, fund-raising, research, and legal assistance. In return, increasing numbers of people of color are assisting mainstream organizations to redefine their limited environmental agendas and expand their outreach by serving on boards, staffs, and advisory councils. Grassroots activists have thus been the most influential activists in placing equity and social justice issues onto the larger environmental agenda and democratizing and diversifying the movement as a whole. Such changes are necessary if the environmental movement is to successfully help spearhead a truly global movement for a just, sustainable, and healthy society and effectively resolve pressing environmental disputes. Environmentalists and civil rights activists of all stripes should welcome the growing movement of African Americans, Latinos, Asians, Pacific Islanders, and Native Americans who are taking up the struggle for environmental justice.

NOTES

1. Stokely Carmichael and Charles V. Hamilton, *Black Power: The Politics of Liberation in America* (New York: Vintage, 1967), pp. 16–17.

2. Robert Blauner, *Racial Oppression in America* (New York: Harper and Row, 1972), p. 26.

3. Michael Omi and Howard Winant, *Racial Formation in the United States: From the 1960s to the 1980s* (New York: Routledge and Kegan Paul, 1986), pp. 76–77.

4. Jane Kay, "Fighting Toxic Racism: LA's Minority Neighborhood Is the 'Dirtiest' in the State," *San Francisco Examiner*, April 7, 1991, A1.

5. Paul Ong and Evelyn Blumenberg, "Race and Environmentalism," paper read at Graduate School of Architecture and Urban Planning, UCLA, March 14, 1990.

6. D. R. Wernette and L. A. Nieves, "Breathing Polluted Air," *EPA Journal* 18 (March/April 1992): 16–17.

7. Joe R. Feagin and Clairece B. Feagin, *Discrimination American Style: Institutional Racism and Sexism* (Malabar, FL: Robert E. Krieger, 1986); Mark Gottdiener and Joe R. Feagin, "The Paradigm Shift in Urban Sociology," *Urban Affairs Quarterly* 24, no. 2 (December 1988): 163–87.

8. Cerrell Associates, Inc., *Political Difficulties Facing Waste-to-Energy Conversion Plant Siting*, California Waste Management Board, Technical Information Series (Los Angeles: Cerrell Associates, 1984), p. 43.

9. Center for Investigative Reporting and Bill Moyers, *Global Dumping Grounds: The International Trade in Hazardous Waste* (Washington, DC: Seven Locks Press, 1990); Roberto Sanchez, "Health and Environmental Risks of the Maquiladora in Mexicali," *Natural Resources Journal* 30 (Winter 1990): 163–86; Jo Ann Zuniga, "Watchdog Keeps Tabs on Politics of Environment along Border," *Houston Chronicle*, May 24, 1992, 22A.

10. Matthew Witt, "An Injury to One Is a Gravio a Todo: The Need for a Mexico-US Health and Safety Movement," *New Solutions: A Journal of Environmental and Occupational Health Policy* 1 (March 1991): 28–33.

11. Working Group on Canada-Mexico Free Trade, "Que Pasa? A Canada-Mexico 'Free' Trade Deal," *New Solutions: A Journal of Environmental and Occupational Health Policy* 2 (January 1991): 10–25.

12. Greenpeace, *The International Trade in Wastes: A Greenpeace Inventory* (Washington, DC: Greenpeace, USA, 1990), p. 3.

13. Bunyan Bryant and Paul Mohai, *Race and the Incidence of Environmental Hazards* (Boulder, CO: Westview Press, 1992).

14. Myrick A. Freeman, "The Distribution of Environmental Quality," in *Environmental Quality Analysis*, ed. Allen V. Kneese and Blair T. Bower (Baltimore: Johns Hopkins University Press for Resources for the Future, 1971); Michel Gelobter, "The Distribution of Air Pollution by Income and Race," paper presented at the Second Symposium on Social Science in Resource Management, Urbana, IL, June 1988; Leonard Gianessi, H. M. Peskin, and E. Wolff, "The Distributional Effects of Uniform Air Pollution Policy in the US," *Quarterly Journal of Economics* (May 1979): 281–301; Wernette and Nieves, "Breathing Polluted Air," pp. 16–17.

15. Pat C. West, M. Fly, and R. Marans, "Minority Anglers and Toxic Fish Consumption: Evidence from a State-Wide Survey of Michigan," *Proceedings of the Michigan Conference on Race and the Incidence of Environmental Hazards*, ed. Bunyan Bryant and Paul Mohai (Ann Arbor, MI: University of Michigan School of Natural Resources, 1989), pp. 108–22.

16. Robert D. Bullard, "Solid Waste Sites and the Black Houston Community," *Sociological Inquiry* 53 (Spring 1983): 273–88; Robert D. Bullard, *Invisible Houston: The Black Experience in Boom and Bust* (College Station: Texas A&M University Press, 1987); Robert D. Bullard, *Dumping in Dixie: Race, Class and Environmental Quality* (Boulder, CO: Westview Press, 1990); Robert D. Bullard, "Environmental Justice for All," *EnviroAction* 9 (November 1991).

17. United Church of Christ Commission for Racial Justice, *The First National People of Color Environmental Leadership Summit: Program Guide* (New York: United Church of Christ, 1992).

18. Agency for Toxic Substances and Disease Registry, *The Nature and Extent of Lead Poisoning in Children in the United States: A Reprint to Congress* (Atlanta: US Department of Health and Human Services, 1988).

19. Ibid., pp. 1–12.

20. Nancy A. Denton and Douglas S. Massey, "Residential Segregation of Blacks,

Hispanics, and Asians by Socioeconomic Class and Generation," *Social Science Quarterly* 69 (1988): 797–817; Gerald D. Jaynes and Robin M. Williams Jr., *A Common Destiny: Blacks and American Society* (Washington, DC: National Academy Press, 1989).

21. Kenneth M. Bachrach and Alex J. Zautra, "Coping with Community Stress: The Threat of a Hazardous Waste Landfill," *Journal of Health and Social Behavior* 26 (June 1985): 127–41; Robert D. Bullard, *Dumping in Dixie*; Robert D. Bullard and Beverly H. Wright, "Blacks and the Environment," *Humboldt Journal of Social Relations* 14 (1987): 165–84; Frederick Buttel and William L. Flinn, "The Structure and Support for the Environmental Movement 1968–70," *Rural Sociology* 39 (1974): 56–69; Riley E. Dunlap, "Public Opinion on the Environment in the Reagan Era: Polls, Pollution, and Politics," *Environment* 29 (1987): 6–11, 31–37; Paul Mohai, "Public Concern and Elite Involvement in Environmental Conservation," *Social Science Quarterly* 66 (December 1985): 820–38; Paul Mohai, "Black Environmentalism," *Social Science Quarterly* 71 (April 1990): 744–65; Denton E. Morrison, "The Soft Cutting Edge of Environmentalism: Why and How the Appropriate Technology Notion Is Changing the Movement," *Natural Resources Journal* 20 (April 1980): 275–98; Denton E. Morrison, "How and Why Environmental Consciousness Has Trickled Down," in *Distributional Conflict in Environmental Resource Policy*, ed. Allan Schnaiberg, Nicholas Watts, and Klaus Zimmermann, 187–220 (New York: St. Martin's Press, 1986).

22. Bryant and Mohai, *Race and the Incidence of Environmental Hazards*; Bullard, *Dumping in Dixie*; Center for Investigative Reporting and Moyers, *Global Dumping Grounds*.

23. Michael H. Brown, *Laying Waste: The Poisoning of America by Toxic Chemicals* (New York: Pantheon Books, 1980); Michael H. Brown, *The Toxic Cloud: The Poisoning of America's Air* (New York: Harper and Row, 1987).

24. William Catton, *Overshoot: The Ecological Basis of Revolutionary Change* (Chicago: University of Illinois Press, 1982).

25. Dana Alston, *We Speak for Ourselves: Social Justice, Race, and Environment* (Washington, DC: Panos Institute, 1990); Bryant and Mohai, *Race and the Incidence of Environmental Hazards*; Bullard, *Dumping in Dixie*; Robert D. Bullard, *Directory of People of Color Environmental Groups 1992* (Riverside: University of California, Department of Sociology, 1992); Catton, *Overshoot*.

26. Bullard and Wright, "Blacks and the Environment," pp. 165–84; Robert D. Bullard and Beverly H. Wright, "The Quest for Environmental Equity: Mobilizing the African American Community for Social Change," *Society and Natural Resources* 3 (1991): 301–11; United Church of Christ Commission for Racial Justice, *People of Color Environmental Leadership Summit*.

27. Robert D. Bullard, "Urban Infrastructure: Social, Environmental, and Health Risks to African Americans," in *The State of Black America 1992*, ed. Billy J. Tidwell (New York: National Urban League, 1992), pp. 183–96.

28. Bill Lann Lee, "Environmental Litigation on Behalf of Poor, Minority Children: *Matthews v. Coye*: A Case Study," paper presented at the annual meeting of the American Association for the Advancement of Science, Chicago, April 1992.

23.

THE ENVIRONMENTAL JUSTICE MOVEMENT

LUKE COLE AND SHEILA FOSTER

Pointing to a particular date or event that launched the environmental justice movement is impossible, as the movement grew organically out of dozens, even hundreds, of local struggles and events and out of a variety of other social movements. Nevertheless, certain incidents loom large in the history of the movement as galvanizing events.

Many observers point to protests by African Americans against a toxic dump in Warren County, North Carolina, in 1982 as the beginning of the movement. The sociologist Robert Bullard points to African American student protests over the drowning death of an eight-year-old girl in a garbage dump in a residential area of Houston in 1967.[1] Others note that the Rev. Dr. Martin Luther King Jr. was traveling to Memphis to support striking garbage workers in what is now considered an environmental justice struggle when he was assassinated in 1968.[2] The United Farm Workers' struggle against pesticide poisoning in the workplace, beginning in the 1960s (and continuing to this day), is the starting point for some. Some Native American activists and others consider the first environmental justice struggles on the North American continent to have taken place five hundred years ago with the initial invasion by Europeans.

Rather than an incident-focused history of the movement, however, we think it more useful to think metaphorically of the movement as a river, fed over time by many tributaries.[3] No one tributary made the river the force that it is today; indeed, it is difficult to point to the headwaters, since so many tributaries have nourished the movement. Particular events can be seen as high-water marks (or

From: *From the Ground Up: Environmental Racism and the Rise of the Environmental Justice Movement* (New York: New York University Press, 2001), pp. 19–33.

perhaps, to push the metaphor, exciting rapids) in each stream, or the main river. With this idea in mind, we discuss here some of the most important tributaries of the river of the environmental justice movement.

FOUNDATIONS OF THE ENVIRONMENTAL JUSTICE MOVEMENT

The Civil Rights Movement

Perhaps the most significant source feeding into today's environmental justice movement is the civil rights movement of the 1950s, 1960s, and 1970s. Through that movement, hundreds of thousands of African Americans and their allies, primarily but not solely in the southern United States, pressed for social change and experienced empowerment through grassroots activism.[4]

The spirit and experience of resistance through the civil rights movement was widespread in the southern United States and in many northern urban areas. The movement was strongly church based; many of its leaders, like the Rev. Dr. Martin Luther King Jr. and the Rev. Ralph Abernathy, were ministers. When the environmental justice movement began building momentum in the early 1980s, it was church-based civil rights leaders, seasoned in the civil rights movement, who were at its fore. The 1982 protests in Warren County, North Carolina, against a PCB dump were led by local church officials and by the Rev. Benjamin Chavis, a longtime civil rights activist and at that time the head of the United Church of Christ's Commission for Racial Justice.[5] The environmental justice movement's roots in civil rights and church-based advocacy is evidenced in the United Church of Christ Commission for Racial Justice's landmark 1987 study, *Toxic Wastes and Race in the United States*. Perhaps the single best-known work documenting the disproportionate impact of environmental hazards on people of color, *Toxic Wastes and Race* galvanized the movement. (Its author, Charles Lee, while working for a church-based civil rights group, also helped organize early meetings of academics to talk about environmental justice issues.)

Environmental justice movement leaders coming out of civil rights organizing include not only those who advocated for the rights of African Americans but also Latino activists. Movement leaders like Jean Gauna and Richard Moore of Albuquerque came out of Chicano political organizing in the Southwest, which involved mass protests against the Vietnam War, police brutality, and racism in housing and education.[6]

Civil rights activists brought three things to the environmental justice movement: a history of, and experience with, direct action, which led to similar exercises of grassroots power by the environmental justice movement; a perspective that recognized that the disproportionate impact of environmental hazards was not random or the result of "neutral" decisions but a product of the same social and economic structure which had produced de jure and de facto segregation and other racial oppression; and the experience of empowerment through political action. The seasoned civil rights leaders recognized environmental racism and set about using the tools and techniques they knew in their effort to combat it. The Warren County protests, for example, in which more than five hundred people were arrested in acts of civil disobedience,[7] directly echoed the sit-ins and civil disobedience of the 1960s. Similarly, marches, a signature of the civil rights movement, have become a fixture in local environmental justice struggles.

Civil rights movement leaders now in positions of power have also lent assistance to the environmental justice movement. For instance, in 1992, Rep. John Lewis of Georgia, a prominent participant in the protests of the 1960s, introduced the Environmental Justice Act.[8] Though the act did not pass Congress, it raised environmental justice issues to a new stature in Washington. Lewis, in speaking about the bill, recognized that "the quest for environmental justice has helped to renew the civil rights movement" through its call for environmental protection as a "right of all, not a privilege for a few."[9]

The Antitoxics Movement

The second major tributary to the river is the grassroots antitoxics movement. Communities have long resisted and organized against hazardous waste facilities, landfills, and incinerators.[10] The grassroots antitoxics movement burst into national prominence in the late 1970s, when President Jimmy Carter declared Love Canal, New York, a disaster area and evacuated residents of a housing development built on a former toxic waste dump.[11] While Love Canal and the subsequent evacuation and relocation of another contaminated community at Times Beach, Missouri, are perhaps the best-known early examples of "grassroots environmentalism," similar stories have taken place across the United States. The proliferation of local actions marked an important shift in environmental activism when it began in the late 1970s: as Andrew Szasz notes, these local environmental conflicts "tended not to be about nature, per se, but about land use, social impact, [and] human health."[12] These local actions and activists also transformed toxic waste from a "nonentity to a full-fledged issue."[13]

The grassroots antitoxics movement grew to prominence after the civil rights movement; in contrast to that movement, its leadership is generally characterized by a lack of political organizing experience before a particular toxic struggle. "I have never been an activist before this fight" is a common story in the antitoxics movement, in which residents, primarily women, are galvanized to action by threats to their health, their families, and their communities. As these grassroots leaders heard about other, similar antitoxics fights in nearby communities, they slowly linked their local struggles together into a larger "movement."[14]

The antitoxics movement became loosely organized under several national umbrella organizations in the 1980s, which helped make its actions more technically sophisticated and strategically coherent. For example, Citizens Clearinghouse for Hazardous Wastes (CCHW), an organization founded by former residents of Love Canal, has assisted grassroots activists nationwide for the past fifteen years, working with more than seven thousand local groups.[15] Thousands of these groups used (and continue to use) direct action protests to effectuate their demands. National groups like CCHW and regional groups that sprang up, such as the Environmental Health Coalition in San Diego, California, also used science and technical information, placing a high premium on demystifying arcane documents such as environmental impact statements and processes such as risk assessment. The antitoxics movement sought to understand, and then restructure, the system of toxic waste production in the United States. Growing out of their concrete experiences in their own communities, antitoxics activists came up with the idea of "pollution prevention"—that is, eliminating the use of toxic chemicals in industrial practices so that the production of toxic waste is stopped as well. Under the force of years of organizing, pollution prevention has moved from being a movement demand to being national policy.[16]

Like the civil rights movement, the grassroots antitoxics movement also brought the experiential base of direct action into the environmental justice movement. It further contributed both the experience of using (and, when need be, discrediting) scientific and technical information and the conceptual framework that pushed pollution prevention and toxics use reduction as policy goals. Antitoxics groups also had built national networks by linking local activists, an experience that they brought to the movement.

The grassroots antitoxics movement also contributed a structural understanding of power, albeit different from civil rights leaders'. Civil rights advocates came, through the process of the civil rights struggles, to understand discrete racial assaults (from epithets to lynchings to segregation laws) as part of a social structure of racial oppression that ultimately had to be dismantled if racial justice was to be achieved. Antitoxics activists, through the process of local fights against polluting facilities, came to understand discrete toxic assaults as part of an economic structure in which, as part of the "natural" functioning of the economy, certain communities would be polluted. Antitoxics leaders thus focused on corporate power and the structure of the United States and the global economies and on strategies for changing that structure. It was when, in the 1980s, civil rights leaders began to embrace the antitoxics movement's economic analysis and the antitoxics leaders embraced the civil rights activists' racial critique that the conceptual fusion took place that helped create the environmental justice movement.[17]

Academics

A third important contributory stream to the environmental justice movement comes from a seemingly unlikely spot: academia. Academics, however, have played a crucial role in both sparking and shaping the environmental justice movement, perhaps a larger one than they have played in any other broad-based social movement in the United States. Beginning in the 1960s, isolated researchers discovered that environmental hazards had a disproportionate impact on people of color and low-income people.[18] Dr. Robert Bullard, studying Houston land-use patterns, found in the late 1970s that garbage dumps had a disproportionate impact on African Americans; this research led to Bullard's pioneering work in the field. In the late 1980s, Bullard did a literature search using the terms "minority" and "environment" and found twelve articles—six of which he had written. At that time, several academics, led by Bunyan Bryant at the University of Michigan, Bullard (then at the University of California–Riverside), and Charles Lee of the United Church of Christ, began to discuss the findings of disparate impact among themselves and held conferences on the subject.

In 1990, a group of academics convened at the University of Michigan to discuss their most recent findings. At that meeting, they decided that the energy and the momentum generated in their weekend together were too exciting to let dissipate in the usual academic papers. Instead, the group wrote letters to Louis Sullivan, the secretary of the US Department of Health and Human Services, and to William Reilly, head of the US Environmental Protection Agency. In the letters, the professors, who came to be known as the Michigan Group, set out some of their findings of disproportionate impact and asked for a meeting with the officials to discuss a government response. As Bullard reports, the group never heard from Secretary Sullivan. William Reilly, however, agreed to meet with the Michigan Group, and, later in 1990, seven professors met with Reilly and EPA staffers in Washington, DC.[19] The result of the Michigan Group's advocacy with Administrator Reilly was EPA's creation of a Work Group on Environmental Equity.[20] Reilly later created an Office of Environmental Equity, which newly appointed EPA administrator Carol Browner renamed the Office of Environmental Justice in 1993.

Beyond lobbying the federal government, the academics researched and wrote (and continue to produce) studies that demonstrate the disproportionate impact of environmental hazards on people of color and on low-income people. These studies, dialectically fueled by and fueling the movement, played a series of roles. For one, the studies sparked and moved forward local struggles. In Los Angeles, for example, a community struggle led by Concerned Citizens of South Central Los Angeles against a giant garbage incinerator received what its leaders call crucial support when, before a key vote of the city council, the UCLA School of Urban Planning released a seven hundred-page critique of the incinerator project and its disproportionate impact on people of color in Los Angeles.

The academics' work also shaped or reaffirmed movement leaders' consciousness about the structural or systemic nature of environmental oppression. "I thought it was just us until I began to hear about the United Church of Christ study and the other studies," says Mary Lou Mares, an activist who has fought for more than ten years against Chemical Waste Management's toxic dump near her Latino community of Kettleman City. "Then I realized we were part of a national pattern."[21]

At other times, the academics have provided expertise to community groups during litigation or administrative advocacy in a local environmental justice struggle. In fact, the career of the most prolific and influential academic, Dr. Robert Bullard, was launched by a court case in Houston in the late 1970s in which his wife needed an expert witness.[22] Bullard and others have since prepared studies and testified for dozens of community groups nationwide. In perhaps the best-known example, Professor Bullard's documentation of racially biased decision-making criteria in the siting of a nuclear waste processing facility in rural Louisiana was directly responsible for the federal government's decision to deny a permit to the facility.[23]

The concrete victories achieved in Los Angeles, Louisiana, and elsewhere were marked by the synergy between community activism and the academic support that played a critical role in each fight. On a local level, the education went both ways: the academics learned from community residents the situation on the ground, while local residents came to understand their community's struggle in the context of a larger regional or national pattern and movement.

Finally, the academics' work provides a basis for policy changes at the local, state, and national levels. In signing the Executive Order on Environmental Justice, for example, President Clinton acknowledged the need to "focus Federal attention on the environmental and human health conditions in minority communities and low-income communities with the goal of achieving environmental justice."[24] Without the previous decade of studies that had established the scope of the environmental injustice in these communities, the problem never would have reached the attention of the White House.

Today, academics continue to play a crucial supporting role through such institutions as the Environmental Justice Resource Center at Clark-Atlanta University, founded and run by Robert Bullard, and the Deep South Center for Environmental Justice at Xavier University in New Orleans, run by Beverly Hendrix Wright. These centers, and others like them, provide crucial research that aids local struggles, as well as train a new generation of professionals of color.

Native American Struggles

A fourth significant stream feeding the environmental justice movement has been organizing by Native Americans. Native Americans have struggled for self-determination in land use decisions since their first encounters with Europeans more than five hundred years ago.[25] Activism by Native Americans in the late 1960s and early 1970s was the precursor to today's organizing around environmental issues by Indians on and off the reservations, organizing that contributes one of the most vibrant and ever-expanding tributaries to the movement. The struggles that led to the creation of the American Indian Movement were often focused around land and environmental exploitation, including such well-known and iconic incidents of Indian resistance as the shootout at Pine

Ridge in 1975, which took place on the very day that the corrupt Pine Ridge Tribal chair Dickie Wilson was in Washington, DC, signing away rights to mineral exploration in the sacred Black Hills to major oil companies.[26]

Native American activists brought to the environmental justice movement the experiences of centuries of struggle for self-determination and resistance to resource-extractive land use. The struggles of the 1870s to protect tribal land honed skills that would be useful later. As the first victims of environmental racism, Native Americans brought a deep understanding of the concept to the environmental justice movement.

The Native American tributary to the movement also helped define one of its central philosophies, the concept of self-determination. The centuries-old Native American idea of sovereignty echoed with, and helped create, the environmental justice movement's credo, "We speak for ourselves." While for some other communities the slogan was an attempt to take back environmental policy decisions from traditional environmental groups, for Native Americans the slogan defined their relationship to state and federal governments.

The significant contributions of Native Americans to the environmental justice movement were institutionalized in the formation of the Indigenous Environmental Network in 1990 . . .

The Labor Movement

Various strands of the labor movement have also contributed to the environmental justice movement. The largest labor tributary has been the historical struggle of farmworkers to gain control over their working conditions. The farmworker movement of the 1960s, led by Cesar Chavez, was perhaps the first nationally known effort by people of color to address an environmental issue. Much of the activity took the form of union organizing drives. For instance, the United Farm Workers (UFW) included in its initial organizing and contractual demands the ban of certain dangerous pesticides, including DDT. Union contracts in the late 1960s prohibited the use of such pesticides, and the lawsuits that ultimately led the US government to ban the chemical outright were brought by migrant farmworkers.[27] Farmworkers' struggle for self-determination in the workplace—for the power to control decisions that affected their lives, such as the use of pesticides—mirrored the struggles by Native Americans and African Americans for political self-determination and by the grassroots antitoxics movement for a role in local decisions. Unionization and protection of farmworkers' health and safety were integrally linked from the earliest days of the farmworker organizing drives, and they continue to be linked today; it was thus natural for farmworkers to become active participants in the movement for environmental justice.

A second, and much less significant, labor tributary, one fed by the public-health activism of the 1970s, is the occupational safety and health movement. The rise of Committees on Occupational Safety and Health (COSHs) across the country in the 1970s and 1980s brought increased attention to the environmental hazards faced by workers in the workplace. The COSHs were active in regions of the country—the South, for example—and in industry sectors—such as textiles and high technology—that traditionally had little or no union representation. COSH activists, such as Mandy Hawes in San Jose, California, became early organizers of and advocates in the environmental justice movement.

A third labor tributary is the increased attention paid to occupational safety and health by industrial unions. Led by sometimes renegade union activists such as Tony Mazzochi, unions such as the Oil, Chemical, and Atomic Workers Union have paid increased attention to environmental justice issues. While industrial unions have often believed that their interests lay with further development or expansion of industrial plants, some visionary union leaders have understood that the push for safer jobs and a cleaner workplace can help build political support for labor from fence-line communities and environmentalists. This awareness has led to an important collaboration between the environmental justice movement and organized labor in the Campaign for a Just Transition; through this campaign, movement and union leaders have been exploring common ground in phasing out the use of dangerous chemicals. This tributary is still a trickle, but it is an exciting addition to the movement.

Traditional Environmentalists

A very small, and late, tributary to the environmental justice movement is the traditional environmental movement. Perhaps it is the history of the traditional environmental movement that has made it such a small contributor to the environmental justice movement.

Two major waves of traditional environmentalism have swept the United States. The first wave began around the turn of the century, when John Muir, Theodore Roosevelt, and other lovers of wilderness advocated the preservation of natural spaces in the United States. Like the second wave, the first wave encompassed two divergent views that even today remain in tension: the preservationists, who advocate preserving wilderness from humans, and the conservationists, who want to preserve nature for human use through wise stewardship.[28]

Modern environmentalism, or the second wave, began after World War II with the rapid expansion in the use of petrochemical products. When the consequences of the shift to petrochemical production began to be felt, a new wave of activism sprang up, fueled by searing critiques of industrial practices such as Rachel Carson's *Silent Spring*. This wave of environmentalism coalesced around Earth Day in 1970 and was institutionalized in the proliferation of legal-scientific groups such as the Natural Resources Defense Council (NRDC), the Sierra Club Legal Defense Fund (SCLDF), and the Environmental Defense Fund (EDF), organizations that currently dominate the national scene.[29] The second wave— what we call the traditional environmental movement[30]—and the body of statutes and case law known today as environmental law grew out of the social ferment of the 1960s. The civil rights movement and the anti-Vietnam War movement, the two movements in which the second wave has its roots, were explicitly oriented toward social justice.

The second-wave environmentalists have moved away from this social justice orientation, however. In particular, they have moved from a participatory strategy based on broad mobilization of the interested public, such as that used in the civil rights and the antiwar movements, to an insider strategy based on litigation, lobbying, and technical evaluation.[31] The movement away from a participatory strategy paralleled the movement away from the social justice issues that dominated the speeches given on Earth Day in 1970.[32] It also coincided with the traditional groups' desire to control the environmental establishment or at least to have power within it; as one commentator observes, "Shedding the radical skin of their amateur past seemed necessary to achieve that goal."[33]

The second wave, made up overwhelmingly of lawyers, focused primarily on legal and scientific approaches to environmental problems.[34] Second-wave lawyers helped write most of the environmental legislation on the books today, from the National Environmental Policy Act (NEPA), to the Clean Air Act (CAA), the Clean Water Act (CWA), the Resource Conservation and Recovery Act (RCRA), the Comprehensive Environmental Response, Compensation, and Liability Act (CERCLA), the Federal Insecticide, Fungicide, Rodenticide Act (FIFRA), the Toxic Substances Control Act (TSCA), and the Superfund Amendment and Reauthorization Act (SARA). These laws created complex administrative processes that exclude most people who do not have training in the field and necessitate specific technical expertise. The laws, while in some cases successful in cleaning up the environment, have also had an unintended consequence—the exclusion of those without expertise from much of environmental decision making.

Having designed and helped implement most of the nation's environmental laws, the second wave has spent the past twenty-five years in court litigating. Lawsuits are now the primary, and sometimes the only, strategy employed by traditional groups.[35] As the executive director of the Sierra Club Legal Defense Fund stated in 1988, "Litigation is the most important thing the environmental movement has done over the past fifteen years."[36]

Until relatively recently, the traditional environmental law community has largely ignored environmental justice issues.[37] In some cases, the lack of attention has been intentional: in a 1971 national membership survey, the Sierra Club asked its members, "Should the Club concern itself with the conservation problems of such special groups as the urban poor and ethnic minorities?" According to the club's *Bulletin*, "[t]he balance of sentiment was against the Club so involving itself" with "58 percent of all members either strongly or somewhat opposed" to the idea.[38]

Racism and other prejudices have historically excluded activists of color and grassroots activists from the traditional environmental movement.[39] In fact, some of these activists regard the traditional environmental groups as obstacles to progress, if not outright enemies.[40] Some in traditional environmental groups have pushed for a greater focus on environmental justice—one hired an environmental justice coordinator, another hired several environmental justice fellows and announced a new focus on such cases—but, for the most part, the environmental justice movement has operated without the input or assistance of the traditional environmental groups, perhaps to its benefit.[41] Given the second wave of environmental activism's roots in the grassroots activism of the 1960s, this disconnect is ironic, and poignant.

Some have described the grassroots movement for environmental justice as the third wave of environmental activism,[42] but we see the environmental justice movement as separate from and as transcending the environmental movement— as a movement based on environmental issues but situated within the history of movements for social justice.

THE SUMMIT

The disparate strands of the environmental justice movement—civil rights, grassroots antitoxics, academic, labor, indigenous—were consciously brought together for the first time in 1991 at the First National People of Color Environmental Leadership Summit. The summit served notice that the environmental justice movement had arrived as a force to be reckoned with on the national level. It was also, in some ways, a declaration of independence from the traditional environmental movement; a telling statement from attendees was, "I don't care to join the environmental movement, I belong to a movement already."

Ironically, the summit grew out of the environmental justice movement's challenge to traditional environmental groups. In early 1990, Richard Moore, of the SouthWest Organizing Project, and Pat Bryant, of the Gulf Coast Tenants Organization, drafted a letter, ultimately signed by more than one hundred community leaders, to the ten largest traditional environmental groups in which they accused the groups of racism in their hiring and policy development processes.[43] An article in the *New York Times* on the letter initiated a media firestorm around the issue,[44] and in an interview on CNN, the Rev. Ben Chavis, one of the signatories of the letter and at the time the head of the United Church of Christ's Commission for Racial Justice, called for an emergency summit of environmental, civil rights, and community groups. "In his mind, he was thinking about a small group of people getting together to negotiate it out," says Charles Lee, who directed the environmental justice program at the Commission for Racial Justice. Lee had other ideas, however, and, as he put together a planning committee, the summit quickly evolved from a small negotiating session into an event at which people of color could actively put forward their own environmental agenda.

The summit was the product of eighteen months of intensive organizing by movement leaders, including Lee, Bryant, Moore, Dana Alston of the Panos Institute, the Indian activist Donna Chavis of North Carolina, and the academic Robert Bullard. Chavis's initial idea of a small gathering ended up a national event that brought together more than three hundred delegates and four hundred observers and supporters for three heady days in Washington, DC. There were speeches by leaders in the national social justice movement, such as Jesse Jackson and Dolores Huerta, strategy sessions on issues such as toxic dumps and legal challenges, and caucuses for delegates organized by region and by race.

Unprecedented alliances were formed at the summit, and participants made conceptual linkages between seemingly different struggles, identifying common themes of racism and economic exploitation of people and land. Many there came to understand their issues in the context of a larger movement, and on a deeper level than before. Latinos saw the racism African Americans had experienced and likened it to their own experiences; foes of toxic waste dumps understood the fights of those who opposed uranium mining. The raised consciousness took place at a variety of levels: "Native Americans stressing a spiritual connection with the environment were able to find a common ground with Christian African Americans and Mexican Americans," recalls Tom Goldtooth, a delegate to the summit who now coordinates the Indigenous Environmental Network.

Some say that from October 21–24, 1991, the environmental movement in the United States changed forever. Certainly, those in attendance went back to their communities across the country with a renewed understanding of the need for environmental justice and with new ideas on how to fight for it; nascent environmental justice networks (such as the Indigenous Environmental Network . . .) gained momentum. Environmental justice as a concept had reached the national stage, and it was not long afterward that two key planners of the summit, Ben Chavis and Robert Bullard, were appointed to president-elect Clinton's transition team. On a more tangible level, perhaps the most important result of the summit were the Principles of Environmental Justice, seventeen principles agreed to by the delegates. [See Conclusion, p. 407.]

ENVIRONMENTAL JUSTICE ACTIVISTS

Despite the many tributaries from which they come to the environmental justice movement, at least three characteristics unite the movement's activists: motives, background, and perspective. With respect to motives, grassroots activists are often fighting for their health and homes. Environmental justice activists usually have an immediate and material stake in solving the environmental problems they confront; they realize the hazards they face affect the communities where they live and may be sickening or even killing them or their children. Because grassroots activists have such a personal stake in the outcome of particular environmental battles, they are often willing to explore a wider range of strategies than other advocates, including traditional environmental advocates.

Second, with respect to background, grassroots environmentalists are largely, though not entirely, poor or working-class people. Many are people of color who come from communities that are disenfranchised from most major societal institutions. Because of their backgrounds, these activists often have a distrust for the law and are often experienced in the use of nonlegal strategies, such as protest and other direct action.

The third trait, perspective, is an outgrowth of the first two. Most environmental justice activists have a social justice orientation, seeing environmental degradation as just one of many ways their communities are under attack. Because of their experiences, grassroots activists often lose faith in government agencies and elected officials,[45] leading those activists to view environmental problems in their communities as connected to larger structural failings—inner-city disinvestment, residential segregation, lack of decent healthcare, joblessness, and poor education. Similarly, many activists also seek remedies that are more fundamental than simply stopping a local polluter or toxic dumper. Instead, many view the need for broader, structural reforms as a way to alleviate many of the problems, including environmental degradation, that their communities endure.

NOTES

1. Robert D. Bullard, "Environmental Justice for All: It's the Right Thing to Do," *Journal of Environmental Law & Litigation* 9 (1994): 281, 285.

2. Ibid., p. 285; Robert D. Bullard, "Environmental Justice for All," in *Unequal Protection: Environmental Justice and Communities of Color*, ed. Robert D. Bullard, 3–4 (San Francisco: Sierra Club Books, 1994).

3. We thank Charles Lee for this metaphor.

4. See, e.g., Charles M. Payne, *I've Got the Light Of Freedom: The Organizing Tradition and the Mississippi Freedom Struggle* (Berkeley: University of California Press, 1995); Aldon Morris, *The Origins of the Civil Rights Movement: Black Communities Organizing for Change* (New York: Free Press, 1984); Clayborne Carson, *In Struggle: SNCC and the Black Awakening of the 1960s* (Cambridge, MA: Harvard University Press, 1981); *The Eyes on the Prize Civil Rights Reader: Documents, Speeches, and Firsthand Accounts from the Black Freedom Struggle, 1954–1990*, ed. Clayborne Carson et al. (New York: Penguin, 1991); Taylor Branch, *Parting the Waters: America in the King Years, 1954–63* (New York: Simon and Schuster, 1988).

5. Chavis helped bring national attention to the Warren County struggle through his national, church-based network. He was a key leader in the environmental justice movement from the early 1980s through the early 1990s. He later went on to head the National Association for the Advancement of Colored People (NAACP). Also prominent in the Warren County struggle was District of Columbia congressional delegate Walter Fauntroy, who was a product of the civil rights movement as well.

6. Elizabeth Martinez, ed., *500 Years of Chicano History in Pictures* (Alburquerque, NM: SouthWest Organizing Project, 1991), pp. 129–37.

7. Bullard, "Environmental Justice for All," pp. 3, 5; Ken Geiser and Gerry Waneck, "PCBs and Warren County," in *Unequal Protection: Environmental Justice and Communities of Color*, ed. Robert D. Bullard, 43, 44 (San Francisco: Sierra Club Books, 1994).

8. See Congressman John Lewis, "Foreword," in *Unequal Protection: Environmental Justice and Communities of Color*, ed. Robert D. Bullard, vii–x (San Francisco: Sierra Club Books, 1994). This act would have targeted the hundred most polluted locations in the United States for federal attention. These areas would have been designated environmental high-impact areas and require assessment of health conditions in communities that have high concentrations of polluting facilities. Ibid., pp. viii–ix.

9. Ibid., p. viii.

10. See, e.g., Andrew Szasz, *Ecopopulism: Toxic Waste and the Movement for Environmental Justice* (Minneapolis: University of Minnesota Press, 1994), pp. 38–102.

11. Nicholas Freudenberg and Carol Steinsapir, "Not in Our Backyards: The Grassroots Environmental Movement," *Society & Natural Resources* 4 (1991): 237; see also Lois M. Gibbs, *Love Canal: My Story* (Albany: State University of New York Press, 1982).

12. Andrew Szasz, *Ecopopulism*, p. 40.

13. Ibid., p. 41.

14. Ibid., pp. 69–102.

15. CCHW is now known as the Center for Health, Environment, and Justice.

16. See, e.g., Freudenberg and Steinsapir, "Not in Our Backyards: The Grassroots Environmental Movement," pp. 235, 242; Barry Commoner, *Making Peace with the Planet* (New York: Pantheon, 1990), pp. 103–40, 178–90; Luke W. Cole, "Empowerment as the Means to Environmental Protection: The Need for Environmental Poverty Law," *Ecology Law Quarterly* 19 (1992): 619, 645.

17. See Richard Moore and Louis Head, "Building a Net That Works: SWOP," in *Unequal Protection: Environmental Justice and Communities of Color*, ed. Robert D. Bullard, 191 (San Francisco: Sierra Club Books, 1994).

18. See generally Paul Mohai and Bunyan Bryant, "Environmental Racism: Reviewing the Evidence," in *Race and the Incidence of Environmental Hazards: A Time for Discourse*, ed. Bunyan Bryant and Paul Mohai, 163 (Boulder, CO: Westview Press, 1992) (reviewing early studies finding distributional inequities by income and race).

19. At the end of the meeting, the seven asked for an informal meeting of only the African American staffers present, without Reilly or any other white EPA staffers; at that post-meeting, EPA staffers informed the Michigan Group about the reality of working as a person of color at the EPA.

20. Robert D. Bullard, "Introduction," in *Unequal Protection: Environmental Justice and Communities of Color*, ed. Robert D. Bullard, vii, xv–xv (San Francisco: Sierra Club Books, 1994).

21. Interview with Mary Lou Mares, Kettleman City, June 15, 1998.

22. "Bean v. Southwestern Waste Management Corp.," 482 F.Supp. 673 (S. D. Tex. 1979). See generally "A Pioneer in Environmental Justice Lawyering: A Conversation with Linda McKeever Bullard," *Race, Poverty & the Environment* 17 (Fall 1994/Winter 1995): 5. Linda Bullard convinced her husband, Bob, to conduct a study of the demographics of those living around garbage dumps in the Houston area as part of the "Bean" case.

23. "In the Matter of Louisiana Energy Services, L. P.," Decision of the Nuclear Regulatory Commission Atomic Safety and Licensing Board, May 1, 1997 (using the testimony of Dr. Robert Bullard to establish that the siting process at issue was biased and that racial considerations were a factor in the site selection process).

24. President William J. Clinton, Memorandum on Environmental Justice, February 11, 1994.

25. Vine Deloria Jr., *Behind the Trail of Broken Treaties: An Indian Declaration of Independence* (Austin: University of Texas Press, 1985), pp. 8–21.

26. Peter Matthiessen, *In the Spirit of Crazy Horse* (1983); Edward Lazarus, *Black Hills, White Justice: The Sioux Nation versus the United States, 1775 to the Present* (Lincoln: University of Nebraska Press, 1991).

27. Cole, "Empowerment as the Means to Environmental Protection," p. 636 n51.

28. For a nice history of the environmental movement and its underlying ideology, see Anna Bramwell, *Ecology in the 20th Century: A History* (New Haven, CT: Yale University Press, 1989).

29. For a provocative indictment of the second-wave environmentalists, see Peter Montague, "What We Must Do: A Grassroots Offensive against Toxics in the '90s," *Workbook* 14 (1989): 90, 92.

30. By the term "traditional environmental group," we mean primarily the "Group of Ten" environmental organizations, which are national in scope, advocacy, and membership. The "Group of Ten" label was first used by these groups—the nation's ten largest traditional environmental groups—in 1985. Robert Gottlieb, "Earth Day Revisited," *Tikkun* (March/April 1990): 55.

31. Robert Bullard and Beverly Hendrix Wright, "Blacks and the Environment," *Humboldt Journal of Social Relations* 14 (Fall/Winter and Winter/Spring 1986–87): 165, 167.

32. Charles Jordan and Donald Snow, "Diversification, Minorities, and the Mainstream Environmental Movement," in *Voices from the Environmental Movement: Perspectives for a New Era*, ed. Donald Snow, 71, 78 (Washington, DC: Island Press, 1991). Many Earth Day speeches in 1970 made the direct connection between environmental issues and social and racial justice. There was, to be sure, tension between the often affluent environmentalists and activists in the civil rights movement, such as that surrounding the burial of a car at an Earth Day event in San Jose, California. While the white environmentalists saw the interment of a brand new car as a statement of antimaterialism, local inner-city residents criticized them for squandering resources that could be put to better use.

33. Ibid., p. 93.

34. For example, by 1983, the heads of the Sierra Club, the Sierra Club Legal Defense Fund, the Natural Resources Defense Council, the Audubon Society, the Environmental Defense Fund, and the Wilderness Society were all attorneys. Christopher Manes, *Green Rage: Radical Environmentalism and the Unmaking of Civilization* (Boston: Little, Brown, 1990), p. 255 n8. For analysis of the second wave and its focus on litigation, see generally Tom Turner, "The Legal Eagle," *Amicus Journal* (Winter 1988): 25; Melia Franklin, "What's Four-Legged and Green All Over? . . . Sorting out the Environmental Movement," *California Tomorrow* (Fall 1988): 14. In a Sierra Club national membership survey in 1972, "lawsuits and lobbying were strongly endorsed as appropriate methods. . . . More than two-thirds of the members, in each case, *strongly* agreed that they were appropriate. Only five percent disapproved." Don Coombs, "The Club Looks at Itself," *Sierra Club Bulletin* (July/August 1972): 38 (emphasis in original).

35. "[T]he legal victories won in the late sixties and early seventies formed the foundation on which the modern environmental movement is built," according to John Adams, the executive director of NRDC. Tom Turner, "The Legal Eagle," *Amicus Journal* (Winter 1988): 27–28. Another pioneer of the environmental law field states, "In no other political and social movement has litigation played such an important and dominant role. Not even close." Ibid., p. 27 (quoting David Sive).

36. Ibid., p. 27 (quoting Frederick Sutherland).

37. There are notable exceptions to this overgeneralization. For example, the Sierra Club worked with the Urban Environment Conference and the National Urban League to put on the City Care conference on the urban environment in 1979. See National Urban League et al., "City Care," in *National Conference on the Urban Environment Proceedings* (1979); Urban Environment Conference, *Resource Book—"Taking Back Our Health": An Institute on Surviving the Toxics Threat to Minority Communities* (January 1985) (conference handbook).

38. Coombs, "The Club Looks at Itself," pp. 35, 37.

39. See Jordan and Snow, *Diversification, Minorities, and the Mainstream Environmental Movement*, pp. 75–78 (detailing racist exclusion of people of color from early conservation clubs and hunting preserves). Several Southern California chapters of the Sierra Club, for example, formerly deliberately excluded blacks and Jews from membership; when the San Francisco chapter tried in 1959 to introduce a policy of inclusion of the "four recognized colors" into the Sierra Club, the resolution failed. Ibid., p. 76; Stephen Fox, *John Muir and His Legacy: The American Conservation Movement* (Boston: Little, Brown, 1981), p. 349.

40. Grassroots leaders have seen traditional environmentalists as enemies when those groups accept contributions from industry and have appointed executives of corporations such as Waste Management to their boards at the same time activists were fighting the companies. As Dana Alston writes about the National Wildlife Federation, those "who are engaged in life and death struggles with Waste Management were hard-pressed to understand why such a corporation is represented on the board of directors of one of the largest and most influential environmental organizations." Dana Alston, "Transforming a Movement," *Race, Poverty & the Environment* (Fall 1991/Winter 1992): 1, 29. As one environmental justice advocate asserted,

> We are not always well served by the Environmentalist establishment in Washington. Perhaps with the best of intentions they have legitimated a system of destruction. Their batteries of lawyers and lobbyists battle over insignificant or irrelevant measures while implicitly recognizing the right of polluters to carry on business as usual. They are caught up in a deadly game, thrilled at the prospect of being "players."

Paul deLeon, "The STP Schools: Education for Environmental Action," *New Solutions* (Summer 1990): 22, 23.

41. See Luke W. Cole, "Foreword: A Jeremiad on Environmental Justice and the Law," *Stanford Environmental Law Journal* 14 (1995): ix, xii–xv.

42. See, e.g., Peter Montague, "What We Must Do: A Grassroots Offensive against Toxics in the '90s." Even one of the authors of this book has referred to the environmental justice movement as the "third wave."

43. Letter from SouthWest Organizing Project to "Group of 10" National Environmental Organizations, February 21, 1990 (letter on behalf of more than one hundred community leaders of color).

44. Philip Shabecoff, "Environmental Groups Told They Are Racist in Hiring," *New York Times*, February 1, 1990, p. A16.

45. Donald Unger et al., "Living near a Hazardous Waste Facility: Coping with Individual and Family Distress," *American Journal of Orthopsychiatry* 62 (1992): 55, 57. Anecdotal evidence shows that even among children living near toxic waste sites, there was a "loss of faith in governmental institutions." Ibid; see also Freudenberg and Steinsapir, "Not in Our Backyards: The Grassroots Environmental Movement," pp. 235, 237, 239 (noting that in one survey of grassroots groups, 45 percent of those responding claimed that government agencies had blocked their access to needed information).

24.

DEVELOPMENT, ECOLOGY, AND WOMEN

VANDANA SHIVA

DEVELOPMENT AS A NEW PROJECT OF WESTERN PATRIARCHY

"Development" was to have been a postcolonial project, a choice for accepting a model of progress in which the entire world remade itself on the model of the colonizing modern West, without having to undergo the subjugation and exploitation that colonialism entailed. The assumption was that Western-style progress was possible for all. Development, as the improved well-being of all, was thus equated with the Westernization of economic categories—of needs, of productivity, of growth. Concepts and categories about economic development and natural resource utilization that had emerged in the specific context of industrialization and capitalist growth in a center of colonial power were raised to the level of universal assumptions and applicability in the entirely different context of basic needs satisfaction for the people of the newly independent third world countries. Yet, as Rosa Luxemburg has pointed out, early industrial development in Western Europe necessitated the permanent occupation of the colonies by the colonial powers and the destruction of the local "natural economy."[1] According to her, colonialism is a constantly necessary condition for capitalist growth: without colonies, capital accumulation would grind to a halt. "Development" as capital accumulation and the commercialization of the economy for the generation of "surplus" and profits thus involved the reproduction not merely of a particular form of creation of wealth, but also of the associated creation of poverty and dispossession. A replication of economic development based on commercialization of resource use for commodity production in the newly independent countries cre-

From: *Staying Alive: Women, Ecology, and Development* (London: Zed Books, 1988), pp. 1–9, 13.

293

ated the internal colonies.[2] Development was thus reduced to a continuation of the process of colonization; it became an extension of the project of wealth creation in modern Western patriarchy's economic vision, which was based on the exploitation or exclusion of women (of the West and non-West), on the exploitation and degradation of nature, and on the exploitation and erosion of other cultures. "Development" could not but entail destruction for women, nature, and subjugated cultures, which is why, throughout the third world, women, peasants, and tribals are struggling for liberation from development just as they earlier struggled for liberation from colonialism.

The UN Decade for Women was based on the assumption that the improvement of women's economic position would automatically flow from an expansion and diffusion of the development process. Yet, by the end of the decade, it was becoming clear that development itself was the problem. Insufficient and inadequate "participation" in "development" was not the cause for women's increasing underdevelopment; it was rather, their enforced but asymmetric participation in it, by which they bore the costs but were excluded from the benefits, that was responsible. Development exclusivity and dispossession aggravated and deepened the colonial processes of ecological degradation and the loss of political control over nature's sustenance base. Economic growth was a new colonialism, draining resources away from those who needed them most. The discontinuity lay in the fact that it was now new national elites, not colonial powers, that masterminded the exploitation on grounds of "national interest" and growing GNPs, and it was accomplished with more powerful technologies of appropriation and destruction.

Ester Boserup has documented how women's impoverishment increased during colonial rule;[3] those rulers who had spent a few centuries in subjugating and crippling their own women into de-skilled, de-intellectualized appendages, disfavored the women of the colonies on matters of access to land, technology, and employment. The economic and political processes of colonial underdevelopment bore the clear mark of modern Western patriarchy, and while large numbers of women and men were impoverished by these processes, women tended to lose more. The privatization of land for revenue generation displaced women more critically, eroding their traditional land-use rights. The expansion of cash crops undermined food production, and women were often left with meager resources to feed and care for children, the aged and the infirm, when men migrated or were conscripted into forced labor by the colonizers. As a collective document by women activists, organizers and researchers stated at the end of the UN Decade for Women, "The almost uniform conclusion of the Decade's research is that with a few exceptions, women's relative access to economic resources, incomes and employment has worsened, their burden of work has increased, and their relative and even absolute health, nutritional and educational status has declined."[4]

The displacement of women from productive activity by the expansion of development was rooted largely in the manner in which development projects appropriated or destroyed the natural resource base for the production of sustenance and survival. It destroyed women's productivity both by removing land, water, and forests from their management and control, as well as through the ecological destruction of soil, water, and vegetation systems so that nature's productivity and renewability were impaired. While gender subordination and patriarchy are the oldest of oppressions, they have taken on new and more violent forms through the project of development. Patriarchal categories which understand destruction as "production" and regeneration of life as "passivity" have generated a crisis of survival. Passivity, as an assumed category of the "nature" of nature and of women, denies the activity of nature and life. Fragmentation and uniformity as assumed categories of progress and development destroy the living forces which arise from relationships within the "web of life" and the diversity in the elements and patterns of these relationships.

The economic biases and values against nature, women, and indigenous peoples are captured in this typical analysis of the "unproductiveness" of traditional natural societies: "Production is achieved through human and animal, rather than mechanical, power. Most agriculture is unproductive; human or animal manure may be used but chemical fertilisers and pesticides are unknown. . . . For the masses, these conditions mean poverty."[5]

The assumptions are evident: nature is unproductive; organic agriculture based on nature's cycles of renewability spells poverty; women and tribal and peasant societies embedded in nature are similarly unproductive, not because it has been demonstrated that in cooperation they produce less goods and services for needs, but because it is assumed that "production" takes place only when mediated by technologies for commodity production, even when such technologies destroy life. A stable and clean river is not a productive resource in this view: it needs to be "developed" with dams in order to become so. Women, sharing the river as a commons to satisfy the water needs of their families and society, are not involved in productive labor: when replaced by the engineering man, water management and water use become productive activities. Natural forests remain unproductive till they are developed into monoculture plantations of commercial species. Development, thus, is equivalent to maldevelopment, a development bereft of the feminine, the conservation, the ecological principle. The neglect of nature's work in renewing herself and women's work in producing sustenance in the form of basic, vital needs is an essential part of the paradigm of maldevelopment, which sees all work that does not produce profits and capital as non- or unproductive work. As Maria Mies has pointed out, this concept of surplus has a patriarchal bias because, from the point of view of nature and women, it is not based on material surplus produced *over and above* the requirements of the community: it is stolen and appropriated through violent modes from nature (who needs a share of her produce to reproduce herself) and from women (who need a share of nature's produce to produce sustenance and ensure survival).[6]

From the perspective of third world women, productivity is a measure of producing life and sustenance; that this kind of productivity has been rendered invisible does not reduce its centrality to survival—it merely reflects the domination of modern patriarchal economic categories which see only profits, not life.

MALDEVELOPMENT AS THE DEATH OF THE FEMININE PRINCIPLE

In this analysis, maldevelopment becomes a new source of male-female inequality. "Modernization" has been associated with the introduction of new forms of dominance. Alice Schlegel has shown that under conditions of subsistence, the interdependence and complementarity of the separate male and female domains of work is the characteristic mode, based on diversity, not inequality.[7] Maldevelopment militates against this equality in diversity, and superimposes the ideologically constructed category of Western technological man as a uniform measure of the worth of classes, cultures, and genders. Dominant modes of perception based on reductionism, duality, and linearity are unable to cope with equality in diversity, with forms and activities that are significant and valid, even though different. The reductionist mind superimposes the roles and forms of power of Western male-oriented concepts on women, all non-Western peoples, and even on nature, rendering all three "deficient," and in need of "development." Diversity, and unity and harmony in diversity, become epistemologically unattainable in the context of maldevelopment, which then becomes synonymous with women's underdevelopment (increasing sexist domination), and nature's depletion (deepening ecological crises). Commodities have grown, but nature has shrunk. The poverty crisis of the South arises from the growing scarcity of water, food, fodder, and fuel, associated with increasing maldevelopment and ecological destruction. This poverty crisis touches women most severely, first because they are the poorest among the poor, and then because, with nature, they are the primary sustainers of society.

Maldevelopment is the violation of the integrity of organic, interconnected, arid interdependent systems, that sets in motion a process of exploitation, inequality, injustice, and violence. It is blind to the fact that a recognition of nature's harmony and action to maintain it are preconditions for distributive justice. This is why Mahatma Gandhi said, "There is enough in the world for everyone's need, but not for some people's greed."

Maldevelopment is maldevelopment in thought and action. In practice, this fragmented, reductionist, dualist perspective violates the integrity and harmony of man in nature, and the harmony between men and women. It ruptures the cooperative unity of masculine and feminine, and places man, shorn of the feminine principle, above nature and women, and separated from both. The violence to nature as symptomatized by the ecological crisis, and the violence to women, as symptomatized by their subjugation and exploitation, arise from this subjugation of the feminine principle. I want to argue that what is currently called development is essentially maldevelopment, based on the introduction or accentuation of the domination of man over nature and women. In it, both are viewed as the "other," the passive non-self. Activity, productivity, creativity, which were associated with the feminine principle, are expropriated as qualities of nature and women, and transformed into the exclusive qualities of man. Nature and women are turned into passive objects, to be used and exploited for the uncontrolled and uncontrollable desires of alienated man. From being the creators and sustainers of life, nature and women are reduced to being "resources" in the fragmented, antilife model of maldevelopment.

TWO KINDS OF GROWTH, TWO KINDS OF PRODUCTIVITY

Maldevelopment is usually called "economic growth," measured by the gross national product. Jonathon Porritt, a leading ecologist, has this to say of GNP:

> *Gross* National Product—for once a word is being used correctly. Even conventional economists admit that the hey-day of GNP is over, for the simple reason that as a measure of progress, it's more or less useless. GNP measures the lot, all the goods and services produced in the money economy. Many of these goods and services are not beneficial to people, but rather a measure of just how much is going wrong; increased spending on crime, on pollution, on the many human casualties of our society, increased spending because of waste or planned obsolescence, increased spending because of growing bureaucracies. It's all counted. . . .[8]

The problem with GNP is that it measures some costs as benefits (e.g., pollution control) and fails to measure other costs completely. Among these hidden costs are the new burdens created by ecological devastation, costs that are invariably heavier for women, both in the North and the South. It is hardly surprising, therefore, that as GNP rises, it does not necessarily mean that either wealth or welfare increase proportionately. I would argue that GNP is becoming, increasingly, a measure of how real wealth—the wealth of nature and that produced by women for sustaining life—is rapidly decreasing. When commodity production as the prime economic activity is introduced as development, it destroys the potential of nature and women to produce life and goods and services for basic needs. More commodities and more cash mean less life—in nature (through ecological destruction) and in society (through denial of basic needs). Women are devalued first, because their work cooperates with nature's processes, and second, because work which satisfies needs and ensures sustenance is devalued in general. Precisely because more growth in maldevelopment has meant less sustenance of life and life-support systems, it is now imperative to recover the feminine principle as the basis for development which conserves and is ecological. Feminism as ecology, and ecology as the revival of *Prakriti*, the source of all life, become the decentered powers of political and economic transformation and restructuring.

This involves, first, a recognition that categories of "productivity" and growth which have been taken to be positive, progressive, and universal are, in reality, restricted patriarchal categories. When viewed from the point of view of nature's productivity and growth, and women's production of sustenance, they are found to be ecologically destructive and a source of gender inequality. It is no accident that the modern, efficient, and productive technologies created within the context of growth in market economic terms are associated with heavy ecological costs, borne largely by women. The resource and energy intensive production processes they give rise to demand ever-increasing resource withdrawals from the ecosystem. These withdrawals disrupt essential ecological processes and convert renewable resources into nonrenewable ones. A forest, for example, provides inexhaustible supplies of diverse biomass over time if its capital stock is maintained and it is harvested on a sustained yield basis. The heavy and uncontrolled demand for industrial and commercial wood, however, requires the continuous overfelling of trees which exceeds the regenerative capacity of the forest ecosystem, and eventually converts the forests into nonrenewable resources. Women's work in the collection of water, fodder, and fuel is thus rendered more energy- and time-consuming. (In Garhwal, for example, I have seen women who originally collected fodder and fuel in a few hours now traveling long distances by truck to collect grass and leaves in a task that might take up to two days.) Sometimes the damage to nature's intrinsic regenerative capacity is impaired not by overexploitation of a particular resource but, indirectly, by damage caused to other related natural resources through ecological processes. Thus the excessive overfelling of trees in the catchment areas of streams and rivers destroys not only forest resources, but also renewable supplies of water, through hydrological destabilization.

Resource intensive industries disrupt essential ecological processes not only by their excessive demands for raw material, but by their pollution of air and water and soil. Often such destruction is caused by the resource demands of non-vital industrial products. In spite of severe ecological crises, this paradigm continues to operate because for the North and for the elites of the South, resources continue to be available, even now. The lack of recognition of nature's processes for survival *as factors in the process of economic development* shrouds the political issues arising from resource transfer and resource destruction, and creates an ideological weapon for increased control over natural resources in the conventionally employed notion of productivity. All other costs of the economic process consequently become invisible. The forces which contribute to the increased "productivity" of a modern farmer or factory worker, for instance, come from the increased use of natural resources. Amory Lovins has described this increase as the amount of "slave" labor presently at work in the world. According to him, each person on Earth, on an average, possesses the equivalent of about fifty slaves, each working a forty-hour week. Man's global energy conversion from all sources (wood, fossil fuel, hydroelectric power, nuclear) is currently approximately 8×10^{12} watts. This is more than twenty times the energy content of the food necessary to feed the present world population at the FAO standard diet of 3,600 cal/day. The "productivity" of the western male compared to women or third world peasants is not intrinsically superior; it is based on inequalities in the distribution of this "slave" labor. The average inhabitant of the United States, for example, has 250 times more "slaves" than the average Nigerian. "If Americans were short of 249 of those 250 'slaves,' one wonders how efficient they would prove themselves to be?"[9]

It is these resource and energy intensive processes of production which divert resources away from survival, and hence from women. What patriarchy sees as productive work, is, in ecological terms, highly destructive production. The second law of thermodynamics predicts that resource intensive and resource wasteful economic development must become a threat to the survival of the human species in the long run. Political struggles based on ecology in industrially advanced countries are rooted in this conflict between *long-term survival options* and *short-term overproduction and overconsumption*. Political struggles of women, peasants, and tribals based on ecology in countries like India are far more acute and urgent since they are rooted in the *immediate threat to the options for survival* for the vast majority of the people, *posed by resource-intensive and resource-wasteful economic growth* for the benefit of a minority.

In the market economy, the organizing principle for natural resource use is the maximization of profits and capital accumulation. Nature and human needs are managed through market mechanisms. Demands for natural resources are restricted to those demands registering on the market; the ideology of development is in large part based on a vision of bringing all natural resources into the market economy for commodity production. When these resources are already being used by nature to maintain her production of renewable resources and by women for sustenance and livelihood, their diversion to the market economy generates a scarcity condition for ecological stability and creates new forms of poverty for women. . . .

The paradox and crisis of development arises from the mistaken identification of culturally perceived poverty with real material poverty, and the mistaken identification of the growth of commodity production as better satisfaction of basic needs. In actual fact, there is less water, less fertile soil, less genetic wealth as a result of the development process. Since these natural resources are the basis of nature's economy and women's survival economy, their scarcity is impoverishing women and marginalized peoples in an unprecedented manner. Their new impoverishment lies in the fact that resources which supported their survival were absorbed into the market economy while they themselves were excluded and displaced by it.

The old assumption that with the development process the availability of goods and services will automatically be increased and poverty will be removed, is now under serious challenge from women's ecology movements in the third world, even while it continues to guide development thinking in centers of patriarchal power. Survival is based on the assumption of the sanctity of life; maldevelopment is based on the assumption of the sacredness of "development." Gustavo Esteva asserts that the sacredness of development has to be refuted because it threatens survival itself. "My people are tired of development," he says, "they just want to live."[10]

The recovery of the feminine principle allows a transcendence and transformation of these patriarchal foundations of maldevelopment. It allows a redefinition of growth and productivity as categories linked to the production, not the destruction, of life. It is thus simultaneously an ecological and a feminist political project which legitimizes the way of knowing and being that create wealth by enhancing life and diversity, and which delegitimizes the knowledge and practice of a culture of death as the basis for capital accumulation.

NOTES

1. Rosa Luxemburg, *The Accumulation of Capital* (London: Routledge and Kegan Paul, 1951).

2. An elaboration of how "development" transfers resources from the poor to the well endowed is contained in Jayanta Bandyopadhyay and Vandana Shiva, "Political Economy of Technological Polarisations," *Economic and Political Weekly* 7, no. 45 (November 6, 1982): 1827–32; and Jayanta Bandyopadhyay and Vandana Shiva, "Political Economy of Ecology Movements," *Economic and Political Weekly* 23, no. 24 (June 11, 1988): 1223–32.

3. Ester Boserup, *Women's Role in Economic Development* (London: Allen and Unwin, 1970).

4. DAWN, *Development Crisis and Alternative Visions: Third World Women's Perspectives* (Bergen: Christian Michelsen Institute, 1985), p. 21.

5. M. George Foster, *Traditional Societies and Technological Change* (Delhi: Allied Publishers, 1973).

6. Maria Mies, *Patriarchy and Accumulation on a World Scale* (London: Zed Books, 1986).

7. Alice Schlegel, ed., *Sexual Stratification: A Cross-Cultural Study* (New York: Columbia University Press, 1977).

8. Jonathon Porritt, *Seeing Green* (Oxford: Blackwell, 1984), p. 121.

9. Amory Lovins, quoted in S. R. Eyre, *The Real Wealth of Nations* (London: Edward Arnold, 1978), p. 133.

10. Gustavo Esteva, remarks made at a Conference of the Society for International Development, Rome, 1985.

25.

CONSERVATION REFUGEES

MARK DOWIE

A low fog envelopes the steep and remote valleys of southwestern Uganda most mornings, as birds found only in this small corner of the continent rise in chorus and the great apes drink from clear streams. Days in the dense montane forest are quiet and steamy. Nights are an exaltation of insects and primate howling. For thousands of years the Batwa people thrived in this soundscape, in such close harmony with the forest that early twentieth-century wildlife biologists who studied the flora and fauna of the region barely noticed their existence. They were, as one naturalist noted, "part of the fauna."

In the 1930s, Ugandan leaders were persuaded by international conservationists that this area was threatened by loggers, miners, and other extractive interests. In response, three forest reserves were created—the Mgahinga, the Echuya, and the Bwindi—all of which overlapped with the Batwa's ancestral territory. For sixty years these reserves simply existed on paper, which kept them off-limits to extractors. And the Batwa stayed on, living as they had for generations, in reciprocity with the diverse biota that first drew conservationists to the region.

However, when the reserves were formally designated as national parks in 1991 and a bureaucracy was created and funded by the World Bank's Global Environment Facility to manage them, a rumor was in circulation that the Batwa were hunting and eating silverback gorillas, which by that time were widely recognized as a threatened species and also, increasingly, as a featured attraction for ecotourists from Europe and America. Gorillas were being disturbed and even poached, the Batwa admitted, but by Bahutu, Batutsi, Bantu, and other tribes

From: "Conservation Refugees: When Protecting Nature Means Kicking People Out," *Orion Magazine* (November/December 2005): 16–27.

301

who invaded the forest from outside villages. The Batwa, who felt a strong kinship with the great apes, adamantly denied killing them. Nonetheless, under pressure from traditional Western conservationists, who had come to believe that wilderness and human community were incompatible, the Batwa were forcibly expelled from their homeland.

These forests are so dense that the Batwa lost perspective when they first came out. Some even stepped in front of moving vehicles. Now they are living in shabby squatter camps on the perimeter of the parks, without running water or sanitation.

Tomas Mtwandi, who was born in the Mgahinga and evicted with his family when he was fourteen, is adapting slowly and reluctantly to modern life. He is employed as an indentured laborer for a local Bantu farmer and is raising a family in a one-room shack near the Bwindi park border. He is regarded as rich by his neighbors because his roof doesn't leak and he has a makeshift metal door on his mud-wall home. As a "registered resource user," Mtwandi is permitted to harvest honey from the Bwindi and pay an occasional visit to the graves of his ancestors in the Mgahinga, but he does so at the risk of being mistaken for a poacher and shot on sight by paid wardens from neighboring tribes. His forest knowledge is waning, and his family's nutrition is poor. In the forest they had meat, roots, fruit, and a balanced diet. Today they have a little money but no meat. In one more generation their forest-based culture—songs, rituals, traditions, and stories—will be gone.

It's no secret that millions of native peoples around the world have been pushed off their land to make room for big oil, big metal, big timber, and big agriculture. But few people realize that the same thing has happened for a much nobler cause: land and wildlife conservation. Today the list of culture-wrecking institutions put forth by tribal leaders on almost every continent includes not only Shell, Texaco, Freeport, and Bechtel, but also more surprising names like Conservation International (CI), The Nature Conservancy (TNC), the World Wildlife Fund (WWF), and the Wildlife Conservation Society (WCS). Even the more culturally sensitive World Conservation Union (IUCN) might get a mention.

In early 2004 a United Nations meeting was convened in New York for the ninth year in a row to push for passage of a resolution protecting the territorial and human rights of indigenous peoples. The UN draft declaration states: "Indigenous peoples shall not be forcibly removed from their lands or territories. No relocation shall take place without the free and informed consent of the indigenous peoples concerned and after agreement on just and fair compensation and, where possible, with the option to return." During the meeting an indigenous delegate who did not identify herself rose to state that while extractive industries were still a serious threat to their welfare and cultural integrity, their new and biggest enemy was "conservation."

Later that spring, at a Vancouver, British Columbia, meeting of the International Forum on Indigenous Mapping, all two hundred delegates signed a declaration stating that the "activities of conservation organizations now represent the single biggest threat to the integrity of indigenous lands." These are rhetorical jabs, of course, but they have shaken the international conservation community, as have a subsequent spate of critical articles and studies, two of them conducted by the Ford Foundation, calling big conservation to task for its historical mistreatment of indigenous peoples.

"We are enemies of conservation," declared Maasai leader Martin Saning'o, standing before a session of the November 2004 World Conservation Congress sponsored by IUCN in Bangkok, Thailand. The nomadic Maasai, who have over the past thirty years lost most of their grazing range to conservation projects throughout eastern Africa, hadn't always felt that way. In fact, Saning'o reminded his audience, "We were the original conservationists." The room was hushed as he quietly explained how pastoral and nomadic cattlemen have traditionally protected their range: "Our ways of farming pollinated diverse seed species and maintained corridors between ecosystems." Then he tried to fathom the strange version of land conservation that has impoverished his people, more than one hundred thousand of whom have been displaced from southern Kenya and the Serengeti Plains of Tanzania. Like the Batwa, the Maasai have not been fairly compensated. Their culture is dissolving and they live in poverty.

"We don't want to be like you," Saning'o told a room of shocked white faces. "We want you to be like us. We are here to change your minds. You cannot accomplish conservation without us."

Although he might not have realized it, Saning'o was speaking for a growing worldwide movement of indigenous peoples who think of themselves as conservation refugees. Not to be confused with ecological refugees—people forced to abandon their homelands as a result of unbearable heat, drought, desertification, flooding, disease, or other consequences of climate chaos—conservation refugees are removed from their lands involuntarily, either forcibly or through a variety of less coercive measures. The gentler, more benign methods are sometimes called "soft eviction" or "voluntary resettlement," though the latter is contestable. Soft or hard, the main complaint heard in the makeshift villages bordering parks and at meetings like the World Conservation Congress in Bangkok is that relocation often occurs with the tacit approval or benign neglect of one of the five big international nongovernmental conservation organizations, or as they have been nicknamed by indigenous leaders, the BINGOs.

The rationale for "internal displacements," as these evictions are officially called, usually involves a perceived threat to the biological diversity of a large geographical area, variously designated by one or more BINGOs as an "ecological hot spot," an "ecoregion," a "vulnerable ecosystem," a "biological corridor," or a "living landscape." The huge parks and reserves that are created often involve a debt-for-nature swap (some of the host country's national debt paid off or retired in exchange for the protection of a parcel of sensitive land) or similar financial incentive provided by the World Bank's Global Environment Facility and one or more of its "executing agencies" (bilateral and multilateral banks). This trade may be paired with an offer made by the funding organization to pay for the management of the park or reserve. Broad rules for human use and habitation of the protected area are set and enforced by the host nation, often following the advice and counsel of a BINGO, which might even be given management powers over the area. Indigenous peoples are often left out of the process entirely.

Curious about this brand of conservation that puts the rights of nature before the rights of people, I set out last autumn to meet the issue face to face. I visited with tribal members on three continents who were grappling with the consequences of Western conservation and found an alarming similarity among the stories I heard.

Khon Noi, matriarch of a remote mountain village, huddles next to an open-pit stove in the loose, brightly colored clothes that identify her as Karen, the most populous of six tribes found in the lush, mountainous reaches of far northern Thailand. Her village of sixty-five families has been in the same wide valley for over two hundred years. She chews betel, spitting its bright red juice into the fire, and speaks softly through black teeth. She tells me I can use her name, as long as I don't identify her village.

"The government has no idea who I am," she says. "The only person in the village they know by name is the 'headman' they appointed to represent us in government negotiations. They were here last week, in military uniforms, to tell us we could no longer practice rotational agriculture in this valley. If they knew that someone here was saying bad things about them they would come back again and move us out."

In a recent outburst of environmental enthusiasm stimulated by generous financial offerings from the Global Environment Facility, the Thai government has been creating national parks as fast as the Royal Forest Department can map them. Ten years ago there was barely a park to be found in Thailand, and because those few that existed were unmarked "paper parks," few Thais even knew they were there. Now there are 114 land parks and 24 marine parks on the map. Almost twenty-five thousand square kilometers, most of which are occupied by hill and fishing tribes, are now managed by the forest department as protected areas.

"Men in uniform just appeared one day, out of nowhere, showing their guns," Kohn Noi recalls, "and telling us that we were now living in a national park. That was the first we knew of it. Our own guns were confiscated . . . no more hunting, no more trapping, no more snaring, and no more 'slash and burn.' That's what they call our agriculture. We call it crop rotation and we've been doing it in this valley for over two hundred years. Soon we will be forced to sell rice to pay for greens and legumes we are no longer allowed to grow here. Hunting we can live without, as we raise chickens, pigs, and buffalo. But rotational farming is our way of life."

A week before our conversation, a short flight south of Noi's village, six thousand conservationists were attending the World Conservation Congress in Bangkok. Lining the hallways of a massive convention center were the display booths of big conservation, adorned with larger-than-life photos of indigenous peoples in splendid tribal attire. At huge plenary sessions praise was heaped on Thailand's beloved Queen Sirkit and her environment minister, who came accompanied by a sizable delegation from the Royal Forest Department.

But if delegates had taken the time to attend smaller panels and workshops, some held outside the convention center in a parking lot, they would have heard Khon Noi's story repeated a dozen times or more by indigenous leaders who came to Bangkok from every continent, at great expense, to lobby conservation biologists and government bureaucrats for fairer treatment. And they would have heard a young Karen father of two boys ask why his country, whose cabinet had ordered its environmental bureaucracy to evict his people from their traditional homeland, was chosen by IUCN to host the largest conservation convention in history.

The response of big conservation, in Bangkok and elsewhere, has been to deny that they are party to the evictions while generating reams of promotional material about their affection for and close relationships with indigenous peoples. "We recognize that indigenous people have perhaps the deepest understanding of the Earth's living resources," says Conservation International chairman and CEO Peter Seligman, adding that "we firmly believe that indigenous people must have ownership, control and title of their lands." Such messages are carefully projected toward major funders of conservation, which in response to the aforementioned Ford Foundation reports and other press have become increasingly sensitive to indigenous peoples and their struggles for cultural survival.

Financial support for international conservation has in recent years expanded well beyond the individuals and family foundations that seeded the movement to include very large foundations like Ford, MacArthur, and Gordon and Betty Moore, as well as the World Bank, its Global Environment Facility, foreign governments, USAID, a host of bilateral and multilateral banks, and transnational corporations. During the 1990s USAID alone pumped almost $300 million into the international conservation movement, which it had come to regard as a vital adjunct to economic prosperity. The five largest conservation organizations, CI, TNC, and WWF among them, absorbed over 70 percent of that expenditure. Indigenous communities received none of it. The Moore Foundation made a singular ten-year commitment of nearly $280 million, the largest environmental grant in history, to just one organization—Conservation International. And all of the BINGOs have become increasingly corporate in recent years, both in orientation and affiliation. The Nature Conservancy now boasts almost two thousand corporate sponsors, while Conservation International has received about $9 million from its 250 corporate "partners."

With that kind of financial and political leverage, as well as chapters in almost every country of the world, millions of loyal members, and nine-figure budgets, CI, WWF, and TNC have undertaken a hugely expanded global push to increase the number of so-called protected areas (PAs)—parks, reserves, wildlife sanctuaries, and corridors created to preserve biological diversity. In 1962, there were some 1,000 official PAs worldwide. Today there are 108,000, with more being added every day. The total area of land now under conservation protection worldwide has doubled since 1990, when the World Parks Commission set a goal of protecting 10 percent of the planet's surface. That goal has been exceeded, as over 12 percent of all land, a total area of 11.75 million square miles, is now protected. That's an area greater than the entire land mass of Africa.

At first glance, so much protected land seems undeniably positive, an enormous achievement of very good people doing the right thing for our planet. But the record is less impressive when the impact upon native people is considered. For example, during the 1990s the African nation of Chad increased the amount of national land under protection from 0.1 to 9.1 percent. All of that land had been previously inhabited by what are now an estimated six hundred thousand conservation refugees. No other country besides India, which officially admits to 1.6 million, is even counting this growing new class of refugees. World estimates offered by the UN, IUCN, and a few anthropologists range from 5 million to tens of millions. Charles Geisler, a sociologist at Cornell University who has studied displacements in Africa, is certain the number on that continent alone exceeds 14 million.

The true worldwide figure, if it were ever known, would depend upon the semantics of words like "eviction," "displacement," and "refugee," over which parties on all sides of the issue argue endlessly. The larger point is that conservation refugees exist on every continent but Antarctica, and by most accounts live far more difficult lives than they once did, banished from lands they thrived on for hundreds, even thousands of years. . . .

In many locations, once a Community Conservation Area (CCA) is established and territorial rights are assured, the founding community invites a BINGO to send its ecologists and wildlife biologists to share in the task of protecting biodiversity by combining Western scientific methodology with indigenous ecological knowledge. And on occasion they will ask for help negotiating with reluctant governments. For example, the Guarani Izoceños people in Bolivia invited the Wildlife Conservation Society to mediate a comanagement agreement with their government, which today allows the tribe to manage and own part of the new Kaa-Iya del Gran Chaco National Park.

Too much hope should probably not be placed in a handful of successful comanagement models or a few field staffs' epiphanies. The unrestrained corporate lust for energy, hardwood, medicines, and strategic metals is still a considerable threat to indigenous communities, arguably a larger threat than conservation. However, the lines between the two are being blurred. Particularly problematic is the fact that international conservation organizations remain comfortable working in close quarters with some of the most aggressive global resource prospectors, such as Boise Cascade, Chevron-Texaco, Mitsubishi, Conoco-Phillips, International Paper, Rio Tinto Mining, Shell, and Weyerhauser, all of whom are members of a CI-created entity called the Center for Environmental Leadership in Business. Of course if the BINGOs were to renounce their corporate partners, they would forfeit millions of dollars in revenue and access to global power without which they sincerely believe they could not be effective.

And there are some respected and influential conservation biologists who still strongly support top-down, centralized "fortress" conservation. Duke University's John Terborgh, for example, author of the classic *Requiem for Nature*, believes that comanagement projects and CCAs are a huge mistake. "My feeling is that a park should be a park and it shouldn't have any resident people in it," he says. He bases his argument on three decades of research in Pew's Manu National Park, where native Machiguenga Indians fish and hunt animals with traditional weapons. Terborgh is concerned that they will acquire motorboats, guns, and chainsaws used by their fellow tribesmen outside the park, and that biodiversity will suffer. Then there's paleontologist Richard Leakey, who at the 2003 World Parks Congress in South Africa set off a firestorm of protest by denying the very existence of indigenous peoples in Kenya, his homeland, and arguing that "the global interest in biodiversity might sometimes trump the rights of local people."

Not all of Leakey's colleagues agree with him. Many conservationists are beginning to realize that most of the areas they have sought to protect are rich in biodiversity precisely because the people who were living there had come to understand the value and mechanisms of biological diversity. Some will even admit that wrecking the lives of 10 million or more poor, powerless people has been an enormous mistake—not only a moral, social, philosophical, and economic mistake, but an ecological one as well. Others have learned from experience that national parks and protected areas surrounded by angry, hungry people who describe themselves as "enemies of conservation" are generally doomed to fail. As Cristina Eghenter of WWF observed after working with communities surrounding the Kayan Mentarang National Park in Borneo, "It is becoming increasingly evident that conservation objectives can rarely be obtained or sustained by imposing policies that produce negative impacts on indigenous peoples."

More and more conservationists seem to be wondering how, after setting aside a "protected" land mass the size of Africa, global biodiversity continues to decline. Might there be something terribly wrong with this plan—particularly after the Convention on Biological Diversity has documented the astounding fact that in Africa, where so many parks and reserves have been created and where indigenous evictions run highest, 90 percent of biodiversity lies outside of protected areas? If we want to preserve biodiversity in the far reaches of the globe, places that are in many cases still occupied by indigenous people living in ways that are ecologically sustainable, history is showing us that the dumbest thing we can do is kick them out.

PART VI

SPIRITUAL ECOLOGY

26.

REINVENTING EDEN

CAROLYN MERCHANT

She has taken up with a snake now. The other animals are glad, for she was always experimenting with them and bothering them; and I am glad, because the snake talks, and this enables me to get a rest. She says the snake advises her to try the fruit of that tree, and says the result will be a great and fine and noble education. . . . I advised her to keep away from the tree. She said she wouldn't. I forsee trouble. Will emigrate.[1]

Mark Twain, *The Diaries of Adam and Eve*

Two grand historical narratives explain how the human species arrived at the present moment in history. Both are Recovery narratives, but the two stories have different plots, one upward, the other downward. The first story is the traditional biblical narrative of the Fall from the Garden of Eden from which humanity can be redeemed through Christianity. But the Garden itself can also be recovered. By the time of the Scientific Revolution of the seventeenth century, the Christian narrative had merged with advances in science, technology, and capitalism to form the mainstream Recovery Narrative. The story begins with a precipitous Fall from Eden followed by a long, slow, upward attempt to recreate the Garden of Eden on Earth. The outcome is a better world for all people. This first story—the mainstream Recovery Narrative—is a story of upward progress in which humanity gains the power to manage and control the earth.

The second story, also a Recovery Narrative, instead depicts a long, slow decline from a prehistoric past in which the world was ecologically more pristine and society was more equitable for all people and for both genders. The decline

From: *Reinventing Eden: The Fate of Nature in Western Culture* (New York: Routledge, 2003), pp. 11–20, 26–38.

continues to the present, but the possibility and, indeed, the absolute necessity of a precipitous, rapid Recovery exists today and could be achieved through a sustainable ecology and an equitable society. This second story is one told by many environmentalists and feminists.

Both stories are enormously compelling and both reflect the beliefs and hopes of many people for achieving a better world. They differ fundamentally, however, on who and what wins out. In the mainstream story, humanity regains its life of ease at the expense of the earth; in the environmental story, the earth is both the victim of exploitation and the beneficiary of restoration. Women play pivotal roles in the two stories, as cause and/or victim of decline and, along with men, as restorers of a reclaimed planet. But, I argue that a third story, one of partnership between humanity and the earth and between women and men, that draws on many of the positive aspects of the two stories is also emerging. Here I develop, compare, and critically assess the roots and broad outlines of these stories.

THE CHRISTIAN NARRATIVE

The Christian story of the Fall and Recovery begins with the Garden of Eden as told in the Bible. The Christian story is marked by a precipitous fall from a pristine past. The initial lapsarian moment, or lapse from innocence, is the decline from garden to desert as the first couple is cast from the light of an ordered paradise into a dark, disorderly wasteland to labor in the earth. Instead of giving fruit readily, the earth now extracts human labor. The blame for the Fall is placed on woman.

The biblical Garden of Eden story has three central chapters: Creation, Temptation, and Expulsion (later referred to as the Fall). A woman, Eve, is the central actress, and the story's plot is declensionist (a decline from Eden) and tragic. The end result is a poorer state of nature and human nature. The valence of woman is bad. The end valence of nature is bad. Men become the agents of transformation. After the Fall, men must labor in the earth, to produce food. They become the earthly saviors who strive, through their own agricultural labor, to re-create the lost garden on Earth, thereby turning the tragedy of the Fall into the comedy of Recovery. The New Testament adds the Resurrection—the time when the earth and all its creatures, especially humans, are reunited with God to re-create the original oneness in a heavenly paradise. The biblical Fall and Recovery story has become the mainstream narrative, shaping and legitimating the course of Western culture.

The Bible offers two versions of the Christian origin story that preceded the Fall. In the Genesis 1 version, God created the land, sea, grass, herbs, and fruit; the stars, sun, and moon; and the birds, whales, cattle, and beasts, after which he made "man in his own image . . . male and female created he them." The couple was instructed "to be fruitful and multiply, replenish the earth, and subdue it," and was given "dominion over the fish of the sea, the fowl of the air, and over every living thing that moveth on the face of the earth." This version of creation is thought to have been contributed by the priestly school of Hebrew scholars in the fifth century BCE. These scholars edited and codified earlier material into the first five books (or Pentateuch) of the Old Testament, adding the first chapter of Genesis.[2]

The alternative Garden of Eden story of creation, temptation, and expulsion (Genesis 2 and 3) derives from an earlier school. Writers in Judah in the ninth century BCE produced a version of the Pentateuch known as the J source, the Book of J, or the Yahwist version (since Yahweh is the Hebrew deity). These writers recorded the oral traditions embodied in songs and folk stories handed down through previous centuries. In addition to the Garden of Eden story, these records include the heroic narratives of Abraham, Jacob, Joseph, and Moses; the escape from Egypt; and the settlement in the promised land of Canaan.[3]

In the Genesis 2 story, God first created "man" from the dust. The name Adam derives from the Hebrew word *adama*, meaning earth or arable land. *Adama* is a feminine noun, meaning an earth that gives birth to plants. God then created the Garden of Eden, the four rivers that flowed from it, and the trees for food (including the tree of life and the tree of the knowledge of good and evil in the center). He put "the man" in the garden "to dress and keep it," formed the birds and beasts from dust, and brought them to Adam to name. Only then did he create "the woman" from Adam's rib: "And Adam said, This is now bone of my bones, and flesh of my flesh: she shall be called Woman, because she was taken out of man."[4]

Biblical scholar Theodore Hiebert argues that the Yahwist's Eden narrative is told from the perspective of an audience outside the garden familiar with the post-Edenic landscape. The use of the word "before" in the phrases that God made "every plant of the field before it was in the earth," and "every herb of the field before it grew," signify the pasturage and field crops of the post-Edenic cultivated land in which the listener is situated. Similarly, the phrases "God had not caused it to rain upon the earth" and "a mist from the earth" that "watered the whole face of the ground" indicate a post-Edenic rain-based agriculture centered on cultivation of the *adama*, or the arable land.[5]

The Garden of Eden described in Genesis 2, however, is a different landscape from that of the post-Edenic *adama*. The garden is filled with spring-fed water out of which the four rivers flow. It contains the "beasts of the field," "fowls of the air," cattle, snakes, and fruit trees, including the fig, as well as humans "to dress and keep it." The image of the garden in which animals, plants, man, and woman live together in peaceful abundance in a well-watered garden is a powerful image. It provides the starting and ending points for both plots of the overarching Recovery Narrative.

Hiebert compares the garden to a desert oasis irrigated by springs. "The term 'garden' (*gan*) is itself the common designation in biblical Hebrew for irrigation supported agriculture." Irrigation agriculture was typified by the river valley civilizations of Mesopotamia and Egypt in which rivers overflowed onto the land

and water was channeled into ditches running to fields. Of the four rivers mentioned in Genesis 2, two are the Tigris (Hiddekel) and Euphrates of Mesopotamia, while the Pison and Gihon "are placed by the Yahwist south of Israel in the area of Arabia and Ethiopia (2:11–13), and have been identified by some as the headwaters of the Nile." The Edenic landscape is thus spring-fed, river-based, and irrigated, whereas the post-Edenic landscape initiated by the temptation is rain-based. Irrigation itself later becomes a technology of humanity's hoped-for return to the garden.[6]

Genesis 3 begins with "the woman's" temptation by the serpent and the consumption of the fruit from the tree of the knowledge of good and evil. (In the Renaissance this fruit became an apple, owing to a play on the Latin word *bad*, or *malum*, which also means apple). The story details the loss of innocence through the couple's discovery of nakedness, followed by God's expulsion from the garden of Adam and his "wife," whom he now calls Eve, because she is to become "the mother of all the living." Adam was condemned to eat bread "in the sweat of thy face," and was "sent forth from the garden of Eden, to till the ground [the *adama*, or arable land] from whence he was taken," the same *adama* to which he would return after death. But because Adam listened to his wife, the *adama* was cursed. Thorns and thistles would henceforth grow in the ground where the "herb of the field" (field crops) must be grown for bread. After the couple's expulsion, God placed "at the east of the garden of Eden," the cherubim and flaming sword to guard the tree of life.[7]

The landscape into which Adam and Eve are expelled is described by Evan Eisenberg in *The Ecology of Eden*. By 1100 BCE the Israelites were farming the hills of Judea and Samaria in Canaan with ox-drawn scratch-plows and were planting wheat, barley, and legumes, such as peas and lentils. They pastured sheep, goats, and cattle and grew grapes in vineyards, olives on hillside groves, and figs, apricots, almonds, and pomegranates in orchards. "Where least disturbed, the landscape was [a] sort of open Mediterranean woodland . . . with evergreen oak, Aleppo pine, and pistachio. Elsewhere this would dwindle to . . . a mix of shrubs and herbs such as rosemary, sage, summer savory, rock rose, and thorny burnet. The settlers cleared a good deal of this forest for pasture and cropland." They captured water in cisterns and terraced the land to retain the rich but shallow red soil for planting, using the drier areas for pasturage. The arid hill country in which arable and pasturage lands was mingled was therefore the landscape that would be inhabited by the descendants of Adam and Eve.[8]

Genesis 4 recounts the fate of Adam and Eve's sons, Abel, "keeper of sheep"—a pastoralist—and Cain, "tiller of the ground"—a farmer. God accepts Abel's lamb as a first fruit, but rejects Cain's offering of the "fruit of the ground," grown on the *adama*. Although the seminomadic pastoralists and farmers of the Near East often existed in mutual support, they also engaged in conflict. Cain's killing of Abel may represent both that conflict and the historical ascendancy of settled farmers over nomadic pastoralists. A second explanation stems from the fact that Israelite farms in the hill country incorporated both farming and pastoralism into a subsistence way of life. According to Hiebert, the elder son was responsible for the tilling of the land, whereas the younger son was the keeper of the sheep. Hiebert argues that God's banishment of Cain after the killing of Abel represents a prohibition against settling disputes through the killing of kin.[9]

When human beings "fell" into a more labor-intensive way of life, their view of nature reflected this decline. Nature acting through God metes out floods, droughts, plagues, and disasters in response to humanity's sins or bountiful harvests in response to obedience. The Christian interpreter Paul "regarded the whole of nature as being in some way involved in the fall and redemption of man. He spoke of nature as 'groaning and travailing' (Romans 8:22)—striving blindly towards the same goal of union with Christ to which the Church is tending, until finally it is re-established in that harmony with man and God which was disrupted by the Fall." While the use of the term the Fall to characterize the expulsion or going forth from Eden is absent from the Bible, it becomes commonplace in the ensuing Christian tradition. Beginning with Saint Augustine, the story is interpreted as a fall that can be undone by a savior.[10]

Before the Fall, nature was an entirely positive presence. The garden, which is the beginning and end points of the Recovery Narrative, is an idealized landscape. The beasts and herbs of Genesis 1 are described as "very good," as are the cattle, fowl, beasts, and trees in the Garden of Eden of Genesis 2. The "dust" of Genesis 2, from which "the man" was formed and which was watered by "a mist from the earth," is positive in valence. The "ground," from which the other creatures are made, is positive as well. But after the couple disobeys God, the ground is "cursed." Adam eats of it in sorrow, and it brings forth thorns and thistles. The serpent changes from being "more subtle" than the other beasts to being "cursed above all cattle and above every beast of the field." In the Christian tradition, the thorns, thistles, and serpent symbolize barren desert and infertile ground, a negative nature from which humanity must recover to regain the garden.[11]

With the Fall from Eden, humanity abandons an original, "untouched" nature and enters into history. Nature is now a fallen world and humans fallen beings. But this fall through the lapsarian moment (or lapse from innocence) sets up the opposite—or the Recovery—moment. The effort to recover Eden henceforth encompasses all of human history. Reattaining the lost garden, its life of ease from labor, and its innocent happiness (and, I would add, the potential for human partnership with the earth) become the primary human endeavor. The Eden narrative is "a story of originary presence which is subsequently usurped by difference; and then of a final presence, reinstituted, sweeping away the unfortunate misadventure."[12]

The Recovery Narrative begins with the Fall from the garden into the desert (and the loss of an original partnership with the land), moves upward to the re-creation of Eden on Earth (the earthly paradise), and culminates with the vision of attainment of a heavenly paradise, a recovered garden. Paradise is defined as heaven, a state of bliss, an enclosed garden or park, an Eden. Derived from a Sumerian word, paradise was once the name of a fertile place that had become dry and barren. The Persian word for park, or enclosure, evolved through Greek and Latin to take on the meaning of garden, so that by the medieval period, Eden was depicted as an enclosed garden. The religious path to a heavenly paradise, practiced throughout the early Christian and medieval periods, incorporated the promise of salvation to atone for the original sin of tasting the forbidden fruit. In the Christian story, time has two poles: beginning and end—creation and salvation.[13]

The resurrection or end drama, heralded in the New Testament, envisions an earth reunited with God when the redeemed earthly garden merges into a higher heavenly paradise. The Second Coming of Christ was to occur either at the outset of the thousand-year period of his reign of peace on Earth foretold in Revelations 20 (the millennium) or at the Last Judgment, when the faithful were reunited with God at the resurrection. Since medieval times, millenarian sects have awaited the advent of Christ on Earth.[14]

The Parousia is the idea of the end of the world, expressed as the hope set forth in the New Testament that "he shall come again to judge both the quick and the dead." It depicts a redeemed Earth and redeemed humans. "The scene of the future consummation is a radically transformed earth." Parousia derives from the Latin *parere*, meaning to produce or bring forth. Hope for Parousia was a motivating force behind the Church's missionary work, both in its early development and in the New World. Christians prepared for this expected age of glory when God would enter history. "The coming of this Kingdom was conceptualized as a sudden catastrophic moment, or as preceded by the Messianic kingdom, during which it was anticipated that progressive work would take place."[15]

THE MODERN NARRATIVE

A second, secular version of Recovery became paramount during the Scientific Revolution of the seventeenth century, one in which the earth itself becomes a new Eden. This is the mainstream narrative of modern Western culture, one which continues to this day—it is *our* story, one so compelling we cannot escape its tentacles. In the 1600s, Europeans and New World colonists began a massive effort to reinvent the whole earth in the image of the Garden of Eden. Aided by the Christian doctrine of redemption and the inventions of science, technology, and capitalism, the long-term goal of the Recovery project has been to turn the entire earth into a vast cultivated garden. The seventeenth-century concept of recovery came to mean more than recovery from the Fall. It also entailed restoration of health, reclamation of land, and recovery of property. The strong interventionist version in Genesis 1 validates Recovery through domination, while the softer Genesis 2 version advocates dressing and keeping the garden through human management (stewardship). Human labor would redeem the souls of men and women, while the earthly wilderness would be redeemed through cultivation and domestication.[16]

The Garden of Eden origin story depicts a comic or happy state of human existence, while the Fall exemplifies a tragic state. Stories and descriptions about nature and human nature told by explorers, colonists, settlers, and developers present images of and movement between comic (positive) or tragic (negative) states. Northrop Frye describes the elements of these two states. In comic stories, the human world is a community and the animal world comprises domesticated flocks and birds of peace. The vegetable world is a garden or park with trees, while the mineral world is a city or temple with precious stones and starlit domes. And the unformed world is depicted by a river. In tragic stories, the human world is an anarchy of individuals and the animal world is filled with birds and beasts of prey (such as wolves, vultures, and serpents). The vegetable world is a wilderness, desert, or sinister forest, the mineral world is filled with rocks and ruins, and the unformed world is a sea or flood. All of these elements are present in the two versions of the Recovery Narrative.[17]

The plot of the tragedy moves from a better or comic state to a worse or tragic state (from the Garden of Eden to a desert wilderness). The comedy, on the other hand, moves from an initial tragic state to a comic outcome (from a desert to a recovered garden). *The primary narrative of Western culture has been a precipitous, tragic Fall from the Garden of Eden, followed by a long, slow, upward Recovery to convert the fallen world of deserts and wilderness into a new earthly Eden.* Tragedy is turned into comedy through human labor in the earth and the Christian faith in redemption. During the Scientific Revolution of the seventeenth century, the Christian and modern stories merged to become the mainstream Recovery Narrative of Western culture.

ENVIRONMENTALIST NARRATIVES

An alternative to the mainstream story of the Fall and Recovery is told by many environmentalists and feminists. This second narrative begins in a Stone Age Garden of Eden and depicts a gradual, rather than precipitous, loss of a pristine condition. It uses archeological, anthropological, and ecological data, along with myth and art, to re-create a story of decline. Both environmental and feminist accounts idealize an Edenic prehistory in which both sexes lived in harmony with each other and nature, but they are nevertheless compelling in their critique of environmental disruption and the subjugation of both women and nature. When viewed critically, both can contribute to a new narrative of sustainable partnership between humanity and nature.

One version of the environmental narrative is exemplified by the work of philosopher Max Oelschlaeger. Paleolithic people, says Oelschlaeger, did not distinguish between nature and culture, but saw "themselves as one with plants and animals, rivers and forests, as part of a larger, encompassing whole." In that deep past, people in gathering-hunting bands lived sustainably and "comfortably in the wilderness," albeit within cycles of want and plenty. Contained within the sacred oneness of the *Magna Mater* (the Great Mother), hunters followed rituals that respected animals and obeyed rules for preparing food and disposing of remains. Cave paintings, for example, reveal human-animal hybrids that suggest identity with the *Magna Mater*, while the cave itself is her womb. Although myth rather than science explained life, Stone Age peoples, argues Oelschlaeger, were just as intelligent as their "modern" counterparts.[18]

Oelschlaeger sees humankind's emergence from the original oneness with the *Magna Mater* as the beginning of a wrenching division, just as birth is a traumatic separation from the human mother.

> No one knows for certain how long prehistoric people existed in an Edenlike condition of hunting-gathering, but 200,000 years or more is not an unreasonable estimate for the hegemony of the Great Hunt. Even while humankind lived the archaic life, clinging conceptually to the bosom of the *Magna Mater*, the course of cultural events contained the seeds of an agricultural revolution, since prehistoric peoples were practicing rudimentary farming and animal husbandry.[19]

Oelschlaeger's narrative is one of gradual decline from the Paleolithic era rather than the precipitous Fall depicted in the Genesis 2 story. Near the end of the last ice age, around 10,000 BCE, changes in climate disrupted Paleolithic ecological relations. Animals and grains were gradually domesticated for herding and cultivation, heralding a change to pastoral and horticultural ways of life, particularly in the Near East. Once humans became agriculturists, Oelschlaeger observes, "the almost paradisiacal character of prehistory was irretrievably lost." Differences between humans and animals, male and female, people and nature became more distinct.[20] Humanity lost the intimacy it once had with the *Magna Mater*: "Western culture was now alienated from the Great Mother of the Paleolithic Mind."[21]

The first environmental problems stemming from large-scale agriculture occurred in Mesopotamia. Canals stretched from the Tigris to the Euphrates, bringing fertility to thousands of square miles of cropland. But, as these irrigation waters evaporated, salts accumulated in the soils and reduced productivity. Oelschlager suggests that agriculture marks a decline from an Edenic past: "If the thesis that agriculture underlies humankind's turn upon the environment, even if out of climatological exigency, is cogent, then the ancient Mediterranean theater is where the 'fall from Paradise' was staged . . ."[22]

In the Near East, the great town-based cultures emerged around 4000 BCE. By about 1000 BCE, the ancient tribes of Yahweh had become a single kingdom, ruled by David, which practiced rain-based agriculture. The God Yahweh above the earth represents a rupture with the *Magna Mater* of the Paleolithic and a legitimization of the settled agriculture and pastoralism of the Neolithic. The

Hebrews rebelled against sacred animals as idols and placed Yahweh as one God above and outside of nature. Time was no longer viewed as a cyclical return, but as a linear history with singular determinative events. As the "chosen people," Hebrew agriculturists and pastoralists became part of a broad-based transition from gathering-hunting to farming-herding.[23]

Ecologically, the Fall from Eden as told in Genesis 2 may reflect the differences between gathering-hunting and farming-herding initiated thousands of years earlier. In the Garden of Eden's age of gathering, Adam and Eve picked the fruits of the trees without having to labor in the earth. The transition from foraging and hunting to settled agriculture took place some nine to ten thousand years ago (7000–8000 BCE) with the domestication of wheat and barley in the oak forests and steppes of the Near East. Around five thousand years ago (3200–3100 BCE), fruits such as the olive, grape, date, pomegranate, and fig were domesticated. By 600 BCE, when the biblical stories were codified, fruit trees were cultivated throughout the Near East. The Genesis 2 story may reflect the state of farming at the time and the labor required for tilling fields as opposed to tending and harvesting fruit trees.[24]

The tilling, planting, harvesting, and storing of wheat and barley represents a form of settled agriculture in which the earth was managed for grain production. "By the time the Genesis stories were composed," states Oelschlaeger, "man had already embarked on the task of transforming nature. In the Genesis stories [he] justifies his actions."[25] In Genesis 1, the anthropocentric God of the Hebrews commands that the earth be subdued. This represents a rupture with the nature gods of the past that occurred during the transition from polytheism to monotheism, and which was codified during the years of Israelite exile in Babylon between 587 and 538 BCE.

During the Iron Age (1200–1000 BCE), the cultures of Israel and Canaan had overlapped. Canaanite mythology included a pantheon of deities: the patriarch, El; his consort and mother-goddess, Asherah; the storm-god, Baal; and his sister/consort, Anat. Although the worship of Yahweh predominated, Israelites also worshipped El, Baal, and Asherah. During the period of the monarchy (ca. 1000–587 BCE), the figure of Yahweh assimilated characteristics of the other deities, and Israel then rejected Baal and Asherah as part of its religion. "By the end of the monarchy," states Mark S. Smith, "much of the spectrum of religious practice had largely disappeared; monolatrous Yahwism was the norm in Israel, setting the stage for the emergence of Israelite monotheism."[26]

Monotheism represented an irrevocable break with the natural world. According to Henri and H. A. Frankfort, the emergence of monotheism represents the highest level of abstraction and constitutes the "emancipation of thought from myth." The two philosophers state: "The dominant tenet of Hebrew thought is the absolute transcendence of God. Yahweh is not in nature. . . . The God of the Hebrews is pure being, unqualified, ineffable. . . . Hence all concrete phenomena are devaluated." Although God had human characteristics, he was not human. Although God had characteristics assimilated from other deities, he was the one God, not one among many gods.[27]

From an ecological perspective, the separation of God from nature constitutes a rupture with nature. God is not nature or of nature. God is unchanging, nature is changing and inconstant. The human relationship to nature was not one of I to thou, not one of subject to subject, nor of a human being to a nature alive with gods and spirits. The intellectual construction of a transcendent God is yet another point in a narrative of decline. The separation of God from nature legitimates humanity's separation from nature and sets up the possibility of human domination and control over nature. In the agricultural communities of the Old Testament, humanity is the link between the soil and God. Humans are of the soil, but separate from and above the soil. They till the land with plows and reap the harvest with scythes. They clear the forests and pollute the rivers. Their goats and sheep devour the hillsides and erode the soil. Over time, the natural landscape is irrevocably transformed. At the same time, however, nature is an unpredictible actor in the story. Noah's flood, plagues of locusts, earthquakes, droughts, and devastating diseases inject uncertainties into the outcome. Efforts to control nature come up against chaotic events that upset the linearity of the storyline and create temporary or permanent setbacks.[28]

The environmentalist narrative of decline initiated by the transition to agriculture continues to the present. Tools and technologies allow people to spread over the entire globe and to subdue the earth. The colonizers denude the earth for ores and build cities and highways across the land. Despite this destruction, however, environmentalists hope for a Recovery that reverses the decline by means of planetary restoration. The Recovery begins with the conservation and preservation movements of the nineteenth century and continues with the environmental movement of the late twentieth century.

FEMINIST NARRATIVES

Many feminists likewise see history as a downward spiral from a utopian past in which women were held in equal or even higher esteem than men. This storyline was developed in the nineteenth century by Marxist philosopher Friedrich Engels, who saw the "worldwide defeat of the female sex" at the dawn of written history, and by anthropologists such as Johann Bachofen, August Bebel, and Robert Briffault. It was elaborated in a series of compelling studies by twentieth-century feminists such as Jane Harrison, Helen Diner, Esther Harding, Elizabeth Gould Davis, Merlin Stone, Adrienne Rich, Françoise d'Eaubonne, Marija Gimbutas, Pamela Berger, Gerda Lerner, Monica Sjöö, Barbara Mor, Riane Eisler, Elinor Gadon, Rosemary Radford Ruether, and a host of other feminists and ecofeminists. Like the environmental story, the feminist story captures the imagination by its symbolic force and its dramatic loss of female power. But like the environmental narrative, it must be critically evaluated for its overly utopian past from which women "fell" and its polarization of the sexes into positive female valences and negative male valences.[29]

In broad outlines the story of the decline of women, goddesses, and female symbolism woven by feminist writers is as follows. Elizabeth Gould Davis in *The First Sex* sets out the storyline:

> When recorded history begins we behold the finale of the long pageant of prehistory. . . . On the stage, firmly entrenched on her ancient throne, appears woman, the heroine of the play. About her, her industrious subjects perform their age-old roles. Peace, Justice, Progress, Equality play their parts with a practiced perfection. . . . Off in the wings, however, we hear a faint rumbling— the . . . jealous complaints of the new men who are no longer satisfied with their secondary role in society. . . . [T]he rebellious males burst onstage, overturn the queen's throne, and take her captive. . . . The queen's subjects—Democracy, Peace, Justice, and the rest—flee the scene in disarray. And man, for the first time in history, stands triumphant, dominating the stage as the curtain falls.[30]

This story of decline from a past dominated by female cultural symbols and powerful female deities into one of female subordination is presented by many feminist writers. The plot is a downward trajectory throughout prehistory and written history in which female power is lost or obscured. Recovery, however, can occur with emancipation, social and economic equality, and the return of powerful cultural icons that validate women's power and promise. Merlin Stone conveys the argument when she writes that in the Neolithic era (ca. 7000 BCE) people worshipped a female creator, a Great Goddess who was overthrown with the advent of newer religions. The loss of paradise, she holds, is the loss of a female deity. The beginnings of this narrative occur in the ancient Near East with the overthrow of goddess worshipping horticulturalists by horse-mounted warriors.[31]

Horticulturists who lived during the period 7000 to 3500 BCE in Old Europe—the area of present-day Greece and the former Yugoslavia—were, according to archeologist Marija Gimbutas, apparently peaceful groups who did not develop destructive weapons. Men and women were buried side by side, indicating equal status. Their lives revolved around fertility rituals based on the female principle. Birth, death, and regeneration were reflected in statues of female deities with large buttocks, pregnant bellies, and cylindrical necks. The concepts of male and female, animal and human, were fused. Nature was venerated. Artifacts show large eggs with snakes wound around them that symbolized the cosmos, while fish, water birds, butterflies, and bees captured the vibrancy of the natural world. Gimbutas's interpretation of grave sites as representing equality and her conjectures about the symbolic meanings of markers on vases and statues have been questioned, but her work is nonetheless compelling in part because the storyline she imposes on the past is one of great power, especially for women.

Between 4400 to 2800 BCE, Gimbutas argues, the apparent oneness with nature and equality between genders was ruptured. She identifies three major waves of horse-mounted Kurgan invaders that conquered Old Europe and introduced hierarchical social relations and sun god worship. Excavated graves from this period reveal male chiefs. They were buried with servants at their feet, and their graves contained weapons of human destruction and material possessions to indicate their high status. Sky gods rather than earth deities appear on pottery, suggesting a new worship of the heavens above rather than animate spirits within nature. This interpretation has likewise undergone scrutiny because it attributes all disruption to external forces and seems to give far less credence to internal social changes and adaptations to external events.[32]

The feminist narrative continues with the overthrow of goddesses in ancient Mesopotamia and Egypt and their replacement by male principles. Throughout the Mediterranean world, as a more settled way of life began, shifting settlements became towns, and civilizations with recorded histories arose. These cultures were rooted in the cyclical return of rains. Sumeria (Mesopotamia) blossomed in the fertile crescent between the Tigris and Euphrates Rivers. Sumerian gods were identified with nature: sky (An), earth (Ki), air (Enlil), and water (Enki). Domesticated animals, such as the bull and cow, symbolized fertility.[33]

A array of powerful female deities existed who were overthrown and replaced by male deities. In Mesopotamia, the Sumerian goddess Ishtar (Inanna) was portrayed with her much smaller son-lover, Tammuz. She renewed life each spring when she descended to the underworld to bring Tammuz back from the dead. Over time, however, Ishtar, faded in importance to Tammuz. Another female deity was the life-giving Tiamat who symbolized the earth. She was slain by her great great grandson, Marduk, who went on to create the heavens and the earth, heralding the rise of patriarchal society. Similarly, the male hero Gilgamesh (second millennium BCE), who slew the forest god Humbaba, symbolized agriculture's encroachment on the ancient forests.[34]

In Egypt, Isis represented the maternal principle. She produced vegetation when impregnated by Osiris, her brother-husband. Every spring her tears overflowed to flood the Nile, which made the soil fertile. In one hand she carried a sistrum, or rattle, to awaken the powers of nature. In the other she held a bucket of Nile water, and her gown was decorated with stars and flowers to symbolize nature.[35] Osiris was the god of the people, who bestowed gifts on humankind. He was killed by his brother Seth and restored to life by Isis, his sister-spouse. Osiris, however, was a male deity, who descended from Atum-Re, the sun god, and was associated with the Egyptian sun kings, or pharoahs, who embodied male power and virility.[36]

Feminists argue that a similar transition in the worship of goddesses to that of gods and a decline in the relative importance of female to male principles also occurred in ancient Greece. The Mycenaeans who worshipped the goddess on the island of Crete at the Palace of Knossos (ca. 1400 BCE) founded cities on mainland Greece bringing with them worship of the mother goddess, which thrived from 1450 to 1100 BCE. Artemis, goddess of the hunt, was worshipped as well as the fertility goddesses Demeter and Persephone. The Achaean invasions of the thirteenth century BCE began to weaken matrilineal traditions, and by the close of the second millennium BCE with the advent of the Dorians, patrilineal succession became established. The goddess Athene was reconfigured as a motherless female, free of maternal desire and labor pains, springing from the head of the male god Zeus. Here the male gives birth to the female, reversing the natural birth process. While the common people continued to worship Artemis, Demeter, and Persephone, the ruling elite set up Olympian gods, including Zeus and Apollo, as a patriarchal, rational, idealized pantheon.[37]

The feminist narrative also reverses the biblical story. It begins with powerful female creative principles. It was the goddess Anat (Eve), mother of all the living, who created Yahweh. And, following the tradition in which goddesses gave birth to sons who then became their spouses, Eve created Adam who then became her consort. Moreover, in the feminist story, Adam was born of Eve's rib, not vice versa. The very idea that Adam should give birth to Eve (as Zeus similarly gave birth to Athena) reverses the biological process in which women give birth to men. States Elizabeth Gould Davis: "[T]he whole intention of the distortion manifested in the Hebrew tale of Adam and Eve is twofold: first, to deny the tradition of a female creator; and second, to deny the original supremacy of the female sex."[38]

The feminist narrative likewise reveals important relationships between Eve and Nature. Eve's mythological connections to the mother goddesses Tiamat, Inanna, Ishtar, Isis, and Demeter are reinforced by her associations with the garden, the serpent, and the tree, all of which were both nature and of nature. First of all, the Garden of Eden itself is nature. It was originally created by the mother goddess and its loss represents the loss of intimacy between woman and nature. Second, the serpent, associated as divine counsel with the mother goddesses and female deities of Mesopotamia (Tiamat, Ishtar), Egypt (Hathor, Maat), Crete (the priestesses of Knossos), and Greece (Athena, Hera, Gaia), was the intimate link between Eve and a nature with which she communicated through speech. Third, the tree symbolized the fertility of nature and Eve's initial ingestion of its fruit initiated sexual consciousness. In the biblical expulsion story, Eve, the serpent, nature, and the body are all relegated, after the Fall, to the lowest levels of being. Merlin Stone sums up the consequences of these ancient associations between Eve and Nature: "[A] woman, listening to the advice of the serpent, eating the forbidden fruit, suggesting that men try it too and join her in sexual consciousness . . . caused the downfall and misery of all humankind."[39]

While many feminists have found evidence for a transition from matriarchy to patriarchy, other writers such Riane Eisler see humanity as taking a five thousand-year detour from a partnership society in prehistory to a dominator society that has existed throughout most of recorded history. She argues that today we have the possibility of reestablishing a partnership society in which men and women are linked as equals rather than ranked as dominant and submissive. Although feminist theologian Rosemary Radford Ruether does not employ the term partnership, in *Gaia and God* she calls for a healing process that will reconfigure the positive features of Western culture and Christianity. She advocates a reordering of social relations that will promote justice in relationships between women and men and among races, classes, and nations. And in "Gender and the Problem of Prehistory," Ruether suggests that "the only way we can, as humans, integrate ourselves into a life-sustaining relationship to nature, is for both of us, males as much as females, to see ourselves as equally rooted in the cycles of life and death, and equally responsible for creating ways of living sustainably together in that relationship."[40]

COMPARING THE NARRATIVES

The mainstream, environmentalist, and feminist Recovery Narratives all have strengths and weaknesses. The mainstream story of the Recovery of Eden through modern science, technology, and capitalism is perhaps the most powerful narrative in Western culture. It has been absorbed consciously and unconsciously by millions of people over several centuries. This story-writ-large is one in which people participate as actors and which they incorporate into their daily lives. As a narrative it is both inspiring and realizable, providing a positive earthly goal and a promise of ultimate salvation. A vast treasury of first-rate scholarship exists on the origins and transmission of the Christian and modern stories and their impacts and implications for history and society.

Yet however comprehensive and positive as a narrative, the mainstream Recovery story is also an ideology of domination over nature and other people. This narrative provides a justification for the takeover of New World lands and peoples and the management and transformation of forests, fields, and deserts. The Christian narrative is based on the belief and assumption that a monotheistic deity exists who has ordained a mode of behavior for humanity and designated roles for men and women. Such beliefs are based on acts of faith rather than credible evidence. Whatever positive ethics of care and stewardship arise from such beliefs, there exists an equal catalog of war and violence against humanity and atrocities against the earth in the name of that deity. The deity can take on any attributes any group wishes to assign to it and becomes a rationale for any actions a particular group wishes to take. As such, God (however defined and by whatever religion or sect) is a social construct that becomes a justification and an ideology for human behavior. The sacred texts that reveal such a deity are humanly constructed stories arising out of specific social, historical, and environmental circumstances.

The environmentalist and feminist narratives likewise have strengths and weaknesses. They use climatological, archeological, anthropological, and historical, as well as mythological evidence to support the storylines. The stories can be criticized, revised, or rejected on the basis of how they use, accept, and organize their evidence. To the extent that they deal with prehistory, their validity depends on how they interpret archeological, anthropological, and mythological evidence and the generalizability of that evidence.

Deciding how an early society behaved toward nature from surviving, non-decomposable artifacts is enormously difficult. Whether a *Magna Mater* or a variety of nature spirits or goddesses existed in prehistory is built on conjecture and extrapolation from later historical documents and anthropological observations. Whether mythologies recorded later in time actually reflect social realities or influence human behavior is problematical. Moreover, of the many statues and images that have survived, some are female, others are male, and still others are male/female or simply anthropomorphic. Some female images are buxom or pregnant with broad buttocks oriented toward the earth, while others are slender with outstretched arms reaching toward the sky, casting doubt on the universality of female fertility symbols. Other problems arise from the causes of transformation from a presumed egalitarian or matriarchal to a patriarchal society. External migrations such as horse-mounted warriors who infused sky-gods into earth-centered egalitarian cultures or invasions of dominant outsiders places too much weight on external as opposed to internal processes, adaptations, and mutual influences. Such critiques undercut the power of the overarching storyline of the environmental and feminist narratives.

Additional problems exist with respect to the very concept of narrative itself. A narrative, whether Christian, environmentalist, or feminist, is an ideal form into which particular bits of content are poured. The form is the organizing principle; the content is the matter. Like Plato's pure forms that explain the changing world of appearances, a narrative is a variant of idealism. What is real is the idea itself. In this sense, a Recovery Narrative is an idealist philosophy. To the extent to which people believe in or absorb the story, it organizes their behavior and hence their perception of the material world. The narrative thus entails an ethic and the ethic gives permission to act in a particular way toward nature and other people.

Narratives, however, are not deterministic. Their plots and ethical implications can be embraced or challenged. Naming the narrative gives people the power to change it, to move outside it, and to reconstruct it. People as material actors living in a real world can organize that world and their behaviors to bring about change and to break out of the confines of a particular storyline.

My own view is that out of the global ecological crisis, a new story or set of stories will emerge, but the new stories will arise out of new forms of production and reproduction as sustainable partnerships with nature are tested and become viable. Revisions of older spiritual traditions may help to create a new story, but spirituality alone cannot bring about a transformation. Nevertheless, probing the meanings of narrative, gender, and ethics embedded in the Bible and other historical narratives is critical for the twenty-first century.

NOTES

1. Mark Twain, "Extracts from Adam's Diary," in *The Diaries of Adam and Eve* (replica of the 1904–1905 first ed.), *The Oxford Mark Twain* (New York: Oxford University Press, 1996), vol. 26, p. 41.

2. Roy B. and Herman Feldman Chamberlin, *The Dartmouth Bible, An Abridgment of the King James Version with Aids to Its Understanding as History and Literature and as a Source of Religious Experience* (Boston, MA: Houghton Mifflin, 1961). See Genesis 1: 26–28 and introduction, pp. 9–10.

3. Chamberlin and Feldman, *Dartmouth Bible*, introduction, pp. 8–9; David Rosenberg and Harold Bloom, *The Book of J*, trans. David Rosenberg (New York: Vintage).

4. Chamberlin and Feldman, *Dartmouth Bible*, Genesis 2:7–22; introduction, pp. 8–9. Everett Fox, ed. *The Five Books of Moses* (New York: Schocken Books, 1995), Genesis 2:23: "She shall be called Woman/Isha, for from Man/Ish she was taken." Adam is named in Genesis 2:19: "God formed every beast of the field, and every fowl of the air; and brought them unto Adam to see what he would call them." The "woman" is created in Genesis 2:21–22 but is not named Eve until after the couple's disobedience and punishment in Genesis 3:20: "And Adam called his wife's name Eve; because she was the mother of all living." The name Eve may have come from the Sumerian name Nin-ti, meaning "lady of the rib" or "lady of life." See W. Gunther Plaut, ed., *The Torah: A Modern Commentary* (New York: Union of Hebrew Congregations, 1981), p. 30n21.

5. Theodore Hiebert, *The Yahwist's Landscape: Nature and Religion in Early Israel* (New York: Oxford University Press, 1996), pp. 32–35.

6. Hiebert, *The Yahwist's Landscape*, pp. 53–55, quotations on pp. 55 and 53.

7. Chamberlin and Feldman, *Dartmouth Bible*, Genesis 3:1–7, 22–4; Bill Moyers, *Genesis: A Living Conversation* (New York: Doubleday, 1996), p. 67; Hiebert, *The Yahwist's Landscape*, pp. 33–35.

8. Evan Eisenberg, *The Ecology of Eden* (New York: Knopf, 1998), pp. 86–89, quotation on p. 87.

9. J. Baird Callicott, "Genesis Revisited: Muirian Musings on the Lynn White Jr. Debate,"*Environmental Review* 14, nos. 1–2 (Spring/Summer) (1990): 65–92, 81; quotation from Bible, Genesis 5:20. See Genesis 1:29–30; Genesis 2:9; Genesis 3:18, 19, 23; Hiebert, *The Yahwist's Landscape*, pp. 40–41.

10. J. L. Russell, "Time in Christian Thought," in *The Voices of Time: A Cooperative Survey of Man's Views of Time as Expressed by the Sciences and Humanities*, ed. J. T. Fraser (Amherst: University of Massachusetts Press, 1981), quoted in Oelschlaeger, *Idea of Wilderness*, p. 67.

11. Genesis 1:31; Genesis 2:6–7; Genesis 3:1, 14, 18.

12. Victor Rotenberg, "The Lapsarian Moment" (unpublished manuscript, University of California, Berkeley, 1993). I thank Victor Rotenberg for sharing his manuscript with me. Henry Goldschmidt, "Rupture Tales: Stories and Politics in and around the Garden of Eden" (unpublished manuscript, University of California, Santa Cruz, 1994), quotations on pp. 8–9. I thank Henry Goldschmidt for sending me his manuscript. As postmodern philosopher Jacques Derrida puts it, the story is "an onto-theology determining the . . . meaning of being as presence, as parousia, as life without difference." See Jacques Derrida, *Of Grammatology*, trans. Gayatri Chakravorty Spivak (Baltimore: Johns Hopkins University Press, 1976), quotation on p. 71.

13. Anonymous, *Oxford English Dictionary*, compact ed., 2 vols. (Oxford: Oxford University Press, 1971), vol. 1, s.v. "Eden"; vol. 2, s.v. "paradise"; Plaut, *The Torah*, p. 29n8. In the Jewish tradition, Eden is the home of the righteous after death. On time in the Christian tradition, see Oelschlaeger, *Idea of Wilderness*, pp. 65–66.

14. Jeffrey L. Sheler, "The Christmas Covenant," *U.S. News & World Report*, December 19, 1994, pp. 62–71, see esp. p. 66. Religious sects differ as to forms of millennialism. Premillennialists, such as fundamentalist and evangelical Christians, believe a catastrophe or final battle of Armageddon will initiate the age of Christ on Earth. Postmillennialists argue for Christ's return only after a golden age of peace on earth brought about by working within the church. Antimillennialists, who include most Protestants and Roman Catholics, do not accept the thousand-year reign of Christ on Earth, but instead believe in a period prior to the final resurrection in which Christ works through the church and individual lives.

15. A. L. Moore, *The Parousia in the New Testament* (Leiden: E. J. Brill, 1966), pp. 2, 3, 5, 16, 17, 20, 21, 25–26, 28. "The divine intervention in history was the manifestation of the Kingdom of God. . . . [T]his would involve a total transformation of the present situation, hence the picture of world renewal enhanced sometimes by the idea of an entirely supernatural realm" (pp. 25–26). "Concerning the central figure in the awaited End-drama there is considerable variation. In some visions the figure of Messiah is entirely absent. In such cases 'the kingdom was always represented as under the immediate sovereignty of God'" (p. 21).

16. The concept of a recovery from the original Fall appears in the seventeenth century. See the *Oxford English Dictionary*, compact ed., vol. 2, p. 2447: "The act of recovering oneself from a mishap, mistake, fall, etc." See Bishop Edward Stillingfleet, *Origines Sacrae* (London, 1662), II, i, sec 1: "The conditions on which fallen man may expect a recovery." William Cowper, *Retirement* (1781), p. 138: "To . . . search the themes, important above all Ourselves, and our recovery from our fall." See also Richard Eden, *The Decades of the Newe Worlde or West India* (1555), p. 168, "The recoverie of the kyngedome of Granata." The term recovery also embraced the idea of regaining a "natural" position after falling and a return to health after sickness. It acquired a legal meaning in the sense of gaining possession of property by a verdict or judgment of the court. In common recovery, an estate was transferred from one party to another. John Cowell, *The Interpreter* (1607), s.v. recoverie: "A true recoverie is an actuall or reall recoverie of anything, or the value thereof by Judgement." Another meaning was the restoration of a person or thing to a healthy or normal condition, or a return to a higher or better state, including the reclamation of land. Anonymous, *Captives Bound in Chains: The Misery of Graceless Sinners and the Hope of Their Recovery by Christ* (1674); Bishop Joseph Butler, *The Analogy of Religion Natural and Revealed* (1736), pt. II, conclusion, p. 295: "Indeed neither Reason nor Analogy would lead us to think . . . that the Interposition of Christ . . . would be of that Efficacy for Recovery of the World, which Scripture teaches us it was." Joseph Gilbert, *The Christian Atonement* (1836), pt. i, p. 24: "A modified system, which shall include the provision of means for recovery from a lapsed state." James Martineau, *Essays, Reviews, and Addresses* (1890–1891), pt. II, p. 310: "He is fitted to be among the prophets of recovery, who may prepare for us a more wholesome future." John Henry Newman, *Historical Sketches* (1872–1873) pt. II, 1, iii, p. 121: "The special work of his reign was the recovery of the soil."

17. On the tragic and comic visions of the human, animal, vegetable, mineral, and

unformed worlds, see Northrup Frye, *Fables of Identity* (New York: Harcourt Brace, 1963), p. 19–20.

18. Max Oelschlaeger, *The Idea of Wilderness: From Prehistory to the Age of Ecology* (New Haven, CT: Yale University Press, 1991), pp. 11–12, 14, 16, 17–18, 20, 23; quotation on pp. 11–12.

19. Ibid., quotation on p. 24.

20. Ibid., pp. 25, 28.

21. Ibid., quotations on pp. 60, 65, 67.

22. Ibid., pp. 39, 31.

23. Ibid., pp. 42, 47–48.

24. Carol Manahan, "The Genesis of Agriculture and the Agriculture of Genesis," manuscript in possession of the author, Richmond, CA. On the domestication of crops and the rise of settled agriculture, see David R. Harris and Gordon C. Hillman, eds., *Foraging and Farming: The Evolution of Plant Exploitation* (Boston: Unwin Hyman, 1989); Daniel Zohary and Pinhas Spiegel-Roy, "Beginnings of Fruit-Growing in the Old World," *Science* 187 (January 31, 1975): 319–27.

25. John Passmore, quoted in Oelschlaeger, *Idea of Wilderness*, p. 46.

26. Mark S. Smith, *The Early History of God: Yahweh and the Other Deities in Ancient Israel* (San Francisco: HarperCollins, 1990), pp. xix–xxvii, quotation on p. xxvii.

27. Henri Frankfort et al., *Before Philosophy* (1946; Baltimore: Penguin, 1949), pp. 241–48, 253, quotations on pp. 241–42.

28. J. Donald Hughes, *Ecology in Ancient Civilizations* (Albuquerque: University of New Mexico Press, 1975), pp. 20–28.

29. Friedrich Engels, *Origins of the Family, Private Property, and the State* in *Selected Works* (New York: International Publishers, 1968); Johann Jacob Bachofen, "Mother Right: An Investigation of the Religious and Juridical Character of Matriarchy in the Ancient World" (1861), in *Myth, Religion, and Mother Right: Selected Writings of J. J. Bachofen*, trans. Ralph Manheim (Princeton, NJ: Princeton University Press, 1967), pp. 69–207; August Bebel, *Woman in the Past, Present, and Future* (San Francisco: G. B. Benham, 1897); Robert Briffault, *The Mothers*, abridged ed. (1927; New York: Atheneum, 1977); Jane Ellen Harrison, *Prolegomena to the Study of Greek Religion* (1903; Cambridge: Cambridge University Press, 1922); Jane Ellen Harrison, *The Religion of Ancient Greece* (London: Archibald Constable, 1905); Jane Ellen Harrison, *Myths of the Social Origins of Greek Religion* (Cambridge: Cambridge University Press, 1912); Jane Ellen Harrison, *Mythology* (1924; New York: Harcourt, Brace and World/Harbinger Books, 1963); Helen Diner, *Mothers and Amazons: The First Feminine History of Culture* (1929; New York: Anchor Press/Doubleday, 1973); M. Esther Harding, *Women's Mysteries, Ancient and Modern* (1955; London: Rider, 1971); Elizabeth Gould Davis, *The First Sex* (1971; Baltimore: Penguin, 1972); Merlin Stone, *When God Was a Woman* (New York: Harcourt Brace Jovanovich, 1976); Adrienne Rich, *Of Woman Born* (New York: Norton, 1976); Françoise d'Eaubonne, *Le Féminisme ou la Mort* (Paris, 1974), see above ch. 16; Marija Gimbutas, *The Goddesses and Gods of Old Europe, 6500–3500 BC* (Berkeley: University of California Press, 1982); Pamela Berger, *The Goddess Obscured: The Transformation of the Grain Protectress from Goddess to Saint* (Boston: Beacon, 1985); Gerda Lerner, *The Creation of Patriarchy* (New York: Oxford University Press, 1986); Monica Sjöö and Barbara Mor, *The Great Cosmic Mother: Rediscovering the Religion of the*

Earth (San Francisco: HarperCollins, 1987); Riane Eisler, *The Chalice and the Blade* (San Francisco: HarperCollins, 1988); Elinor Gadon, *The Once and Future Goddess* (San Francisco: HarperCollins, 1989); Rosemary Radford Ruether, *Gaia and God: An Ecofeminist Theology of Earth Healing* (San Francisco: HarperCollins, 1992).

30. Davis, *The First Sex*, pp. 16–17.

31. Stone, *When God Was a Woman*, pp. xii–xiii: "Archaeological, mythological and historical evidence all reveal that the female religion, far from naturally fading away, was the victim of centuries of continual persecution and suppression by the advocates of the newer religions which held male deities as supreme. And from these new religions came the creation myth of Adam and Eve and the tale of the loss of Paradise."

32. Gimbutas, *The Goddesses and Gods of Old Europe*.

33. Rich, *Of Woman Born*, p. 56: "A prehistoric civilization [was] centered around the female, both as mother and head of family, and as deity—the Great Goddess who appears throughout early mythology, as Tiamat, Rhea, Isis, Ishtar, Astarte, Cybele, Demeter, Diana of Ephesus, and by many other names: the eternal giver of life and embodiment of the natural order, including death."

34. Lerner, *The Creation of Patriarchy*, p. 153: "The young god who slays Tiamat in the epic is Marduk, the god worshipped in the city of Babylon. Marduk first emerges during the time of Hammurabi of Babylon, who has made his city-state dominant in the Mesopotamian region."

35. Stone, *When God Was a Woman*, pp. 10–11, 139–44.

36. Lerner, *Creation of Patriarchy*, p. 154: "The changing position of the Mother-Goddess, her dethroning, takes place in many cultures and at different times, but usually it is associated with the same historical processes. . . . In Egypt, where the male God early predominates, we can also find traces of a still earlier predominance of the Goddess. Isis . . . [was] 'the prototype of the life-giving mother and faithful wife.'"

37. Stone, *When God Was a Woman* pp. 51–53; Sjöö and Mor, *The Great Cosmic Mother*, pp. 235–37, quotation on p. 235: According to Sjöö and Mor: "The Olympian god . . . is not born from woman, or earth, or matter, but from his own absolute will. He represents a static perfection, in human form, incapable of transformation or ecstatic change; as a God, he is an intellectual concept."

38. Davis, *The First Sex*, pp. 142–44, quotation on p. 144.

39. Stone, *When God Was a Woman*, pp. 198–223, quotation on p. 223.

40. Eisler, *The Chalice and the Blade*, pp. xvii, 105, 185–203; Ruether, *Gaia and God*, pp. 2–3; Rosemary Radford Ruether, "Gender and the Problem of Prehistory," in *Goddesses and the Divine Feminine: A Western Religious History* (Berkeley: University of California Press, 2005), p. 40.

27.

TOWARD A HEALING
OF SELF AND WORLD

JOANNA MACY

A new paradigm is emerging in our time. Through its lens we see reality structured in such a way that all life-forms affect and sustain each other in a web of radical interdependence. This organic interconnectedness is what we call our Deep Ecology.

"Deep Ecology" is a term coined by Norwegian philosopher Arne Naess, to contrast with "shallow environmentalism," a Band-Aid approach applying piecemeal technological fixes for short-term goals. Deep Ecology teaches us that we humans are neither the rulers nor the center of the universe, but are embedded in a vast living matrix and subject to its laws of reciprocity. Deep Ecology represents a basic shift in ways of seeing and valuing, a shift beyond anthropocentrism:

> Anthropocentrism means human chauvinism. Similar to sexism, but substitute human race for man and *all other species* for woman. [It's about the human race being oppressive to other species and the environment.]
>
> When humans investigate and see through their layers of anthropocentric self-cherishing, a most profound change in consciousness begins to take place. Alienation subsides. The human is no longer an outsider, apart. . . .
>
> What a relief then! The thousands of years of imagined separation are over and we begin to recall our true nature. That is, the change is a spiritual one . . . sometimes referred to as deep ecology (John Seed).

There are, of course, manifold ways of evoking or provoking this change in perspective. Methods of inspiring the experience of Deep Ecology range from

From: "Deep Ecology Work: Toward the Healing of Self and World," *Human Potential Magazine* 17, no. 1 (Spring 1992): 10–13, 29–31.

prayer to poetry, from wilderness vision quests to the induction of altered states of consciousness. The most reliable is direct action in defense of earth—and it is spreading today in many forms.

THE GREENING OF THE SELF

Something important is happening in our world that is not reported in the newspapers. I consider it the most fascinating and hopeful development of our time, and it is one of the reasons I am so glad to be alive today. It has to do with what is occurring to the notion of the self.

The self is the hypothetical piece of turf on which we construct our strategies for survival, the notion around which we focus our instincts for self-preservation, or need for self-approval, and the boundaries for our self-interest. The conventional notion of the self with which we have been raised and to which we have been conditioned by mainstream culture is being undermined. What Alan Watts called "the skin-encapsulated ego" and Gregory Bateson referred to as "the epistemological error of Occidental civilization" is being unhinged, peeled off. It is being replaced by concepts of identity and self-interest which are much wider than the conventional ego . . . by what you might call the ecological self, coextensive with other beings and the life of our planet.

At a recent lecture on a college campus, I gave the students examples of activities which are currently being undertaken in defense of life on Earth— actions in which people risk their comfort and even their lives to protect other species. In the Chipko, or tree-hugging movement in northern India, for example, villagers fight the deforestation of their remaining woodlands. On the open seas, Greenpeace activists are intervening to protect marine mammals from slaughter. After that talk, I received a letter from a student I'll call Michael. He wrote:

> I think of the tree-huggers hugging my trunk, blocking the chainsaws with their bodies. I feel their fingers digging into my bark to stop the steel and let me breathe. I hear the bodhisattvas [note: a Buddhist term for an enlightened, compassionate being] in their rubber boats as they put themselves between the harpoons and me, so I can escape to the depths of the sea. I give thanks for your life and mine, and for life itself. I give thanks for realizing that I too have the powers of the tree-huggers and the bodhisattvas.

What is striking about Michael's words is the shift in identification. Michael is able to extend his sense of self to encompass the self of the tree and of the whale. Tree and whale are no longer removed, separate, disposable objects pertaining to a world "out there" (outside of humans and inside the environment); they are intrinsic to his own vitality. Through the power of his caring, his experience of self is expanded far beyond the skin-encapsulated ego. I quote Michael's words not because they are unusual, but, to the contrary, because they express a desire and a capacity that is being released from the prison cell of old constructs of self. This desire and capacity are arising in more and more people today as, out of deep concern for what is happening to our world, they begin to speak and act on its behalf.

Among those who are shedding these old constructs of self, like old skin or a confining shell, is John Seed, director of the Rainforest Information Centre in Australia. One day we were walking through the rain forest in New South Wales, where he has his office, and I asked him, "You talk about the struggle against the lumbering interests and politicians to save the remaining rain forest in Australia. How do you deal with the despair?" He replied, "I try to remember that it's not me, John Seed, trying to protect the rain forest. Rather I'm part of the rain forest protecting myself. I am that part of the rain forest recently emerged into human thinking." This is what I mean by the greening of the self. It involves a combining of the mystical with the practical and the pragmatic, transcending separateness, alienation, and fragmentation. It is a shift that Seed himself calls "a spiritual change," generating a sense of profound interconnectedness with all life.

This is hardly new to our species. In the past, poets and mystics have been speaking and writing about these ideas, but not people on the barricades agitating for social change. Now the sense of an encompassing self, the deep identity with the wider reaches of life, is a motivation for action. It is a source of courage that helps us stand up to the powers that are still, through force of inertia, destroying the fabric of life. I am convinced that this expanded sense of self is the only basis for adequate and effective action.

Three developments converge in our time to call forth the ecological self. They are: (1) the psychological and spiritual pressure exerted by current dangers of mass annihilation, (2) the emergence in science of the systems view of the world, and (3) a renaissance of nondualistic forms of spirituality.

CURRENT DANGERS OF MASS ANNIHILATION: PAIN FOR THE WORLD

The move to a wider ecological sense of self is in large part a function of the dangers that are threatening to overwhelm us. We are confronted by social breakdown, wars, nuclear proliferation, and the progressive destruction of our biosphere. Polls show that people today are aware that the world, as they know it, may come to an end. This loss of certainty that there will be a future is the pivotal psychological reality of our time.

Over the past twelve years my colleagues and I have worked with tens of thousands of people in North America, Europe, Asia, and Australia, helping them confront and explore what they know and feel about what is happening to their world. The purpose of this work, which was first known as *"Despair and Empowerment Work,"* is to overcome the numbing and powerlessness that result from suppression of painful responses to massively painful realities. As their grief and fear for the world is allowed to be expressed without apology or argument and validated as a wholesome, life-preserving response, people break through their avoidance mechanisms, break through their sense of futility and isolation. Generally what they break through into is a larger sense of identity. It is as if the pressure of their acknowledged awareness of the suffering of our world stretches or collapses the culturally defined boundaries of the self.

It becomes clear, for example, that the grief and fear experienced for our world and our common future are categorically different from similar sentiments relating to one's personal welfare. This pain cannot be equated with dread of one's own individual demise. Its source lies less in concerns for personal survival than in apprehensions of collective suffering—of what looms for human life and other species and unborn generations to come. Its nature is akin to the original meaning of compassion—"suffering with." It is the distress we feel on behalf of the larger whole of which we are a part. And, when it is so defined, it serves as a trigger or getaway to a more encompassing sense of identity, inseparable from the web of life in which we are as intricately interconnected as cells in a larger body.

This shift in consciousness is an appropriate, adaptive response. For the crisis that threatens our planet, be it seen in its military, ecological, or social aspects, derives from a dysfunctional and pathogenic notion of the self. It is a mistake about our place in the order of things. It is the delusion that the self is so separate and fragile that we must delineate and defend its boundaries, that it is so small and needy that we must endlessly acquire and endlessly consume, that it is so aloof that we can—as individuals, corporations, nation-states, or as a species—be immune to what we do to other beings.

This view of human nature is not new, of course. Many have felt the imperative to extend self-interest to embrace the whole. What is notable in our situation is that this extension of identity can come not through an effort to be noble or good or altruistic, but simply to be present and own our pain. That is why this shift in the sense of self is credible to people. As the poet Theodore Roethke said, "I believe my pain."

SCIENCE AND THE SYSTEMS VIEW: CYBERNETICS OF THE SELF

The findings of twentieth-century science undermine the notion of a separate self, distinct from the world it observes and acts upon. As Einstein showed, the self's perceptions are shaped by its changing position in relation to other phenomena. And these phenomena are affected not only by location but, as Heisenberg demonstrated, by the very act of observation. Now contemporary systems science and systems cybernetics go yet further in challenging old assumptions about a distinct, separate, continuous self.

We are open, self-organizing systems; our very breathing, acting, and thinking arise in interaction with our shared world through the currents of matter, energy, and information that flow through us. In the web of relationships that sustain these activities, there are no clear lines demarcating a separate self. As systems theorists aver, there is no categorical "I" set over and against a categorical "you" or "it."

One of the clearer expositions of this is offered by Gregory Bateson, whom I earlier quoted as saying that the abstraction of a separate "I" is "the epistemological fallacy of Western civilization." He says that the process that decides and acts cannot be neatly identified with the isolated subjectivity of the individual or located within the confines of the skin. He contends that "the total self-corrective unit that processes information is a system whose boundaries do not at all coincide with the boundaries either of the body or what is popularly called 'self' or 'consciousness.'" He goes on to say, "The self is ordinarily understood as only a small part of a much larger trial-and-error system which does the thinking, acting, and deciding."

Bateson uses the example of a woodcutter, about to fell a tree. His hands grip the handle of the axe. Whump, he makes a cut, and then whump, another cut. What is the feedback circuit, where is the information that is guiding that cutting down of the tree? That is the self-correcting unit, that is what is doing the chopping down of the tree. In another illustration, a blind person with a cane is walking along the sidewalk. Tap, tap, whoops, there's a fire hydrant, there's a curb. What is doing the walking? Where is the self then of the blind person? What is doing the perceiving and deciding? That self-corrective feedback circuit is the arm, the hand, the cane, the curb, the ear. At that moment that is the self that is walking. Bateson's point is that the self as we usually define it is an improperly delimited part of a much larger field of interlocking processes. And he maintains that

> this false reification of the self is basic to the planetary ecological crisis in which we find ourselves. We have imagined that we are a unit of survival and we have to see to our own survival, and we imagine that the unit of survival is the separate individual or a separate species, whereas in reality through the history of evolution, it is the individual plus the environment, the species plus the environment, for they are essentially symbiotic.

The self is a metaphor. We can decide to limit it to our skin, our person, our family, our organization, or our species. We can select its boundaries in objective reality. As the systems theorists see it, our consciousness illuminates a small arc in the wider currents and loops of knowing that interconnect us. It is just as plausible to conceive of mind as coexistent with these larger circuits, the entire "pattern that connects" as Bateson said. Do not think that to broaden the construct of self this way involves an eclipse of one's distinctiveness. Do not think that you will lose your identity like a drop in the ocean merging into the oneness of Brahman. From the systems perspective, this interaction, creating larger wholes and patterns, fosters and even requires diversity. You become more yourself. Integration and differentiation go hand in hand.

NONDUALISTIC SPIRITUALITY: THE BOUNDLESS HEART OF THE BODHISATTVA

A third factor that nourishes deep ecological consciousness in our world today is the resurgence of nondualistic spirituality. We find it in many realms—in Sufism in Islam, Creation Spirituality in Christianity, and in Buddhism's historic coming to the West. Buddhism is distinctive in its clarity and sophistication about the dynamics of self. In much the same way as systems theory does, Buddhism undermines categorical distinctions between self and other. It then goes further than systems theory in showing the pathogenic character of any reifications of the self, and in offering methods for transcending these difficulties and healing this suffering. What the Buddha woke up to under the Bodhi tree was *paticca samuppada*, the dependent co-arising of phenomena, in which you cannot isolate a separate, continuous self.

We think, "What do we do with the self, this clamorous 'I,' always wanting attention, always wanting its goodies? Do we crucify it, sacrifice it, mortify it, punish it, or do we make it noble?" Upon awaking we realize, "It's just a convention!" When you take it too seriously, when you suppose that it is something enduring which you have to defend and promote, it becomes the foundation of delusion, the motive behind our attachments and our aversions. Consider the Tibetan portrayal of the wheel of life, that mythically depicts all the realms of being. At the very center of that wheel of suffering are three figures: the pig, the rooster, and the snake—they represent delusion, greed, and hatred—and they just chase one another around and around. The linchpin of all the pain is the notion of our self, the notion that we have to protect that self or conquer on its behalf—or do something with it.

The point of Buddhism, and, I think, of Deep Ecology too, is that we do not need to be doomed to the perpetual rat race. The vicious circle can be broken. It can be broken by wisdom, meditation, and morality—that is, when we pay attention to our experience and our actions and discover that they do not have to be in bondage to a separate self. The sense of interconnectedness that can then arise is imaged—one of the most beautiful images coming out of the Mahayana—as the jeweled net of Indra. It is a vision of reality structured very much like the holographic view of the universe, so that each being is a jewel at each node of the net, and each jewel reflects all the others, reflecting back and catching the reflection, just as systems theory sees that the part contains the whole.

The awakening to our true self is the awakening to that entirety, breaking out of the prison-self of separate ego. The one who perceives this is the bodhisattva—and we are all bodhisattvas because we are all capable of experiencing that—it is our true nature. We are profoundly interconnected and therefore we are all able to recognize and act upon our deep, intricate, and intimate inter-existence with one another and all beings. That true nature of ours is already present in our pain for the world. When we turn our eyes away from that homeless figure, are we indifferent, or is the pain of seeing him or her too great? Do not be easily duped by the apparent indifference of those around you. What looks like apathy is really the fear of suffering. But the bodhisattva knows that to experience the pain of all beings it is necessary to experience their joy. It says in the Lotus Sutra that the bodhisattva hears the music of the spheres, and understands the language of the birds, while hearing the cries in the deepest levels of hell.

One of the things I like best about the ecological self that is arising in our time is that it is making moral exhortation irrelevant. Sermonizing is both boring and ineffective. As Arne Naess says, "The extensive moralizing within the eco-logical movement has given the public the false impression that they are being asked to make a sacrifice to show more responsibility, more concern, and a nicer moral standard. But all of that would flow naturally and easily if the self were widened and deepened so that the protection of nature was felt and perceived as protection of our very selves." Please note this important point: virtue is not required for the greening of the self or the emergence of the ecological self. The shift in identification at this point in our history is required precisely because moral exhortation doesn't work, and because sermons seldom hinder us from fol-lowing our self-interest as we conceive it.

The obvious choice, then, is to extend our notions of self-interest. For example, it would not occur to me to plead with you, "Oh, don't saw off your leg. That would be an act of violence." It wouldn't occur to me, or you, because your leg is part of your body. Well, so are the trees in the Amazon rain basin. They are our external lungs. And we are beginning to realize that the world is our body. This ecological self, like any notion of selfhood, is a metaphoric construct and a dynamic one. It involves choice: choices can be made to identify at different moments with different dimensions or aspects of our systemically interrelated existence, be they hunted whales or homeless humans or the planet itself. In doing this, the extended self brings into play wider resources—courage, endurance, ingenuity—like a nerve cell in a neural net opening to the charge of the other neurons.

There is the sense of being acted through and sustained by those very beings on whose behalf one acts. This is very close to the religious concept of grace. In systems language we can talk about it as a synergy. But with this extension, this greening of the self, we can find a sense of buoyancy and resilience that comes from letting flow through us strengths and resources that come to us with con-tinuous surprise and sense of blessing.

28

THE SPIRITUAL DIMENSION OF GREEN POLITICS

CHARLENE SPRETNAK

HOW SHALL WE RELATE TO OUR CONTEXT, THE ENVIRONMENT?

In 1967 Lynn White Jr., a professor of history at UCLA, published in *Science* "The Historical Roots of our Ecologic Crisis," a critical analysis of the attitudes Western religion has encouraged toward our environment. Since then ecologists often point to the injunctions in Genesis that humans should attempt to "subdue" the earth and have "dominion" over all the creatures of the earth as being bad advice with disastrous results. (Many of those critiques, however, have lacked a full sense of the Hebrew words.) Bill Devall, coauthor of *Deep Ecology*, spoke for many activists when he declared in August 1984, "Unless major changes occur in churches, ecologists and all those working in ecology movements will feel very uncomfortable sitting in the pews of most American churches."

The disparity between Judeo-Christian religion and ecological wisdom is illustrated by the experience of a friend of mine who once lived in a seminary overlooking Lake Erie and says he spent two years contemplating the sufferings of Christ without ever noticing that Lake Erie was dying.[1] Even when Catholic clergy speaks today of St. Francis of Assisi, whom Lynn White nominated as the patron saint of ecologists, they often take pains to insist that he was not some "nature mystic,"[2] which, of course, would taint him with "paganism."[3] Religion that sets itself in opposition to nature and vehemently resists the resacralizing of the natural world on the grounds that it would be "pagan" to do so is not sustainable over time.

The cultural historian Thomas Berry has declared that we are entering a new

From: *The Spiritual Dimension of Green Politics* (Santa Fe: Bear and Co., 1986), pp. 52–69.

era of human history, the Ecological Age.[4] How could our religion reflect ecological wisdom and aid the desperately needed transformation of culture? First,
I suggest that Judaism and Christianity should stop being ashamed of their
"pagan" inheritance, *which is substantial*, and should proudly proclaim their
many inherent ties to nature. How many of us realize that the church sets Easter
on the first Sunday after the first *full moon* after the *vernal equinox* and that most
of the Jewish holy days are determined by a lunar calendar?[5] Numerous symbols,
rituals, and names in Jewish and Christian holy days have roots directly in the
nature-revering Old Religion. The list is a long one and should be cause for
self-congratulation and celebration among Christians and Jews.

Second, I hope the stewardship movement, which is gaining momentum in
Christian and Jewish circles, will continue to deepen its analyses and its field of
action. Those people are performing a valuable service by reinterpreting the
overall biblical teachings about the natural world and finding ecological wisdom
that balances or outweighs the "dominance" message. Virtually all spokespersons for the stewardship movement emphasize that nature is to be honored as
God's creation.[6] In fact, that position is firmly rooted in the work of several noted
theologians whose orientation is known as "creation spirituality." They emphasize the interrelatedness of all creation, the understanding that humans do not
occupy the central position in the cosmic creation but have a responsible role to
play, and the transformation of society in directions that will further the continuation of life. Hence peace is a central issue for creation theologians, as is justice. Nearly all of them give greater importance to the female dimension of
creation than do other theologians. Among the Catholic, Protestant, and Jewish
theologians of creation spirituality are Bernhard W. Anderson, Thomas Berry,
Walter Bruggemann, Martin Buber, Marie-Dominique Chenu, Matthew Fox,
Abraham Heschel, Jurgen Moltmann, Paul Santmire, Edward Schillebeeckx,
Odil Hannes Steck, Pierre Teilhard de Chardin, and Samuel Terrien.[7]

The experience of knowing the divine through communication with nature
has been a recurrent theme in art. Alice Walker described a theologically sophisticated, elementally spiritual experience in her Pulitzer Prize–winning novel, *The
Color Purple*, when one black woman in rural Georgia explains to another that
God "ain't a he or a she, but a It":

> It ain't a picture show. It ain't something you can look at apart from anything
> else, including yourself. I believe God is everything, say Shug. Everything that
> is or ever was or ever will be. And when you can feel that, and be happy to feel
> that, you've found it. . . . My first step away from the old white man was trees.
> Then air. Then birds. Then other people. But one day when I was sitting quiet
> and feeling like a motherless child, which I was, it come to me: that feeling of
> being part of everything, not separate at all. I knew that if I cut a tree, my arm
> would bleed. And I laughed and I cried and I run all around the house. I knew
> just what it was. In fact, when it happen, you can't miss it. It sort of like you
> know what, she say, grinning and rubbing high up on my thigh.[8]

I am encouraged that a religion-based respect for nature is showing up in numerous articles and books, especially books like *The Spirit of the Earth* (1984), in which John Hart urges study of and respect for Native American religious perspectives on nature because that is the indigenous tradition of our land and suggests compatibility between their religion and the Judeo-Christian tradition. Yet why is it that attention to loving and caring for nature rarely makes it into the liturgy today? . . . Harold Gilliam in the *San Francisco Chronicle* describ[ed] a magnificent ecological service that spanned twenty-four hours, beginning at sunrise on the autumnal equinox, and took place in the gothic cathedral on Nob Hill in San Francisco, Grace Cathedral. At the sound of a bell and a conch shell, the Episcopal bishop of California opened the service:

> We are gathered here at sunrise to express our love and concern for the living waters of the Central Valley of California and for the burrowing owls, white-tailed kites, great blue herons, migratory waterfowl, willow trees, cord grass, water lilies, beaver, possum, striped bass, anchovies, and women, children, and men of the Great Family who derive their life and spiritual sustenance from these waters. Today we offer our concerns and prayers for the ascending health and spirit of these phenomena of life and their interwoven habitats and rights.

Poets, spiritual teachers, musicians, and ecologists all participated in the service, which included whale and wolf calls emanating from various corners of the cathedral's sound system, as well as the projection of nature photography onto the walls and pillars. Gary Snyder and his family read his "Prayer for the Great Family," which is based on a Mohawk prayer.[9] The celebrants poured water from all the rivers of California into the baptismal font. They committed themselves to changing our society and our environment into "a truly Great Family," and they assigned to each US senator a totemic animal or plant from his or her region in order to accentuate the rights of our nonhuman family members. I read the account with awe and then noticed with sadness that it was dated October 17, 1971. (No subsequent ecological services took place in that church because a few influential members of the congregation pronounced it paganism.) How many species have been lost since then, how many tons of topsoil washed away, how many aquifers polluted—while we have failed to include nature in our religion?

Knowledge of nature must precede respect and love for it. We could urge that ecological wisdom regarding God's creation be incorporated in Sunday school as well as in sermons and prayer. We could suggest practices such as the planting of trees on certain holy days. We could mention in the church bulletin ecological issues that are crucial to our community.[10] There is no end to what we *could* do to focus spiritually based awareness and action on saving the great web of life.

HOW SHALL WE RELATE TO OTHER PEOPLE?

This last basic question has two parts: distinction by gender and then by other groups. Our lives are shaped to a great extent not by the differences between the sexes, but by the cultural response to those differences. There is no need to belabor the point that in patriarchal cultures the male is considered the norm and the female is considered "the Other." For our purposes here, however, it is relevant to note that Judeo-Christian religion has played a central role in constructing the subordinate role for women in Western culture. Suffice it to say that the eminent mythologist Joseph Campbell once remarked that in all his decades of studying religious texts worldwide he had never encountered a more relentlessly misogynist book than the Old Testament. Numerous Christian saints and theologians have continued the tradition.

The results for traditional society of denying women education and opportunity have been an inestimable loss of talent, intelligence, and creativity. For women it has meant both structural and direct violence. Of the former, Virginia Woolf observed that women under patriarchy are uncomfortable with themselves because they know society holds them in low esteem. The structural violence of forced dependency sometimes provides the conditions for physical violence, that is, battering. Finally, patriarchal culture usurps control over a woman's body from the woman herself, often inflicting torturous pain. . . . Some people accuse the Greens of being hypocritical in calling themselves a "party of life" and adopting a "pro-choice" stance on abortion. . . . The quality of the debate in [European] green parties over abortion has more integrity than that currently being waged in American politics precisely because all aspects of the issue are considered. In our country half of the debate often seems to be missing: women's suffering. The issue is obviously complex, and there are people of good conscience on both sides. I offer my views merely as personal ones, not official positions of American Greens.

Church leaders of many varieties are demanding an end to all legal (that is, medically safe) abortion. I suspect they can maintain a position demanding the criminalization of abortion only because they have never witnessed a woman going through pregnancy, labor, and delivery—or else they believe the biblical injunction that woman is *supposed* to suffer. Sometimes birth is textbook simple, but usually it is not. Some men say they remember their wife's screams for months. Many men say the birth experience made them "pro-choice" on the abortion issue because they would never want to force any woman to go through such an ordeal against her will.

. . . Where is the compassion for the lonely teenager from an unnurturing family situation who tried to find affection and love where she could? Where is the compassion for the innocent victims of rape, including incestual rape and the increasing frequency of the "date rape" and "acquaintance rape"? Where is the compassion for *any* woman who discovers that she is "in trouble"? The number of abortions needed in this country would plummet if the problematic conditions were addressed effectively: disintegrated families, widespread pornography depicting violence against women, culturally approved hyper-macho behavior on dates, what has been called "patriarchy's dirty little secret" (the shocking statistics on sexual abuse by male relatives), selfishness and lack of spiritual grounding on the part of men who emotionally coerce their girlfriends, and lack of self-confidence and spiritual grounding in their own being among young women.

There comes a time at the end of many lives when life is not viable without machinery, and most people say they would like the machinery turned off if it came to that. Similarly, there is a time at the beginning of life when a fertilized egg and then a fetus is not viable life *unless* the woman is willing to give over her body and accept the suffering. To force a woman either to give birth or to abort is violence against the person. Most men and women know this in their hearts. They also know that countless women do not have the financial and other resources for the twenty-year task of raising a child. . . . Male fears of women controlling their own sexuality are deeply rooted in patriarchal culture; during the Renaissance, for example, peasant healers were burned as "witches" for providing women with contraception and abortion. So the debate we are embroiled in is a very old one, and the campaign to "save the embryo; damn the woman" has been mounted many times before.

Men, too, suffer under patriarchal culture. Because woman is regarded as the denigrated Other, men are pressured to react and continually prove themselves very *un*like the female. This dynamic results in what some men have called "the male machine." It has also skewed much of our behavioral and cognitive science since thousands of careers and volumes of commentary on "sex differences" have been funded but no recognized field of "sex similarities" exists.[11] That would be too unnerving. The most serious effect of men under patriarchy needing to prove themselves *very different* from women is the function of military combat as an initiation into true manhood and full citizenship. This deeply rooted belief surfaced as an unexpected element in the struggle to pass the Equal Rights Amendment, for instance. Feminist lobbyists in state legislatures throughout the 1970s were repeatedly informed, "When you ladies are ready to fight in a *war*, we'll be ready to discuss equal rights!"[12] Such an orientation is not sustainable in the nuclear age.

What role could religion play in removing the cultural insistence on women as Other and men as godlike and hence inherently superior? How could religion further the green principle of postpatriarchal consciousness? We know the answers because they are already being tried: women must have equal participation in ritual (as ministers, rabbis, and priests), language in sermons and translations must be inclusive, and the godhead must be considered female as well as male. These solutions are not new, but neither are they very effective, because so many people do not take either the need or the means seriously. Instead, they resent these efforts and feel silly and somewhat embarrassed with the notion of a female God. Being forced to say "God the Mother" once in a while is pointless if people have in mind Yahweh-with-a-skirt. We must first understand who She is: She is not in the sky; She is earth. Here is Her manifestation in the oldest creation story in Western culture:

The Myth of Gaia

Free of birth or destruction, of time or space, of form or condition, is the Void. From the eternal Void, Gaia danced forth and rolled Herself into a spinning ball. She molded mountains along Her spine, valleys in the hollows of Her flesh. A rhythm of hills and stretching plains followed Her contours. From Her warm moisture She bore a flow of gentle rain that fed Her surface and brought life. Wriggling creatures spawned in tidal pools, while tiny green shoots pushed upward through Her pores. She filled oceans and ponds and set rivers flowing through deep furrows. Gaia watched Her plants and animals grow. In time She brought forth from Her womb six women and six men. . . .

Unceasingly the Earth-Mother manifested gifts on Her surface and accepted the dead into her body. In return She was revered by all mortals. Offerings to Gaia of honey and barley cake were left in a small hole in the earth before plants were gathered. Many of Her temples were built near deep chasms where yearly the mortals offered sweet cakes into her womb. From within the darkness of Her secrets, Gaia received their gifts.[13]

Having addressed the self, nature, and gender, we now come to the last half of the last basic question, *How shall we relate to groups and other individuals?* There are, of course, a multiplicity of groups in society at the levels of family, community, region, state, nation, and planet. The following are merely some general considerations.

We must first analyze how our own mode of living affects others in the great family: Does the nature of our existence impose suffering on others—or does it support and assist those who are less privileged than we? Here we can enjoy the convergence of spiritual growth and political responsibility in the spiritual practice of cultivating moment-to-moment awareness, being fully "awake" and focused on our actions—a simple-sounding yet demanding task. There is a story in Zen of a student who studied very hard to master certain religious texts and then went before his spiritual teacher to be questioned. The *roshi* asked simply, "On which side of the umbrella did you place your shoes?" The student was defeated; he had lost awareness (or "spaced out," as we might say).

We can begin our day by focusing mindfulness on our every act. Turning on the water in the bathroom. Where does it come from? Is our town recklessly pumping water from the receding water table instead of calling for conservation measures? Where does our wastewater go when it leaves the sink? What happens after it is treated? Later we are in the kitchen, making breakfast. Where does our coffee come from? A worker-owned cooperative in the third world or an exploitative multinational corporation? Obviously, it is exhausting to continue this practice very long unless one is adept. (It *is* difficult—so much so that a friend of mine has added an amendment to a popular spiritual saying: "Be here now—or now and then.") But everyone can practice *some* mindfulness.

If we analyze our own situation, we may discover that we are benefiting from the suffering of others—and that we ourselves are uncomfortable with the structural systems in which we work. When one thinks of religious people working for economic or social change, the "liberation theology" movement probably comes to mind because of its size in Latin America. . . . In that movement, grassroots Catholic groups (base communities) meet frequently to discuss the teachings in the Gospels and applications of Marxist analysis.

But there is another way: a religion-based movement for social change is beginning to flourish that is completely in keeping with green principles of private ownership and cooperative economics, decentralization, grassroots democracy, nonviolence, social responsibility, global awareness, and the spiritual truth of oneness. This type of call for economic and social change is gaining momentum in Catholic, Protestant, and Jewish communities. We see it, for example, in the statement issued by the Catholic Bishops of Appalachia, *This Land Is My Home: A Pastoral Letter on Powerlessness in Appalachia*, which calls for worker-owned businesses and community-based economics. We see it in *Strangers and Guests: Toward Community in the Heartland* by the Catholic bishops of the Heartland (Midwest) and in *The Land: God's Giving, Our Caring* by the American Lutheran Church, a statement which was then echoed by the Presbyterian Church. Both of these statements address ecological use of the land, and *Strangers and Guests* calls for small-is-beautiful *land reform* as the only sustainable course for rural America. Developing the applications of such principles as "the land should be distributed equitably" and "the land's workers should be able to become the land's owners," the Heartland bishops discuss elimination of capital-gains tax laws which favor "wealthy investors and speculators" and disfavor "small and low-income farm families," taxation of agricultural land "according to its productive value rather than its speculative value," "taxing land progressively at a higher rate according to increases in size and quality of holdings" (a proposal in the Jeffersonian tradition), and low-interest loans to aspiring farmers as well as tax incentives for farmers with large holdings to sell land to them.

We see green-oriented economic and social change now promoted in the Jewish periodical *Menorah* and by the Protestant multidenominational association, Joint Strategy and Action Committee. The lead article in a 1984 issue of the JSAC newsletter began: "If you want to know what ecojustice is, read the Psalms. The dual theme of justice in the social order and integrity in the natural order is pervasive and prominent. The Book is, in large part, a celebration of interrelationships, the interaction, the mutuality, the organic oneness and wholeness of it all that is, that is to say, the Creator and the creation, human and nonhuman."

The green-oriented Jewish and Protestant leaders seek to locate justice and ecological wisdom in the Old Testament. Green-oriented Catholics usually turn to the papal encyclicals, especially Pope Pius XI's 1931 encyclical *Quadragesimo anno* (Forty Years After),[14] which established three cardinal principles: *personalism* (the goal of society is to develop and enrich the individual human person), *subsidiarity* (no organization should be bigger than necessary and nothing should be done by a large and higher social unit than can be done effectively by a lower and smaller unit), and *pluralism* (that a healthy society is characterized by a wide variety of intermediate groups freely flourishing between the individual and the state).[15] . . . Andrew Greeley argues in *No Bigger Than Necessary* that Catholic social theory is firmly rooted in the communitarian, decentralist tradition and that Catholics who drifted into Marxism in recent decades are simply unaware that their own tradition contains a better solution. Joe Holland, a Catholic activist with the Center of Concern in Washington, DC, argues, however, that Left-oriented Catholics have never embraced "scientific Marxism" and the model of a machinelike centralized government and economy. They are attracted, rather, by communitarian ideals and are uncomfortable with the modernity of many socialist assumptions.[16] Hence, we may assume, and I believe Joe Holland would agree, that many of these lukewarm leftists in Catholic circles would readily become green.

The possibilities for locating and working with green-oriented activists in mainline religions have never been better. . . . Within our own green political organizations, however, the question remains of how much religious content is proper in pluralistic meetings and publications. I myself am uncertain about how much overt spirituality the "market will bear" in green conferences and statements, and I am often dissatisfied afterward because I and other Greens have held back too much on spirituality so as not to exclude anyone in the group. . . .

Surely, no Green, whatever his or her spiritual orientation, could object to our structuring our groups according to the Deep Ecology principles of diversity, interdependence, openness, and adaptability—as well as the spiritual principles of cultivating wisdom and compassion. These can be our guidelines as we evolve the ever-changing forms of green politics.

NOTES

1. Paul Ryan, "Relationships," *Talking Wood* 1, no. 4 (1980).

2. One example, although by no means the only one, is Murray Bodo, OFM, *The Way of St. Francis* (Garden City, NY: Doubleday, 1984).

3. "Pagan" is from the Latin word for "country people," *pagani*. It has nothing to do with Satan worship.

4. For collections of Thomas Berry's papers, see Brian Swimme, *The Universe Is a Green Dragon* (Santa Fe: Bear and Co., 1984).

5. Arthur Waskow, *Seasons of Our Joy: A Celebration of Modern Jewish Renewal* (New York: Bantam Books, 1982).

6. See, for example, Mary Evelyn Jegen and Bruno V. Manno, eds., *The Earth Is the Lord's: Essays on Stewardship* (New York: Paulist Press, 1978); Wesley Granberg-Michaelson, *A Worldly Spirituality: The Call to Take Care of the Earth* (New York: Harper and Row, 1984); John Hart, *The Spirit of the Earth: A Theology of the Land* (New York: Paulist Press, 1984); John Carmody, *Ecology and Religion: Toward a New Christian Theology of Nature* (New York: Paulist Press, 1983); and Ian G. Barbour, ed., *Earth Might Be Fair: Reflections on Ethics, Religion, and Ecology* (Englewood Cliffs, NJ: Prentice-Hall, 1972).

7. In addition to the scores of books by the creation theologians cited in the text, there is a relevant anthology, Philip N. Joranson and Ken Butigan, eds., *Cry of the Environment: Rebuilding the Christian Creation Tradition* (Santa Fe: Bear and Co., 1984). A partial "family tree of creation-centered spirituality" may be found in Matthew Fox, *Original Blessing* (Santa Fe: Bear and Co., 1983).

8. Alice Walker, *The Color Purple* (New York: Harcourt Brace Jovanovich, 1982), p. 167.

9. "Prayer for the Great Family" may be found in Gary Snyder's Pulitzer Prize–winning volume of poetry, *Turtle Island* (New York: New Directions Books, 1974).

10. See Byron Kennard, "Mixing Religion and Politics," Ecopinion, *Audubon* 86, no. 2 (March 1984): 14–19.

11. Ruth Bleier, *Science and Gender* (New York: Pergamon Press, 1984).

12. See Charlene Spretnak, "Naming the Cultural Forces That Push Us toward War," *Journal of Humanistic Psychology* 23, no. 1 (Summer, 1983): 104–14; also in *Nuclear Strategy and the Code of the Warrior: Face of Mars and Shiva in the Crisis of Human Survival*, ed. Richard Grossinger and Lindy Hough (Berkeley, CA: Atlantic Books, 1984). Also see the chapters on "The Soldier" and "War" in Mark Gerzon, *A Choice of Heroes* (Boston: Houghton Mifflin, 1982).

13. Charlene Spretnak, *Lost Goddesses of Early Greece: A Collection of Pre-Hellenic Myths* (Boston: Beacon Press, 1981). For some useful compromise positions on the Great Mother, see also Virginia Ramey Mollenkott, *The Divine Female: The Biblical Imagery of God as Female* (New York: Crossroad, 1984).

14. *Quadragesimo anno* was a commemoration and expansion of Pope Leo XIII's 1891 encyclical *Rerum Novarum* (Of the New Situation of the Working Class).

15. Andrew M. Greeley, *No Bigger Than Necessary* (New York: New American Library, 1977), p. 10.

16. Joe Holland, *The Postmodern Paradigm Implicit in the Church's Shift to the Left* (Washington, DC: Center of Concern, 1984).

29

ECOLOGY AND PROCESS THEOLOGY

JOHN COBB JR.

People have become aware of the dangers to the human future resulting from exploitation of the environment. This exploitation has been consistent with both the dominant economic theories and the dominant theologies of the nineteenth and twentieth centuries. Ideally these theories called for treatment of all human beings as ends rather than as means, but the power of the dominant theories has been such that their objectifying categories are readily extended to human beings. People, too, become resources, and the term human resources has become prevalent. In practice powerless human beings and powerless societies have been treated as resources for exploitation by those who have the economic and political power to establish goals and to pursue them. In response to this situation, the task cannot be simply to improve practice in light of existing theory. It must be to change the theory. And because our theory, both in economics and theology, has both shaped and expressed our dominant perceptions and sensibility, it is necessary to change our vision of reality as well.

Process thought has been protesting for some time against some of the features of our dominant practice, theory, and sensibility which are now more widely recognized as damaging. In particular, process thought has offered an alternative to the dominant dualisms of soul and body, spirit and nature, mind and matter, self and other. It should be able to speak with some relevance to the contemporary situation. The theology which has appropriated these contributions of process thought is often called "process theology."

The dominant thinker behind the distinctive approach of process theology is

From: "Process Theology and an Ecological Model," *Pacific Theological Review* 15, no. 2 (Winter 1982): 24–27, 28.

the English mathematician-philosopher Alfred North Whitehead (1861–1947). He taught at Harvard University during his later years and has had his largest following in the United States. In his most important book, his Gifford Lectures of 1927–1928, entitled *Process and Reality*, he proposed a "philosophy of organism." By this he meant that the actual entities of which the world is ultimately made up are better thought of as organisms than as material or mental substances.

The main significance of the idea of organism here is that each entity exists only in its relation to its environment. One cannot think first of an entity and then, incidentally, of its relations to the rest of the world. On the contrary, every entity is relational in its most fundamental nature. It is constituted by its relations. Even in thought it cannot be abstracted from them.

Professor L. Charles Birch, an Australian biologist active in the World Council of Churches, and I have authored a book in which we develop this point. We call ours an ecological model of living things, and we show that this model fits the evidence of biological sciences better than the substantialist, the materialist, and the mechanist models that have dominated their development. The ecological model is also more appropriate to field, relativity, and quantum theory in physics. There are a number of converging trends in the natural sciences that support a Whiteheadian, ecological understanding of nature.

If process thought accepted a dualism between humanity and nature, a changed view of nature would not seem very important theologically. But those who follow Whitehead understand human beings as part of nature. The ecological model applies to us as well. Human experience, too, is constituted by its relations with the body, with other people, and with nonhuman creatures.

Three further clarifications of the meaning of these assertions for theology, as well as for the natural sciences, are needed. First, the ecological model depicts the actual entities as events rather than as objects that exist through time. An event occurs in its fullness and is over. Whereas our previous models have usually explained events by the motions of atoms and particles, the ecological model explains atoms and particles as well as tables and mountains, in terms of events. The explanation is finally sought at the level of the ultimate, indivisible unit events into which larger events can be analyzed. These include subatomic events but also momentary human experiences. Whitehead called these unit events "actual occasions" or "occasions of experience." The world is a vast field of actual occasions within which there emerge relatively enduring patterns of many varieties. The atom, the subatomic particle, the table, and the mountain are all societies of actual occasions with relatively enduring patterns. A human person is too.

Second, our usual ideas of such dualities as mental and physical or subjective and objective are derived from reflection about societies of actual occasions, such as atoms and mountains, wrongly supposed to be substances. These ideas cannot be applied to events or occasions. Mind and matter, as conceived according to these usual ideas, do not exist. But understood more loosely, "mental" and "physical" aspects can be found in every occasion whatsoever.

Every occasion is a synthesis of features derived from its environing world. In this sense it is physical. Because this synthesis is not merely the mechanical product of the forces that impinge upon it, because it includes an element of selectivity and self-determination, we can also say that the occasion has a mental aspect. There are no purely mental occasions and no purely physical ones. There are only occasions in which the element of self-origination is relatively important and others in which it is negligible.

Similarly, when we think of subjects and objects in the usual way, we cannot say that an event or occasion is a subject or that it is an object. As it takes place it is the locus of subjectivity, that is, it receives the activity of the past, and it, in its turn, acts. It is subject to the action of other events, and it is an agent determining its response to these influences and, thereby, its effects upon the world. It actively takes account of its world. When it has taken place, it becomes part of the objective world in which other occasions occur. All occasions are "subjects" in the moment of their occurrence and "objects" for other occasions as soon as they are over.

Third, every occasion in some measure transcends its world. It transcends it, first, simply by being a new occasion. No occasion can recur. However similar an occasion may be to antecedent ones, it is a different entity. In addition, no occasion is qualitatively identical with any other. It cannot be, because no two occasions have the same spatio-temporal locus, and every locus defines a different environment or world. Since this world enters constitutively into the occasion, the occasion must be qualitatively unique. Further, as already noted, no occasion is the mere product of its world. However much the world of the occasion determines its character, the exact form of the occasion is decided only when it happens. Every occasion transcends its world by determining just how it constitutes itself out of that world. And, finally, many occasions, indeed, all living occasions, constitute themselves not only out of the world of past occasions but also out of the world of possibility or the Not Yet. They transcend their world by incorporating possibilities derived, not from it, but from God. This need not involve conscious choice, but it can, and in the human case it often does. This is the context in which we can speak meaningfully of freedom.

Those who are accustomed to contrast material or merely phenomenal nature with human existence as spiritual may be shocked by the insistence of process thought that human beings are part of an inclusive ecosystem. But it is well to recall that the dualism of the human and the natural is a modern one. For the Bible there is the one world of creatures of which human beings are a part. The welfare of humanity and the welfare of other creatures are interdependent. Process theologians rejoice that Christians are beginning to recover the sense of the unity of creation against the modern philosophical dualism of spirit and nature.

Neither for the Bible nor for process theologians does the insistence that human beings are part of the world of creatures mean that we are not distinctive. Human experiences are the most valuable of all the events on this planet, but they are not the only loci of value. In Genesis God declares all other creatures good even before and apart from the creation of human beings, and process theologians share this biblical conviction. But human beings are created in the image of God, and process theologians agree that we share in God's power of creation in quite unique ways. As Genesis indicates, we exercise a unique power over all the other creatures. Hence we are responsible for the welfare of the whole created order on this planet as no other creature could be. That we are exercising that power so extensively for the degradation of the biosphere as well as the oppression of our powerless sisters and brothers expresses the depth of our betrayal of God's trust.

The doctrine of God to which this brings us is at the center of process theology. It is above all Charles Hartshorne who has developed this aspect of process thought. Process theology has been very critical of some features of classical theism. That theism often pictures God as an immutable substance, whereas process theologians see God as the most perfect exemplification of the ecological model. Divine perfection does not consist in being totally self-contained and unaffected by creaturely joy and suffering but in being totally open, totally receptive, and perfectly responsive. God, too, is constituted by relations to all things. In the case of God, these relations are complete and express the perfection of love. Through this perfection of relations, what is perpetually perishing in the world attains everlastingness in the divine life. The threat to meaning that is inherent in transience is thereby overcome.

Too often, also, classical theism has spoken of God's power in a way that suggests the control of a tyrant or a dictator. It is thought that everything that happens in the world, even an earthquake or a war, must express the will of God. Many have turned against this God, and process theologians do so, too. But process theologians do not go to the extreme of asserting that God is powerless and is to be found only in suffering. God is certainly found there, but God is found also in a child's enjoyment of play, in the joy of a happy marriage, and in the creative ecstasy of an artist. Process theology teaches that God's power is perfect and that perfect power is not coercive. Following the biblical image of parental power, we see that coercion expresses the failure of power, not its perfection. God's power lies in the gift to every occasion of some measure of transcendence or freedom and in the call to employ that freedom in love. It is this gift and call which have brought into being all that is good in our world; indeed, they have created our world. But creatures, and especially human creatures, resist the gift of freedom and abuse it. God does not will the evil which we thereby bring about.

The language of process theology about God often leads readers to suppose that process theologians are optimistic, and it is true that they are hopeful. They hope because the giver of freedom and direction is always surprising us. The future cannot be foreseen. It is not just the unfolding of the past. It is the place where the new can be received if we will accept it. However bleak the projections of past trends into the future, we hope for a new heaven and a new earth.

Process theology developed chiefly at the Divinity School of the University of Chicago. During the first three decades of [the twentieth] century that school was committed to sociohistorical research in the service of the social gospel. During the 1930s it became more philosophical, and the influence of Alfred North Whitehead grew. The concerns for political freedom, peace, and justice did not disappear, but they ceased to control the theological program. The dominant emphasis among the theologians was an empirical, rational, and speculative approach to understanding God and God's relationship to human beings. The label "process theology" came to be used during the sixties to emphasize the rejection of static and substantial modes of thought.

When the 1960s brought renewed concern for human liberation as the focus of Christian theology, most process theologians recognized that their work had become too theoretical and abstract. During the [1970s] they made efforts to display the relevance of their cosmological vision to the practical needs of the church. Thus far success has been greatest in relation to interfaith dialogue, ecology, feminism, and pastoral work. Much remains to be done, especially in relation to the world of politics, although a beginning has been made there as well.

30

RECOVERING THE SACRED

WINONA LaDUKE

How does a community heal itself from the ravages of the past? . . . I found an answer in the multifaceted process of recovering that which is "sacred." This complex and intergenerational process is essential to our vitality as Indigenous peoples and ultimately as individuals. . . .

What qualifies something as sacred? That is a question asked in courtrooms and city council meetings across the country. Under consideration is the preservation or destruction of places like the Valley of the Chiefs in what is now eastern Montana and Medicine Lake in northern California, as well as the fate of skeletons and other artifacts mummified by collectors and held in museums against the will of their rightful inheritors. Debates on how the past is understood and what the future might bring have bearing on genetic research, reclamation of mining sites, reparations for broken treaties, and reconciliation between descendants of murderers and their victims. At stake is nothing less than the ecological integrity of the land base and the physical and social health of Native Americans throughout the continent. In the end there is no absence of irony: the integrity of what is sacred to Native Americans will be determined by the government that has been responsible for doing everything in its power to destroy Native American cultures.

Xenophobia and a deep fear of Native spiritual practices came to the Americas with the first Europeans. Papal law was the foundation of colonialism; the Church served as handmaiden to military, economic, and spiritual genocide and domination. Centuries of papal bulls posited the supremacy of Christendom over all other beliefs, sanctified manifest destiny, and authorized even the most brutal practices of colonialism. Some of the most virulent and disgraceful manifesta-

From: *Recovering the Sacred: The Power of Naming and Claiming* (Cambridge, MA: South End Press), pp. 11–15, 241–43, 251–53.

tions of Christian dominance found expression in the conquest and colonization of the Americas.

Religious dominance became the centerpiece of early reservation policy as Native religious expression was outlawed in this country. To practice a traditional form of worship was to risk a death sentence for many peoples. The Wounded Knee Massacre of 1890 occurred in large part because of the fear of the Ghost Dance Religion, which had spread throughout the American West. Hundreds of Native spiritual leaders were sent to the Hiawatha Asylum for Insane Indians for their spiritual beliefs.[1]

The history of religious colonialism, including the genocide perpetrated by the Catholic Church (particularly in Latin America), is a wound from which Native communities have not yet healed. The notion that non-Christian spiritual practices could have validity was entirely ignored or actively suppressed for centuries. So it was by necessity that Native spiritual practitioners went deep into the woods or into the heartland of their territory to keep up their traditions, always knowing that their job was to keep alive their teachers' instructions, and, hence, their way of life.

Native spiritual practices and Judeo-Christian traditions are based on very different paradigms. Native American rituals are frequently based on the reaffirmation of the relationship of humans to the Creation. Many of our oral traditions tell of the place of the "little brother" (the humans) in the larger Creation. Our gratitude for our part in Creation and for the gifts given to us by the Creator is continuously reinforced in Midewiwin lodges, Sundance ceremonies, world renewal ceremonies, and many others. Understanding the complexity of these belief systems is central to understanding the societies built on those spiritual foundations—the relationship of peoples to their sacred lands, to relatives with fins or hooves, to the plant and animal foods that anchor a way of life.[2]

Chris Peters, a Pohik-la from northern California and president of the Seventh Generation Fund, broadly defines Native spiritual practices as affirmation-based and characterizes Judeo-Christian faiths as commemorative.[3] Judeo-Christian teachings and events frequently commemorate a set of historical events: Easter, Christmas, Passover, and Hannukah are examples. Vine Deloria Jr., echoes this distinction:

> Unlike the Mass or the Passover which both commemorate past historical religious events and which believers understand as also occurring in a timeless setting beyond the reach of the corruption of temporal processes, Native American religious practitioners are seeking to introduce a sense of order into the chaotic physical present as a prelude to experiencing the universal moment of complete fulfillment.[4]

The difference in the paradigms of these spiritual practices has, over time, become a source of great conflict in the Americas. Some two hundred years after the US Constitution guaranteed freedom of religion for most Americans, Congress passed the American Indian Religious Freedom Act in 1978 and President Carter signed it into law. Although the act contains worthy language that seems to reflect the founders' concepts of religious liberty, it has but a few teeth. The act states:

> It shall be the policy of the United States to protect and preserve for American Indians their inherent right of freedom to believe, express, and exercise the traditional religions of the American Indian, Eskimo, Aleut and native Hawaiians, including but not limited to access to sites, use and possession of sacred objects, and the freedom to worship through ceremonial and traditional rites.[5]

While the law ensured that Native people could hold many of their ceremonies, it did not protect the places where many of these rituals take place or the relatives and elements central to these ceremonies, such as salt from the sacred Salt Mother for the Zuni or salmon for the Nez Perce. The Religious Freedom Act was amplified by President Clinton's 1996 Executive Order 13007, for preservation of sacred sites: "In managing Federal lands, each executive branch agency with statutory or administrative responsibility for the management of Federal lands shall . . . avoid adversely affecting the physical integrity of such sacred sites."[6]

Those protections were applied to lands held by the federal government, not by private interests, although many sacred sites advocates have urged compliance by other landholders to the spirit and intent of the law. The Bush administration, however, has by and large ignored that executive order. Today, increasing numbers of sacred sites and all that embodies the sacred are threatened.

While Judeo-Christian sacred sites such as "the Holy Land" are recognized, the existence of other holy lands has been denied. There is a place on the shore of Lake Superior, or Gichi Gummi, where the Giant laid down to sleep. There is a place in Zuni's alpine prairie to which the Salt Woman moved and hoped to rest. There is a place in the heart of Lakota territory where the people go to vision quest and remember the children who ascended from there to the sky to become the Pleiades. There is a place known as the Falls of a Woman's Hair that is the epicenter of a salmon culture. And there is a mountain upon which the Anishinaabeg rested during their migration and from where they looked back to find their prophesized destination. The concept of "holy land" cannot be exclusive in a multicultural and multispiritual society, yet indeed it has been treated as such.

We have a problem of two separate spiritual paradigms and one dominant culture—make that a dominant culture with an immense appetite for natural resources. The animals, the trees, and other plants, even the minerals under the ground and the water from the lakes and streams, all have been expropriated from Native American territories. Land taken from Native peoples either by force or the colonists' law was the basis for an industrial infrastructure and now a standard of living that consumes a third of the world's resources.

By the 1930s, Native territories had been reduced to about 4 percent of our original land base. More than 75 percent of our sacred sites have been removed from our care and jurisdiction.[7] Native people must now request permission to use their own sacred sites and, more often than not, find that those sites are in danger of being desecrated or obliterated.

The challenge of attempting to maintain your spiritual practice in a new millennium is complicated by the destruction of that which you need for your ceremonial practice. The annihilation of 50 million buffalo in the Great Plains region by the beginning of the twentieth century caused immense hardship for traditional spiritual practices of the region, especially since the *Pte Oyate*, the Buffalo nation, is considered the older brother of the Lakota nation and of many other Indigenous cultures of the region. Similarly, the decimation of the salmon in Northwest rivers like the Columbia and the Klamath, caused by dam projects, overfishing, and water diversion, has resulted in great emotional, social, and spiritual devastation to the Yakama, Wasco, Umatilla, Nez Perce, and other peoples of the region. New efforts to domesticate, patent, and genetically modify wild rice similarly concern the Anishinaabeg people of the Great Lakes.

It is more than five hundred years since the European invasion of North America and more than two hundred years since the formation of the United States. Despite these centuries of spiritual challenges, Native people continue, as we have for centuries, to always express our thankfulness to Creation—in our prayers, our songs, and our understanding of the sacredness of the land.

Dr. Henrietta Mann is a Northern Cheyenne woman and chair of the Native American Studies Department at Montana State University. She reiterates the significance of the natural world to Native spiritual teaching:

> Over the time we have been here, we have built cultural ways on and about this land. We have our own respected versions of how we came to be. These origin stories—that we emerged or fell from the sky or were brought forth—connect us to this land and establish our realities, our belief systems. We have spiritual responsibilities to renew the Earth and we do this through our ceremonies so that our Mother, the Earth, can continue to support us. Mutuality and respect are part of our tradition—give and take. Somewhere along the way, I hope people will learn that you can't just take, that you have to give back to the land.[8] . . .

TATÉ: THE WIND IS WAKAN

. . . The Lakota are looking to harness *Taté* [the winds] and play their part in moving us from combusting the finite leftovers of the Paleozoic into an era of renewable energy. It is all part of a strategy by Native people across the Great Plains to power their communities in an environmentally sound manner and to, over time, build an export economy based on green power. . . .

"We believe the wind is *wakan*, a holy or great power," explains Pat Spears, from his home on the Lower Brule Reservation in South Dakota. Pat is a big guy with a broad smile, the president of Intertribal COUP [Council on Utility Policy], and a member of the Lower Brule Tribe. "Our grandmothers and grandfathers have always talked about it, and we recognize that this is sacred and this is the future."[9] Indeed, the Lakota, like other Native peoples, have made peace with the wind, recognizing its power in change, historically and perhaps today.

Alex White Plume echoes Spears's words, talking about *Taté* as the power of motion and transformation, a "messenger for the prayers of the Lakota people."[10] The power of the transformation is growing stronger these days and tribal nations want a return to, as Debbie White Plume puts it, "the power the Creator gave us," not the power doled out by electric utilities and energy corporations.[11] The wind is as constant on the Pine Ridge Reservation as anything, bringing the remembrances of ancestors, the smell of new seasons, and a constant reminder of human insignificance in the face of the immensity of Creation. . . .

"We can either give you coal or we can give you wind," Bob Gough quips. The theory of Intertribal COUP and its partners is that if we put up more wind power, those utility companies won't have to buy coal or other fossil fuels or nuclear power. That's the larger strategy. To get there is a bit more of a challenge; the power lines are literally clogged with coal, so we continue to work on decommissioning coal plants. One strategy that's been put into place in several states is the establishment of state laws that require local energy companies to purchase a set amount of energy from renewable or efficient sources.[12] . . .

The tribal wind program is also an opportunity to bring the wealthiest Native communities (those on the East and West coasts) into a partnership with some of the largest landholding, wind-rich tribes in the country. This is not only about sharing wealth, it is about restoring trade relations between Indigenous nations, and, in some ways, allowing Native people an opportunity to recover land and culturally based traditions in the context of a new set of technologies and a new millennium. Speaking with some of the largest casino tribes at the United Southern and Eastern Tribes meeting in February 2003, Bob Gough laid out the potential for tribal investment, income, and environmental protection through new partnerships. "We don't just want to be there when the blue-haired ladies put quarters into the machines. We want to be there any time a light switch goes on."[13]

The Mohegans of Connecticut have taken some of the first steps in adopting alternative energy. The tribe recently purchased a collection of hydrogen fuel cells to power its casino complex. While traditional generating systems create as much as twenty-five pounds of pollutants to generate one thousand kilowatt-hours of power, the same energy produced by fuel cells equates to less than one ounce of pollutants. The Mohegans have been among the first to show an interest in investing in tribal wind power with the cash-strapped, wind-rich tribes in the West. A number of us hope this interest develops further.

The Lakota are also taking advantage of a new sort of politics on the wind.

In March 2004, the mayors of 150 cities joined with Intertribal COUP, Honor the Earth, and other organizations in pledging to voluntarily meet the provisions of the Kyoto Accord, which the United States refused to sign. Taking things a step further, mayors of cities ranging from Denver and Duluth to Santa Monica and Dallas reiterated their commitment to make changes by issuing a Declaration of Energy Independence: "In the course of human events it has now become necessary for all people to face the threat of catastrophic global climate change and so we do hereby declare our commitment to a renewable energy future."[14] The mayors, joining with tribes, urged a commitment to renewable energy and called on Washington to support the transition.

In summer 2004, a coalition of churches joined in the effort, specifically supporting tribal wind generation as a centerpiece of energy justice and part of these churches' calling to be "stewards of God's creation."[15] Michigan Interfaith Power and Light, a collection of one hundred congregations working to both mitigate global climate change and look toward energy and environmental justice, signed an agreement with *Native*Energy to buy "green tags," a market mechanism to reward green power producers, from the Rosebud expansion project and, over time, from other Native wind turbines. As with *Native*Energy's support of the first Rosebud turbine, the green tags will be purchased on an up-front basis to help finance new projects.[16]

Intertribal COUP and groups such as the Apollo Alliance, representing a host of environmental groups and twelve labor unions, are looking toward renewable energy component manufacturing as a way to create jobs here in the United States while mitigating climate change. The Apollo Project has called for a $300 billion federal investment into renewable energy. That investment, according to the project, would add 3.3 million new jobs and stimulate an estimated $1.4 trillion in new gross domestic product.[17] The tribes' goal is to turn some of these investments into jobs in Indian country.

So it is that new wind projects are planned for the reservations at Fort Berthold, Northern Cheyenne (Montana), Makah (Washington), and White Earth (northern Minnesota). The economics makes sense to tribes like the Assiniboine and Sioux of Fort Peck that hope to bring online a 660-kilowatt turbine. The Poplar, Montana, wind turbine would produce enough energy to reduce the tribal electric bill by $134,000 annually and help finance other programs through savings.

While remote tribal communities on the Great Plains and elsewhere wrangle through the "white tape" created by the fossil fuel industry, they are clear about their commitment to ending the unsustainable exploitation of their natural resources. It is as if they were saying, "That was then; this is now. We will not be cheated or stolen from."

The fossil fuel century has been incredibly destructive to the ecological structures—the air, earth, water, and plant and animal life—that keep planet Earth habitable for humans. Whether human populations will continue to flourish a hundred years from now will depend on the choices we make today. Oliver Red

Cloud, a traditional Oglala headman, reminds us of our spiritual agreements with the Creator when he says, "The *Takoche* [grandchild] generation is coming. We've got to take care of all of this for them."[18] Native American communities are creating momentum for change and providing some critical leadership in the face of global climate change and the energy crisis to come. By democratizing power production, Native nations are providing the solutions that all of us will need in order to survive into the next millennium.

NOTES

1. Bradley and Jennifer Soule, "Death at the Hiawatha Asylum for Insane Indians," *Native Voice* (Rapid City, SD) 2, no. 3 (February 2003): B3. Indian agents, appointed as US governors over local Indian tribes, often identified Native individuals who adhered to their traditions as "malcontents," thus consigning them to the asylum. From there very few would ever return.

2. For more on this topic, see Vina Deloria Jr., *God Is Red: A Native View of Religion, 30th Anniversary Edition* (Golden, CO: Fulcrum Publishing, 2003).

3. Chris Peters, author's interview, June 8, 1992.

4. Vine Deloria Jr., "Sacredness among Native Americans," in *Native American Sacred Sites and the Department of Defense*, ed. Vine Deloria Jr. and Richard W. Stoffle (US Department of Defense, Legacy Resource Management Program, 1998), available at: https://www.denix.osd.mil/denix/Public/ES-Programs/Conservation/Legacy/Sacred/ch2.html.

5. Public Law 95-341, 42 U.S.C. 1996 and 1996a, "Protection and Preservation of Traditional Religions of Native Americans," August 11, 1978.

6. Executive Order 13007, "Protection and Accommodation of Access to 'Indian Sacred Sites,'" May 24, 1996, available at: http://www.gsa.gov/gsa/ep/contentView.do?noc=T&contentType=GSA_BASIC&contentId 16912.

7. See http://resourcescommittee.house.gov/democrats/pr2002/20020718Native AmericanSacredLandsActIntro.html.

8. Valerie Taliman, "Sacred Landscapes," *Sierra Magazine*, November/December 2002, p. 3.

9. Pat Spears, author's interview, December 10, 2002.

10. Alex White Plume, author's interview, February 18, 2003.

11. Debbie White Plume, author's interview, February 18, 2003.

12. States which have adopted a Renewable Energy Portfolio include Massachusetts, Maine, New Jersey, Colorado, Iowa, Minnesota, and eight others.

13. Bob Gough at United Southern and Eastern Tribes meeting, Washington, DC, February 4, 2003.

14. More information on the "Declaration of Energy Independence" is available from the International Council for Local Environmental Initiatives, http://www.iclei.org/us/ and from Honor the Earth, http://www.honorearth.org/intiatives/energy/thdependenceday.html

15. Michigan Interfaith Power and Light Web site, http://www.miipl.org.

16. To learn more about what green tags are and how they are retired, go to http://www.ems.org/renewables/green_tags.html.

17. The Institute for America's Future, the Center on Wisconsin Strategy, and the Perryman Group, "The Apollo Jobs Report: For Good Jobs & Energy Independence," Apollo Alliance, January 2004, p. 7, available at http://www.apolloalliance.org/docUploads/ApolloReport%2Epdf.

18. Oliver Red Cloud, author's interview, September 19, 2003.

PART VII
POSTMODERN SCIENCE

31

SYSTEMS THEORY AND THE NEW PARADIGM

FRITJOF CAPRA

CRISIS AND TRANSFORMATION IN SCIENCE AND SOCIETY

The dramatic change in concepts and ideas that happened in physics during the first three decades of [the twentieth] century has been widely discussed by physicists and philosophers for more than seventy years. It led Thomas Kuhn to the notion of a scientific paradigm, a constellation of achievements—concepts, values, techniques, and so on—shared by a scientific community and used by that community to define legitimate problems and solutions. Changes of paradigms, according to Kuhn, occur in discontinuous, revolutionary breaks called paradigm shifts.[1]

Today, [four decades] after Kuhn's analysis, we recognize paradigm shifts in physics as an integral part of a much larger cultural transformation.[2] The intellectual crisis of quantum physicists in the 1920s is mirrored today by a similar but much broader cultural crisis. The major problems of our time—the growing threat of nuclear war, terrorism, the devastation of our natural environment, our inability to deal with poverty and starvation around the world, to name just the most urgent ones—are all different facets of one single crisis, which is essentially a crisis of perception. Like the crisis in quantum physics, it derives from the fact that most of us, and especially our large social institutions, subscribe to the concepts of an outdated worldview, inadequate for dealing with the problems of our overpopulated, globally interconnected world. At the same time, researchers in several sci-

From: "Physics and the Current Change of Paradigms," in *The World View of Contemporary Physics: Does It Need a New Metaphysics?* ed. Richard F. Kitchener (Albany: State University of New York Press, 1988), pp. 144–52.

entific disciplines, various social movements, and numerous alternative organizations and networks are developing a new vision of reality that will form the basis of our future technologies, economic systems, and social institutions.

What we are seeing today is a shift of paradigms not only within science but also in the larger social arena. To analyze that cultural transformation, I have generalized Kuhn's account of a scientific paradigm to that of a *social paradigm*, which I define as "a constellation of concepts, values, perceptions, and practices shared by a community, which form a particular vision of reality that is the basis of the way the community organizes itself."[3]

The social paradigm now receding has dominated our culture for several hundred years, during which it has shaped our modern Western society and has significantly influenced the rest of the world. This paradigm consists of a number of ideas and values, among them the view of the universe as a mechanical system composed of elementary building blocks, the view of the human body as a machine, the view of life in a society as a competitive struggle for existence, the belief in unlimited material progress to be achieved through economic and technological growth and—last but not least—the belief that a society, in which the female is everywhere subsumed under the male, is one that follows from some basic law of nature. During recent decades, all of these assumptions have been found severely limited and in need of radical revision.

Indeed, such a revision is now taking place. The emerging new paradigm may be called a holistic, or an *ecological*, worldview, using the term ecological here in a much broader and deeper sense than it is commonly used. Ecological awareness, in that deep sense, recognizes the fundamental interdependence of all phenomena and the embeddedness of individuals and societies in the cyclical processes of nature.

Ultimately, deep ecological awareness is spiritual or religious awareness. When the concept of the human spirit is understood as the mode of consciousness in which the individual feels connected to the cosmos as a whole, which is the root meaning of the word *religion* (from the Latin *religare*, meaning "to bind strongly"), it becomes clear that ecological awareness is spiritual in its deepest essence. It is, therefore, not surprising that the emerging new vision of reality, based on deep ecological awareness, is consistent with the "perennial philosophy" of spiritual traditions, for example, that of Eastern spiritual traditions, the spirituality of Christian mystics, or with the philosophy and cosmology underlying the Native American traditions.[4]

THE SYSTEMS APPROACH

In science, the language of systems theory, and especially the theory of living systems, seems to provide the most appropriate formulation of the new ecological paradigm.[5] Since living systems cover such a wide range of phenomena—

individual organisms, social systems, and ecosystems—the theory provides a common framework and language for biology, psychology, medicine, economics, ecology, and many other sciences, a framework in which the so urgently needed ecological perspective is explicitly manifest.

The conceptual framework of contemporary physics, and especially those aspects [suggesting a new metaphysics is needed], may be seen as a special case of the systems approach, dealing with nonliving systems and exploring the interface between nonliving and living systems. It is important to recognize, I believe, that in the new paradigm physics is no longer the model and source of metaphors for the other sciences. Even though the paradigm shift in physics is still of special interest, since it was the first to occur in modern science, physics has now lost its role as the science providing the most fundamental description of reality.

I would now like to specify what I mean by the systems approach. To do so, I shall identify five criteria of systems thinking that, I claim, hold for all the sciences—the natural sciences, the humanities, and the social sciences. I shall formulate each criterion in terms of the shift from the old to the new paradigm, and I will illustrate the five criteria with examples from contemporary physics. However, since the criteria hold for all the sciences, I could equally well illustrate them with examples from biology, psychology, or economics.

1. *Shift from the part to the whole.* In the old paradigm, it is believed that in any complex system the dynamics of the whole can be understood from the properties of the parts. The parts themselves cannot be analyzed any further, except by reducing them to still smaller parts. Indeed, physics has been progressing in that way, and at each step there has been a level of fundamental constituents that could not be analyzed any further.

In the new paradigm, the relationship between the parts and the whole is reversed. The properties of the parts can be understood only from the dynamics of the whole. In fact, ultimately there are no parts at all. What we call a part is merely a pattern in an inseparable web of relationships. The shift from the part to the whole was the central aspect of the conceptual revolution of quantum physics in the 1920s. Werner Heisenberg was so impressed by this aspect that he entitled his autobiography *Der Teil und das Ganze* (The Part and the Whole).[6] More recently, the view of physical reality as a web of relationships has been emphasized by Henry Stapp, who showed how this view is embodied in S-matrix theory.[7]

2. *Shift from structure to process.* In the old paradigm, there are fundamental structures, and then there are forces and mechanisms through which these interact, thus giving rise to processes. In the new paradigm, every structure is seen as the manifestation of an underlying process. The entire web of relationships is intrinsically dynamic. The shift from structure to process is evident, for example, when we remember that mass in contemporary physics is no longer seen as measuring a fundamental substance but rather as a form of energy, that is, as measuring activity or processes. The shift from structure to process is also apparent in the work of Ilya Prigogine, who entitled his classic book *From Being to Becoming.*[8]

3. *Shift from objective to "epistemic" science.* In the old paradigm, scientific descriptions are believed to be objective, that is, independent of the human observer and the process of knowing. In the new paradigm, it is believed that epistemology—the understanding of the process of knowledge—has to be included explicitly in the description of natural phenomena. This recognition entered into physics with Heisenberg and is closely related to the view of physical reality as a web of relationships. Whenever we isolate a pattern in this network and define it as a part, or an object, we do so by cutting through some of its connections to the rest of the network, and this may be done in different ways. As Heisenberg put it, "What we observe is not nature itself, but nature exposed to our method of questioning."[9]

This method of questioning, in other words epistemology, inevitably becomes part of the theory. At present, there is no consensus about what is the proper epistemology, but there is an emerging consensus that epistemology will have to be an integral part of every scientific theory.

4. *Shift from "building" to "network" as metaphor of knowledge.* The metaphor of knowledge as a building has been used in Western science and philosophy for thousands of years. There are fundamental laws, fundamental principles, basic building blocks, and so on. The edifice of science must be built on firm foundations. During periods of paradigm shift, it was always felt that the foundations of knowledge were shifting, or even crumbling, and that feeling induced great anxiety. Einstein (1949), for example, wrote in his autobiography about the early days of quantum mechanics: "All my attempts to adapt the theoretical foundations of physics to this (new type) of knowledge failed completely. It was as if the ground had been pulled out from under one, with no firm foundation to be seen anywhere, upon which one could have built."[10]

In the new paradigm, the metaphor of knowledge as a building is being replaced by that of the network. Since we perceive reality as a network of relationships, our descriptions, too, form an interconnected network of concepts and models in which there are no foundations. For most scientists this metaphor of knowledge as a network with no firm foundations is extremely uncomfortable. It is explicitly expressed in physics in Geoffrey Chew's bootstrap theory of particles.[11] According to Chew, nature cannot be reduced to any fundamental entities, but has to be understood entirely through self-consistency. There are no fundamental equations or fundamental symmetries in the bootstrap theory. Physical reality is seen as a dynamic web of interrelated events. Things exist by virtue of their mutually consistent relationships, and all of physics has to follow uniquely from the requirement that its components be consistent with one another and with themselves. This approach is so foreign to our traditional scientific ways of thinking that it is pursued today only by a small minority of physicists.

When the notion of scientific knowledge as a network of concepts and models, in which no part is any more fundamental than the others, is applied to science as a whole, it implies that physics can no longer be seen as the most fundamental level of science. Since there are no foundations in the network, the phenomena described by physics are not any more fundamental than those described, for example, by biology or psychology. They belong to different systems levels, but none of those levels is any more fundamental than the others.

5. *Shift from truth to approximate descriptions.* The four criteria of systems thinking presented so far are all interdependent. Nature is seen as an interconnected, dynamic web of relationships, in which the identification of specific patterns as "objects" depends on the human observer and the process of knowledge. This web of relationships is described in terms of a corresponding network of concepts and models, none of which is any more fundamental than the others.

This new approach immediately raises an important question: If everything is connected to everything else, how can you ever hope to understand anything? Since all natural phenomena are ultimately interconnected, in order to explain any one of them we need to understand all the others, which is obviously impossible.

What makes it possible to turn the systems approach into a scientific theory is the fact that there is such a thing as approximate knowledge. This insight is crucial to all of modern science. The old paradigm is based on the Cartesian belief in the certainty of scientific knowledge. In the new paradigm, it is recognized that all scientific concepts and theories are limited and approximate. Science can never provide any complete and definitive understanding. Scientists do not deal with truth in the sense of a precise correspondence between the description and the described phenomena. They deal with limited and approximate descriptions of reality. Heisenberg often pointed out that important fact. For example, he wrote in *Physics and Philosophy,* "The often discussed lesson that has been learned from modern physics [is] that every word or concept, clear as it may seem to be, has only a limited range of applicability."[12]

SELF-ORGANIZING SYSTEMS

The broadest implications of the systems approach are found today in a new theory of living systems, which originated in cybernetics in the 1940s and emerged in its main outlines over the last [forty] years.[13] As I mentioned before, living systems include individual organisms, social systems, and ecosystems, and thus the new theory can provide a common framework and language for a wide range of disciplines—biology, psychology, medicine, economics, ecology, and many others.

The central concept of the new theory is that of self-organization. A living system is defined as a self-organizing system, which means that its order is not imposed by the environment but is established by the system itself. In other words, self-organizing systems exhibit a certain degree of autonomy. This does not mean that living systems are isolated from their environment; on the contrary, they interact with it continually, but this interaction does not determine their organization.

In this essay, I can give only a brief sketch of the theory of self-organizing systems. To do so, let me distinguish three aspects of self-organization:

1. *Pattern of organization*: the totality of relationships that define the system as an integrated whole
2. *Structure*: the physical realization of the pattern of organization in space and time
3. *Organizing activity*: the activity involved in realizing the pattern of organization

For self-organizing systems, the pattern of organization is characterized by a mutual dependency of the system's parts, which is necessary and sufficient to understand the parts. This is quite similar to the pattern of relationships between subatomic particles in Chew's bootstrap theory. However, the pattern of self-organization has the additional property that gives the whole system an individual identity.

The pattern of self-organization has been studied extensively and described precisely by Humberto Maturana and Francisco Varela, who have called it *autopoiesis*, which means literally self-production. Sometimes it is also called operational closure.[14]

An important aspect of the theory is the fact that the description of the pattern of self-organization does not use any physical parameters, such as energy or entropy, nor does it use the concepts of space and time. It is an abstract mathematical description of a pattern of relationships. This pattern can be realized in space and time in different physical structures, which are then described in terms of the concepts of physics and chemistry. But such a description alone will fail to capture the biological phenomenon of self-organization. In other words, physics and chemistry are not enough to understand life; we also need to understand the pattern of self-organization, which is independent of physical and chemical parameters.

The structure of self-organizing systems has been studied extensively by Ilya Prigogine, who has called it a dissipative structure.[15] The two main characteristics of a dissipative structure are (1) that it is an open system, maintaining its pattern of organization through continuous exchange of energy and matter with its environment; and (2) that it operates far from thermodynamic equilibrium and thus cannot be described in terms of classical thermodynamics. One of Prigogine's greatest contributions has been to create a new thermodynamics to describe living systems.

The organizing activity of living, self-organizing systems, finally, is cognition, or mental activity. This implies a radically new concept of mind, which was first proposed by Gregory Bateson.[16] Mental process is defined as the organizing activity of life. This means that all interactions of a living system with its environment are cognitive, or mental interactions. With this new concept of mind, life and cognition become inseparably connected. Mind, or more accurately, mental process is seen as being immanent in matter at all levels of life.

I have taken some time to outline the emerging theory of self-organizing systems because it is today the broadest scientific formulation of the ecological paradigm with the most wide-ranging implications. The worldview of contemporary physics, in my view, will have to be understood within that broader framework. In particular, any speculation about human consciousness and its relation to the phenomena described by physics will have to take into account the notion of mental process as the self-organizing activity of life.

SCIENCE AND ETHICS

A further reason why I find the theory of self-organizing systems so important is that it seems to provide the ideal scientific framework for an ecologically oriented ethics.[17] Such a system of ethics is urgently needed, since most of what scientists are doing today is not life-furthering and life-preserving but life-destroying. With physicists designing nuclear weapons that threaten to wipe out all life on the planet, with chemists contaminating our environment, with biologists releasing new and unknown types of microorganisms into the environment without really knowing what the consequences are, with psychologists and other scientists torturing animals in the name of scientific progress, with all these activities occurring, it seems that it is most urgent to introduce ethical standards into modern science.

It is generally not recognized in our culture that values are not peripheral to science and technology but constitute their very basis and driving force. During the scientific revolution in the seventeenth century, values were separated from facts, and since that time we have tended to believe that scientific facts are independent of what we do and, therefore, independent of our values. In reality, scientific facts emerge out of an entire constellation of human perceptions, values, and actions—in a word, out of a paradigm—from which they cannot be separated. Although much of the detailed research may not depend explicitly on the scientist's value system, the larger paradigm within which this research is pursued will never be value free. Scientists, therefore, are responsible for their research not only intellectually but also morally.

One of the most important insights of the new systems theory of life is that life and cognition are inseparable. The process of knowledge is also the process of self-organization, that is, the process of life. Our conventional model of knowledge is one of a representation or an image of independently existing facts, which is the model derived from classical physics. From the new systems point of view, knowledge is part of the process of life, of a dialogue between object and subject.

Knowledge and life, then, are inseparable, and, therefore, facts are inseparable from values. Thus, the fundamental split that made it impossible to include ethical considerations in our scientific worldview has now been healed. At present, nobody has yet established a system of ethics that expresses the same ecological awareness on which the systems view of life is based, but I believe that this is now possible. I also believe that it is one of the most important tasks for scientists and philosophers today.

NOTES

1. Thomas Kuhn, *The Structure of Scientific Revolutions* (Chicago: University of Chicago Press, 1970). The definition quoted is my own synthesis of several definitions given by Kuhn.

2. See Fritjof Capra, *The Turning Point* (New York: Bantam Books, 1983).

3. Fritjof Capra, "The Concept of Paradigm and Paradigm Shift" and "New Paradigm Thinking in Science," *ReVision* 9, no. 1 (Summer/Fall 1986): 11.

4. Fritjof Capra, *The Tao of Physics*, 2nd ed. (New York: Bantam Books, 1984).

5. Capra, *The Turning Point*, ch. 9.

6. Werner Heisenberg, *Der Teil und das Ganze* (Piper, 1970); published in the United States as *Physics and Beyond* (New York: Harper and Row, 1971).

7. Henry P. Stapp, "S-Matrix Interpretation of Quantum Theory," *Physical Review D, Particles and Fields* 3, no. 6 (March 15 1971): 1303–20; Henry P. Stapp, "The Copenhagen Interpretation," *American Journal of Physics* 40, no. 7 (July 1972): 1098–1116.

8. Ilya Prigogine, *From Being to Becoming* (San Francisco: W. H. Freeman, 1980).

9. Werner Heisenberg, *Physics and Philosophy* (New York: Harper and Row, 1971), p. 58.

10. Albert Einstein in *Albert Einstein: Philosopher-Scientist*, ed. P. A. Schilpp (Evanston, IL: Library of Living Philosophers, 1949), p. 45.

11. See Fritjof Capra, "Bootstrap Physics: A Conversation with Geoffrey Chew," in *A Passion for Physics: Essays in Honor of Geoffrey Chew*, including an interview with Chew, ed. C. De Tar, J. Finkelstein, and Chung-I Tang (Philadelphia: World Scientific, 1985), pp. 247–86.

12. Heisenberg, *Physics and Philosophy*, p. 125.

13. Capra, *The Turning Point*, ch. 9.

14. Humberto R. Maturana and Francisco J. Varela, *Autopoiesis and Cognition: The Realization of the Living* (Boston: Reidel, 1980).

15. Prigogine, *From Being to Becoming*.

16. Gregory Bateson, *Mind and Nature* (New York: Dutton, 1979).

17. Fritjof Capra, ed., "Science and Ethics," Elmwood Discussion transcript no. 1, Elmwood Institute, Berkeley, CA.

32

THE ECOLOGY OF ORDER AND CHAOS

DONALD WORSTER

The science of ecology has had a popular impact unlike that of any other academic field of research. Consider the extraordinary ubiquity of the word itself: it has appeared in the most everyday places and the most astonishing, on day-glo T-shirts, in corporate advertising, and on bridge abutments. It has changed the language of politics and philosophy—springing up in a number of countries are political groups that are self-identified as "ecology parties." Yet who ever proposed forming a political party named after comparative linguistics or advanced paleontology? On several continents we have a philosophical movement termed "Deep Ecology," but nowhere has anyone announced a movement for "Deep Entomology" or "Deep Polish Literature." Why has this funny little word, ecology, coined by an obscure nineteenth-century German scientist, acquired so powerful a cultural resonance, so widespread a following?

Behind the persistent enthusiasm for ecology, I believe, lies the hope that this science can offer a great deal more than a pile of data. It is supposed to offer a pathway to a kind of moral enlightenment that we can call, for the purposes of simplicity, "conservation." The expectation did not originate with the public but first appeared among eminent scientists within the field. For instance, in his 1935 book, *Deserts on the March*, the noted University of Oklahoma, and later Yale, botanist Paul Sears urged Americans to take ecology seriously, promoting it in their universities and making it part of their governing process. "In Great Britain," he pointed out,

From: *Environmental History Review* 14, nos. 1–2 (Spring/Summer, 1990): 4–16.

the ecologists are being consulted at every step in planning the proper utilization of those parts of the Empire not yet settled, thus . . . ending the era of haphazard exploitation. There are hopeful, but all too few signs that our own national government realizes the part which ecology must play in a permanent program.[1]

Sears recommended that the United States hire a few thousand ecologists at the county level to advise citizens on questions of land use and thereby bring an end to environmental degradation; such a brigade, he thought, would put the whole nation on a biologically and economically sustainable basis.

In a 1947 addendum to his text, Sears added that ecologists, acting in the public interest, would instill in the American mind that "body of knowledge," that "point of view, which peculiarly implies all that is meant by conservation."[2] In other words, by the time of the 1930s and '40s, ecology was being hailed as a much-needed guide to a future motivated by an ethic of conservation. And conservation for Sears meant restoring the biological order, maintaining the health of the land and thereby the well-being of the nation, pursuing by both moral and technical means a lasting equilibrium with nature.

While we have not taken to heart all of Sears's suggestions—have not yet put any ecologists on county payrolls, with an office next door to the tax collector and sheriff—we have taken a surprisingly long step in his direction. Every day in some part of the nation, an ecologist is at work writing an environmental impact report or monitoring a human disturbance of the landscape or testifying at a hearing.

[In 1978] I published a history, going back to the eighteenth century, of this scientific discipline and its ideas about nature.[3] The conclusions in that book still strike me as being, on the whole, sensible and valid: that this science has come to be a major influence on our perception of nature in modern times; that its ideas, on the other hand, have been reflections of ourselves as much as objective apprehensions of nature; that scientific analysis cannot take the place of moral reasoning; that science, including the science of ecology, promotes, at least in some of its manifestations, a few of our darker ambitions toward nature and therefore itself needs to be morally examined and critiqued from time to time. Ecology, I argued, should never be taken as an all-wise, always trustworthy guide. We must be willing to challenge this authority, and indeed challenge the authority of science in general; not be quick to scorn or vilify or behead, but simply, now and then, to question.

During the period since my book was published, there has accumulated a considerable body of new thinking and new research in ecology. In this essay I mean to survey some of that recent thinking, contrasting it with its predecessors, and to raise a few of the same questions I did before. Part of my argument will be that Paul Sears would be astonished, and perhaps dismayed, to hear the kind of advice that ecological experts have to give these days. Less and less do they offer, or even promise to offer, what he would consider to be a program of moral enlightenment—of "conservation" in the sense of a restored equilibrium between humans and nature. There is a clear reason for that outcome, I will argue, and it has to do with drastic changes in the ideas that ecologists hold about the structure and function of the natural world. In Sears's day ecology was basically a study of equilibrium, harmony, and order; it had been so from its beginnings. Today, however, in many circles of scientific research, it has become a study of disturbance, disharmony, and chaos, and coincidentally or not, conservation is often not even a remote concern.

At the time *Deserts on the March* appeared in print, and through the time of its second and even third edition, the dominant name in the field of American ecology was that of Frederic L. Clements, who more than any other individual introduced scientific ecology into our national academic life. He called his approach "dynamic ecology," meaning it was concerned with change and evolution in the landscape. At its heart Clements's ecology dealt with the process of vegetational succession—the sequence of plant communities that appear on a piece of soil, newly made or disturbed, beginning with the first pioneer communities that invade and get a foothold.[4] Here is how I have defined the essence of the Clementsian paradigm:

> Change upon change became the inescapable principle of Clements's science.
> Yet he also insisted stubbornly and vigorously on the notion that the natural
> landscape must eventually reach a vaguely final climax stage. Nature's course,
> he contended, is not an aimless wandering to and fro but a steady flow toward
> stability that can be exactly plotted by the scientist.[5]

Most interestingly, Clements referred to that final climax stage as a "superorganism," implying that the assemblage of plants had achieved the close integration of parts, the self-organizing capability, of a single animal or plant. In some unique sense, it had become a live, coherent thing, not a mere collection of atomistic individuals, and exercised some control over the nonliving world around it, as organisms do.

Until well after World War II Clements's climax theory dominated ecological thought in this country.[6] Pick up almost any textbook in the field written [forty] years ago, and you will likely find mention of the climax. It was this theory that Paul Sears had studied and took to be the core lesson of ecology that his county ecologists should teach their fellow citizens: that nature tends toward a climax state and that, as far as practicable, they should learn to respect and preserve it. Sears wrote that the chief work of the scientist ought to be to show "the unbalance which man has produced on this continent" and to lead people back to some approximation of nature's original health and stability.[7]

But then, beginning in the 1940s, while Clements and his ideas were still in the ascendent, a few scientists began trying to speak a new vocabulary. Words like "energy flow," "trophic levels," and "ecosystem" appeared in the leading journals, and they indicated a view of nature shaped more by physics than botany. Within another decade or two nature came to be widely seen as a flow of energy and nutrients through a physical or thermodynamic system. The early figures prominent in shaping this new view included Chancy Juday, Raymond Lindeman, and G. Evelyn Hutchinson. But perhaps its most influential exponent was Eugene P. Odum, hailing from North Carolina and Georgia, discovering in his southern saltwater marshes, tidal estuaries, and abandoned cotton fields the animating, pulsating force of the sun, the global flux of energy. In 1953 Odum published the first edition of his famous textbook, *The Fundamentals of Ecology*.[8] In 1966 he became president of the Ecological Society of America.

By now anyone in the United States who regularly reads a newspaper or magazine has come to know at least a few of Odum's ideas, for they furnish the main themes in our popular understanding of ecology, beginning with the sovereign idea of the ecosystem. Odum defined the ecosystem as "any unit that includes all of the organisms (i.e., the 'community') in a given area interacting with the physical environment so that a flow of energy leads to clearly defined trophic structure, biotic diversity, and material cycles (i.e., exchange of materials between living and nonliving parts) within the system."[9] The whole earth, he argued, is organized into an interlocking series of such "ecosystems," ranging in size from a small pond to so vast an expanse as the Brazilian rainforest.

What all those ecosystems have in common is a "strategy of development," a kind of game plan that gives nature an overall direction. That strategy is, in Odum's words, "directed toward achieving as large and diverse an organic structure as is possible within the limits set by the available energy input and the prevailing physical conditions of existence."[10] Every single ecosystem, he believed, is either moving toward or has already achieved that goal. It is a clear, coherent, and easily observable strategy; and it ends in the happy state of order.

Nature's strategy, Odum added, leads finally to a world of mutualism and cooperation among the organisms inhabiting an area. From an early stage of competing against one another, they evolve toward a more symbiotic relationship. They learn, as it were, to work together to control their surrounding environment, making it more and more suitable as a habitat, until at last they have the power to protect themselves from its stressful cycles of drought and flood, winter and summer, cold and heat. Odum called that point "homeostasis." To achieve it, the living components of an ecosystem must evolve a structure of interrelatedness and cooperation that can, to some extent, manage the physical world— manage it for maximum efficiency and mutual benefit.

I have described this set of ideas as a break from the past, but that is misleading. Odum may have used different terms than Clements, may even have had a radically different vision of nature at times; but he did not repudiate Clements's notion that nature moves toward order and harmony. In the place of the theory of the "climax" stage he put the theory of the "mature ecosystem." His nature may

have appeared more as an automated factory than as a Clementsian superorganism, but like its predecessor it tends toward order.

The theory of the ecosystem presented a very clear set of standards as to
what constituted order and disorder, which Odum set forth in the form of a "tab-
ular model of ecological succession." When the ecosystem reaches its end point
of homeostasis, his table shows, it expends less energy on increasing production
and more on furnishing protection from external vicissitudes: that is, the biomass
in an area reaches a steady level, neither increasing nor decreasing, and the
emphasis in the system is on keeping it that way—on maintaining a kind of
no-growth economy. Then the little, aggressive, weedy organisms common at an
early stage in development (the r-selected species) give way to larger, steadier
creatures (K-selected species), who may have less potential for fast growth and
explosive reproduction but also better talents at surviving in dense settlements
and keeping the place on an even keel.[11] At that point there is supposed to be
more diversity in the community—i.e., a greater array of species. And there is
less loss of nutrients to the outside; nitrogen, phosphorous, and calcium all stay
in circulation within the ecosystem rather than leaking out. Those are some of the
key indicators of ecological order, all of them susceptible to precise measurement. The suggestion was implicit but clear that if one interfered too much with
nature's strategy of development, the effects might be costly: a serious loss of
nutrients, a decline in species diversity, an end to biomass stability. In short, the
ecosystem would be damaged.

The most likely source of that damage was no mystery to Odum: it was
human beings trying to force up the production of useful commodities and stu-
pidly risking the destruction of their life-support system.

> Man has generally been preoccupied with obtaining as much "production" from
> the landscape as possible, by developing and maintaining early successional
> types of ecosystems, usually monocultures. But, of course, man does not live by
> food and fiber alone; he also needs a balanced CO_2-O_2 atmosphere, the climatic
> buffer provided by oceans and masses of vegetation, and clean (that is, unpro-
> ductive) water for cultural and industrial uses. Many essential life-cycle
> resources, not to mention recreational and esthetic needs, are best provided man
> by the less "productive" landscapes. In other words, the landscape is not just a
> supply depot but is also the *oikos*—the home—in which we must live.[12]

Odum's view of nature as a series of balanced ecosystems, achieved or in the
making, led him to take a strong stand in favor of preserving the landscape in as
nearly natural a condition as possible. He suggested the need for substantial
restraint on human activity—for environmental planning "on a rational and scientific basis." For him as for Paul Sears, ecology must be taught to the public and
made the foundation of education, economics, and politics; America and other
countries must be "ecologized."

Of course not everyone who adopted the ecosystem approach to ecology ended up where Odum did. Quite the contrary; many found the ecosystem idea a wonderful instrument for promoting global technocracy. Experts familiar with the ecosystem and skilled in its manipulation, it was hoped in some quarters, could manage the entire planet for improved efficiency. "Governing" all of nature with the aid of rational science was the dream of these ecosystem technocrats.[13] But technocratic management was not the chief lesson, I believe, the public learned in Professor Odum's classroom; most came away devoted, as he was, to preserving large parts of nature in an unmanaged state and sure that they had been given a strong scientific rationale, as well as knowledge base, to do it. We must defend the world's endangered ecosystems, they insisted. We must safeguard the integrity of the Greater Yellowstone ecosystem, the Chesapeake Bay ecosystem, the Serengeti ecosystem. We must protect species diversity, biomass stability, and calcium recycling. We must make the world safe for K-species.[14]

That was the rallying cry of environmentalists and ecologists alike in the 1960s and early 1970s, when it seemed that the great coming struggle would be between what was left of pristine nature, delicately balanced in Odum's beautifully rational ecosystems, and a human race bent on mindless, greedy destruction. A decade or two later the situation . . . changed considerably. There are still environmental threats around, to be sure, and they are more dangerous than ever. The newspapers inform of us of continuing disasters [such as devastating hurricanes and global climate change], and reporters persist in using words like "ecosystem" and "balance" and "fragility" to describe such disasters. So do many scientists, who continue to acknowledge their theoretical indebtedness to Odum.[15] . . . But all the same, and despite the persistence of environmental problems, Odum's ecosystem is no longer the main theme in research or teaching in the science. A survey of . . . ecology textbooks shows that the concept is not even mentioned in one leading work and has a much-diminished place in the others.[16]

Ecology is not the same as it was. A rather drastic change has been going on in this science of late—a radical shifting away from the thinking of Eugene Odum's generation, away from its assumptions of order and predictability, a shifting toward what we might call a new *ecology of chaos*.

In July 1973, the *Journal of the Arnold Arboretum* published an article by two scientists associated with the Massachusetts Audubon Society, William Drury and Ian Nisbet, and it challenged Odum's ecology fundamentally. The title of the article was simply "Succession," indicating that old subject of observed sequences in plant and animal associations. With both Frederic Clements and Eugene Odum, succession had been taken to be the straight and narrow road to equilibrium. Drury and Nisbet disagreed completely with that assumption. Their observations, drawn particularly from northeastern temperate forests, strongly suggested that the process of ecological succession does not lead anywhere. Change is without any determinable direction and goes on forever, never reaching a point of stability. They found no evidence of any progressive development in nature: no progressive increase over time in biomass stabilization, no progressive diversification of species, no progressive movement toward a greater cohesiveness in plant and animal communities, nor toward a greater success in regulating the environment. Indeed, they found none of the criteria Odum had posited for mature ecosystems. The forest, they insisted, no matter what its age, is nothing but an erratic, shifting mosaic of trees and other plants. In their words, "Most of the phenomena of succession should be understood as resulting from the differential growth, differential survival, and perhaps differential dispersal of species adapted to grow at different points on stress gradients."[17] In other words, they could see lots of individual species, each doing its thing, but they could locate no emergent collectivity, nor any strategy to achieve one.

Prominent among their authorities supporting this view was the nearly forgotten name of Henry A. Gleason, a taxonomist who, in 1926, had challenged Frederic Clements and his organismic theory of the climax in an article entitled, "The Individualistic Concept of the Plant Association." Gleason had argued that we live in a world of constant flux and impermanence, not one tending toward Clements's climaxes. There is no such thing, he argued, as balance or equilibrium or steady-state. Each and every plant association is nothing but a temporary gathering of strangers, a clustering of species unrelated to one another, here for a brief while today, on their way somewhere else tomorrow. "Each . . . species of plant is a law unto itself," he wrote.[18] We look for cooperation in nature and we find only competition. We look for organized wholes, and we can discover only loose atoms and fragments. We hope for order and discern only a mishmash of conjoining species, all seeking their own advantage in utter disregard of others.

Thanks in part to Drury and Nisbet, this "individualistic" view was reborn in the mid-1970s and, during the past decade, it became the core idea of what some scientists hailed as a new, revolutionary paradigm in ecology. To promote it, they attacked the traditional notion of succession; for to reject that notion was to reject the larger idea that organic nature tends toward order. In 1977 two more biologists, Joseph Connell and Ralph Slatyer, continued the attack, denying the old claim that an invading community of pioneering species, the first stage in Clements's sequence, works to prepare the ground for its successors, like a group of Daniel Boones blazing the trail for civilization. The firstcomers, Connell and Slatyer maintained, manage in most cases to stake out their claims and successfully defend them; they do not give way to a later, superior group of colonists. Only when the pioneers die or are damaged by natural disturbances, thus releasing the resources they have monopolized, can latecomers find a foothold and get established.[19]

As this assault on the old thinking gathered momentum, the word "disturbance" began to appear more frequently in the scientific literature and be taken far more seriously. "Disturbance" was not a common subject in Odum's heyday, and it almost never appeared in combination with the adjective "natural." Now, however, it was as though scientists were out looking strenuously for signs of disturbance in nature—especially signs of disturbance that were not caused by humans—and they were finding it everywhere. During the past decade those new ecologists succeeded in leaving little tranquility in primitive nature. Fire is one of the most common disturbances they noted. So is wind, especially in the form of violent hurricanes and tornados. So are invading populations of microorganisms and pests and predators. And volcanic eruptions. And invading ice sheets of the Quaternary Period. And devastating droughts like that of the 1930s in the American West. Above all, it is these last sorts of disturbances, caused by the restlessness of climate, that the new generation of ecologists have emphasized. As one of the most influential of them, Professor Margaret Davis of the University of Minnesota, has written: "For the last 50 years or 500 or 1,000—as long as anyone would claim for 'ecological time'—there has never been an interval when temperature was in a steady state with symmetrical fluctuations about a mean. . . . Only on the longest time scale, 100,000 years, is there a tendency toward cyclical variation, and the cycles are asymmetrical, with a mean much different from today."[20]

One of the most provocative and impressive expressions of the new post-Odum ecology is a book of essays edited by S. T. A. Pickett and P. S. White, *The Ecology of Natural Disturbance and Patch Dynamics* (published in 1985). I submit it as symptomatic of much of the thinking going on . . . in the field. Though the final section of the book does deal with ecosystems, the word has lost much of its former meaning and implications. Two of the authors in fact open their contribution with a complaint that many scientists assume that "homogeneous ecosystems are a reality," when in truth "virtually all naturally occurring and man-disturbed ecosystems are mosaics of environmental conditions." "Historically," they write, "ecologists have been slow to recognize the importance of disturbances and the heterogeneity they generate." The reason for this slowness? "The majority of both theoretical and empirical work has been dominated by an equilibrium perspective."[21] Repudiating that perspective, these authors take us to the tropical forests of South and Central America and to the Everglades of Florida, showing us instability on every hand: a wet, green world of continual disturbance—or as they prefer to say, "of perturbations." Even the grasslands of North America, which inspired Frederic Clements's theory of the climax, appear in this collection as regularly disturbed environments. One paper describes them as a "dynamic, fine-textured mosaic" that is constantly kept in upheaval by the workings of badgers, pocket gophers, and mound-building ants, along with fire, drought, and eroding wind and water.[22] The message in all these papers is consistent. The climax notion is dead, the ecosystem has receded in usefulness, and in their place we have the idea of the lowly "patch." Nature should be regarded as a landscape of patches, big and little, patches of all textures and colors, a patchwork quilt of living things, changing continually through time and space, responding to an unceasing barrage of perturbations. The stitches in that quilt never hold for long. . . .

. . . Nature, many have begun to believe, is *fundamentally* erratic, discontinuous, and unpredictable. It is full of seemingly random events that elude our models of how things are supposed to work. As a result, the unexpected keeps hitting us in the face. Clouds collect and disperse, rain falls or doesn't fall, disregarding our careful weather predictions, and we cannot explain why. Cars suddenly bunch up on the freeway, and the traffic controllers fly into a frenzy. A man's heart beats regularly year after year, then abruptly begins to skip a beat now and then. A ping pong ball bounces off the table in an unexpected direction. Each little snowflake falling out of the sky turns out to be completely unlike any other. Those are ways in which nature seems, in contrast to all our previous theories and methods, to be chaotic. If the ultimate test of any body of scientific knowledge is its ability to predict events, then all the sciences and pseudosciences—physics, chemistry, climatology, economics, ecology—fail the test regularly. They all have been announcing laws, designing models, predicting what an individual atom or person is supposed to do; and now, increasingly, they are beginning to confess that the world never quite behaves the way it is supposed to do.

Making sense of this situation is the task of an altogether new kind of inquiry calling itself the science of chaos. Some say it portends a revolution in thinking equivalent to quantum mechanics or relativity. Like those other twentieth-century revolutions, the science of chaos rejects tenets going back as far as the days of Sir Isaac Newton. In fact, what is occurring may be not two or three separate revolutions but a single revolution against all the principles, laws, models, and applications of classical science, the science ushered in by the great Scientific Revolution of the seventeenth century.[23] For centuries we have assumed that nature, despite a few appearances to the contrary, is a perfectly predictable system of linear, rational order. Give us an adequate number of facts, scientists have said, and we can describe that order in complete detail—can plot the lines along which everything moves and the speed of that movement and the collisions that will occur. Even Darwin's theory of evolution, which in the last century challenged much of the Newtonian worldview, left intact many people's confidence that order would prevail at last in the evolution of life; that out of the tangled history of competitive struggle would come progress, harmony, and stability. Now that traditional assumption may have broken down irretrievably. For whatever reason, whether because empirical data suggests it or because extrascientific cultural trends do—the experience of so much rapid social change in our daily lives—scientists are beginning to focus on what they had long managed to avoid seeing. The world is more complex than we ever imagined, they say, and indeed, some would add, ever can imagine.[24]

Despite the obvious complexity of their subject matter, ecologists have been among the slowest to join the cross-disciplinary science of chaos. I suspect that the influence of Clements and Odum, lingering well into the 1970s, worked against the new perspective, encouraging faith in linear regularities and equilibrium in the interaction of species. Nonetheless, eventually there arrived a day of conversion. In 1974 the Princeton mathematical ecologist Robert May published a paper with the title, "Biological Populations with Nonoverlapping Generations: Stable Points, Stable Cycles, and Chaos."[25] In it he admitted that the mathematical models he and others had constructed were inadequate approximations of the ragged life histories of organisms. They did not fully explain, for example, the aperiodic outbreaks of gypsy moths in eastern hardwood forests or the Canadian lynx cycles in the subarctic. Wildlife populations do not follow some simple Malthusian pattern of increase, saturation, and crash.

More and more ecologists . . . followed May and [began] to try to bring their subject into line with chaotic theory. William Schaefer is one of them; though a student of Robert MacArthur, a leader of the old equilibrium school, he [was] struck by the same anomaly of unpredictable fluctuations in populations as May and others. Though taught to believe in "the so-called 'Balance of Nature,'" he writes, ". . . the idea that populations are at or close to equilibrium," things now are beginning to look very different.[26] He describes himself as having to reach far across the disciplines, to make connections with concepts of chaos in the other natural sciences, in order to free himself from his field's restrictive past.

The entire study of chaos began in 1961, with efforts to simulate weather and climate patterns on a computer at MIT. There, meteorologist Edward Lorenz came up with his now-famous "Butterfly Effect," the notion that a butterfly stirring the air today in a Beijing park can transform storm systems next month in New York City. Scientists call this phenomenon "sensitive dependence on initial conditions." What it means is that tiny differences in input can quickly become substantial differences in output. A corollary is that we cannot know, even with all our artificial intelligence apparatus, every one of the tiny differences that have occurred or are occurring at any place or point in time; nor can we know which tiny differences will produce which substantial differences in output. Beyond a short range, say, of two or three days from now, our predictions are not worth the paper they are written on.

The implications of this "Butterfly Effect" for ecology are profound. If a single flap of an insect's wings in China can lead to a torrential downpour in New York, then what might it do to the Greater Yellowstone ecosystem? What can ecologists possibly know about all the forces impinging on, or about to impinge on, any piece of land? What can they safely ignore and what must they pay attention to? What distant, invisible, minuscule events may even now be happening that will change the organization of plant and animal life in our backyards? This is the predicament, and the challenge, presented by the science of chaos, and it is altering the imagination of ecologists dramatically.

John Muir once declared, "When we try to pick out anything by itself, we find it hitched to everything else in the universe."[27] For him, that was a manifestation of an infinitely wise plan in which everything functioned with perfect harmony. The new ecology of chaos, though impressed like Muir with interdependency, does not share his view of "an infinitely wise plan" that controls and shapes everything into order. There is no plan, . . . scientists say, no harmony apparent in the events of nature. If there is order in the universe—and there will no longer be any science if all faith in order vanishes—it is going to be much more difficult to locate and describe than we thought.

For Muir, the clear lesson of cosmic complexity was that humans ought to love and preserve nature just as it is. The lessons of the new ecology, in contrast, are not at all clear. Does it promote, in Ilya Prigogine and Isabelle Stenger's words, "a renewal of nature," a less hierarchical view of life, and a set of "new relations between man and nature and between man and man"?[28] Or does it increase our alienation from the world, our withdrawal into postmodernist doubt and self-consciousness? What is there to love or preserve in a universe of chaos? How are people supposed to behave in such a universe? If such is the kind of place we inhabit, why not go ahead with all our private ambitions, free of any fear that we may be doing special damage? What, after all, does the phrase "environmental damage" mean in a world of so much natural chaos? Does the tradition of environmentalism to which Muir belonged, along with so many other nature writers and ecologists of the past—people like Paul Sears, Eugene Odum, Aldo Leopold, and Rachel Carson—make sense any longer? I have no space here to attempt to answer those questions or to make predictions, but only to issue a warning that they are too important to be left for scientists alone to answer. Ecology today, no more than in the past, can be assumed to be all-knowing or all-wise or eternally true.

Whether they are true or false, permanent or passingly fashionable, it does seem entirely possible that these changes in scientific thinking toward an emphasis on chaos will not produce any easing of the environmentalist's concern. Though words like ecosystem or climax may fade away and some new vocabulary take their place, the fear of risk and danger will likely become greater than ever. Most of us are intuitively aware, whether we can put our fears into mathematical formulae or not, that the technological power we have accumulated is *destructively* chaotic; not irrationally, we fear it and fear what it can to do us as well as the rest of nature.[29] It may be that we moderns, after absorbing the lessons of today's science, find we cannot love nature quite so easily as Muir did; but it may also be that we have discovered more reason than ever to respect it—to respect its baffling complexity, its inherent unpredictability, its daily turbulence. And to flap our own wings in it a little more gently.

NOTES

1. Paul Sears, *Deserts on the March*, 3rd ed. (Norman: University of Oklahoma Press, 1959), p. 162.

2. Ibid., p. 177.

3. Donald Worster, *Nature's Economy: A History of Ecological Ideas* (New York: Cambridge University Press, 1977).

4. This is the theme in particular of Clements's book *Plant Succession* (Washington: Carnegie Institution, 1916).

5. Worster, *Nature's Economy*, p. 210

6. Clements's major rival for influence in the United States was Henry Chandler Cowles of the University of Chicago, whose first paper on ecological succession appeared in 1899. The best study of Cowles's ideas is J. Ronald Engel, *Sacred Sands: The Struggle for Community in the Indiana Dunes* (Middletown, CT: Wesleyan University Press, 1983), pp. 137–59. Engel describes him as having a less deterministic, more pluralistic notion of succession, one that "opened the way to a more creative role for human beings in nature's evolutionary adventure" (p. 150). See also Ronald C. Tobey, *Saving the Prairies: The Life Cycle of the Founding School of American Plant Ecology, 1895–1955* (Berkeley: University of California, 1981).

7. Sears, *Deserts on the March*, p. 142.

8. This book was coauthored with his brother Howard T. Odum, and it went through six more editions. Island Press has the most recent, a 1988 edition.

9. Eugene P. Odum, *Fundamentals of Ecology* (Philadelphia: W. B. Saunders, 1971), p. 8.

10. Eugene P. Odum, "The Strategy of Ecosystem Development," *Science* 164 (April 18, 1969): 266.

11. The terms "K-species" and "r-species" came from Robert MacArthur and Edward O. Wilson, *Theory of Island Biogeography* (Princeton, NJ: Princeton University Press, 1967). Along with Odum, MacArthur was the leading spokesman during the 1950s and '60s for the view of nature as a series of thermodynamically balanced ecosystems.

12. Odum, "Strategy of Ecosystem Development," p. 266. See also Odum, "Trends Expected in Stressed Ecosystems," *BioScience* 35 (July/August 1995): 419–422.

13. A book of that title was published by Earl F. Murphy, *Governing Nature* (Chicago: Quadrangle Books, 1967). From time to time, Eugene Odum himself seems to have caught that ambition or lent his support to it, and it was certainly central to the work of his brother, Howard T. Odum. On this theme see Peter J. Taylor, "Technocratic Optimism, H. T. Odum, and the Partial Transformation of Ecological Metaphor after World War II," *Journal of the History of Biology* 21 (Summer 1988): 213–44.

14. A very influential popularization of Odum's view of nature (though he is never actually referred to in it) is Barry Commoner's *The Closing Circle: Nature, Man, and Technology* (New York: Knopf, 1971). See in particular the discussion of the four "laws" of ecology, pp. 33–46.

15. Communication from Malcolm Cherrett, *Ecology* 70 (March 1989): 41–42.

16. See Michael Begon, John L. Harper, and Colin R. Townsend, *Ecology: Individuals, Populations, and Communities* (Sunderland, MA: Sinauer, 1986). In another textbook, Odum's views are presented critically as the traditional approach: R. J. Putnam and S. D. Wratten, *Principles of Ecology* (Berkeley: University of California Press, 1984). More loyal to the ecosystem model are Paul Ehrlich and Jonathan Roughgarden, *The Science of Ecology* (New York: Macmillan, 1987); and Robert Leo Smith, *Elements of Ecology*, 2nd ed. (New York: Harper and Row, 1986), although the latter admits that he has shifted from an "ecosystem approach" to more of an "evolutionary approach" (p. xiii).

17. William H. Drury and Ian C. T. Nisbet, "Succession," *Journal of the Arnold Arboretum* 54 (July 1973): 360.

18. H. A. Gleason, "The Individualistic Concept of the Plant Association," *Bulletin of the Torrey Botanical Club* 53 (1926): 25. A later version of the same article appeared in *American Midland Naturalist* 21 (1939): 92–110.

19. Joseph H. Connell and Ralph O. Slatyer, "Mechanisms of Succession in Natural Communities and Their Role in Community Stability and Organization," *American Naturalist* 111 (November/December 1977): 1119–44.

20. Margaret Bryan Davis, "Climatic Instability, Time Lags, and Community Disequilibrium," in *Community Ecology*, ed. Jared Diamond and Ted J. Case, 269 (New York: Harper and Row, 1986).

21. James R. Karr and Kathryn E. Freemark, "Disturbance and Vertebrates: An Integrative Perspective," in *The Ecology of Natural Disturbance and Patch Dynamics*, ed. S. T. A. Pickett and P. S. White, 154–55 (Orlando, FL: Academic Press, 1985). The Odum school of thought is, however, by no means silent. Another recent compilation has been put together in his honor, and many of its authors express a continuing support for his ideas: L. R. Pomeroy and J. J. Albert, eds., *Concepts of Ecosystem Ecology: A Comparative View* (New York: Springer-Verlag, 1988).

22. Orie L. Loucks, Mary L. Plumb-Mentjes, and Deborah Rogers, "Gap Processes and Large-Scale Disturbances in Sand Prairies," ibid., pp. 72–85.

23. This argument is made with great intellectual force by Ilya Prigogine and Isabelle Stengers, *Order Out of Chaos: Man's New Dialogue with Nature* (Boulder: Shambala/New Science Library, 1984). Prigogine won the Nobel Prize in 1977 for his work on the thermodynamics of nonequilibrium systems.

24. An excellent account of the change in thinking is James Gleick, *Chaos: The Making of a New Science* (New York: Viking, 1987). I have drawn on his explanation extensively here. What Gleick does not explore are the striking intellectual parallels between chaotic theory in science and postmodern discourse in literature and philosophy. Postmodernism is a sensibility that has abandoned the historic search for unity and order in nature, taking an ironic view of existence and debunking all established faiths. According to Todd Gitlin, "Post-Modernism reflects the fact that a new moral structure has not yet been built and our culture has not yet found a language for articulating the new understandings we are trying, haltingly, to live with. It objects to all principles, all commitments, all crusades—in the name of an unconscientious evasion." On the other hand, and more positively, the new sensibility leads to emphasis on democratic coexistence: "a new 'moral ecology'—that in the preservation of the other is a condition for the preservation of the self." Gitlin, "Post-Modernism: The Stenography of Surfaces," *New Perspectives Quarterly* 6 (Spring 1989): 57, 59.

25. The paper was published in *Science* 186 (1974): 645–47. See also Robert M. May, "Simple Mathematical Models with Very Complicated Dynamics," *Nature* 261 (1976): 459–67. Gleick discusses May's work in *Chaos*, pp. 69–80.

26. W. M. Schaeffer, "Chaos in Ecology and Epidemiology," in *Chaos in Biological Systems*, ed. H. Degan, A. V. Holden, and L. F. Olsen (New York: Plenum Press, 1987), p. 233. See also Schaeffer, "Order and Chaos in Ecological Systems," *Ecology* 66 (February 1985): 93–106.

27. John Muir, *My First Summer in the Sierra* (1911; Boston: Houghton Mifflin, 1944), p. 157.

28. Prigogine and Stengers, *Order Out of Chaos*, pp. 312–13.

29. Much of the alarm that Sears and Odum, among others, expressed has shifted to a global perspective, and the older equilibrium thinking has been taken up by scientists concerned about the geo- and biochemical condition of the planet as a whole and about human threats, particularly from the burning of fossil fuels, to its stability. One of the most influential texts in this new development is James Lovelock's *Gaia: A New Look at Life on Earth* (Oxford: Oxford University Press, 1979). See also Edward Goldsmith, "Gaia: Some Implications for Theoretical Ecology," *Ecologist* 18, nos. 2–3 (1988): 64–74.

33

POSTMODERN SCIENCE AND A POSTMODERN WORLD

DAVID BOHM

MODERN PHYSICS AND THE MODERN WORLD

With the coming of the modern era, human beings' view of their world and themselves underwent a fundamental change. The earlier, basically religious approach to life was replaced by a secular approach. This approach has assumed that nature could be thoroughly understood and eventually brought under control by means of the systematic development of scientific knowledge through observation, experiment, and rational thought. This idea became powerful in the seventeenth and eighteenth centuries. In fact, the great seal of the United States has as part of its motto "the new secular order," showing the way the founders of the country were thinking. The main focus of attention was on discerning the order of the universe as it manifests itself in the laws of nature. The principal path to human happiness was to be in the discovery of these laws, in complying with them, in utilizing them wherever possible for the benefit of humankind.

So great is the change in the whole context of thought thereby brought about that Huston Smith and some others have described it as the onset of the modern mind.[1] This mind is in contrast with the mind of the medieval period, in which it was generally supposed that the order of nature was beyond human comprehension and in which human happiness consisted in being aware of the revealed knowledge of God and carrying out the divine commandments. A total revolution occurred in the way people were aiming to live.

The modern mind went from one triumph to another for several centuries

From: David Ray Griffin, ed., *The Reenchantment of Science: Postmodern Proposals* (Albany: State University of New York Press, 1988), pp. 57–58, 60–66, 68.

through science, technology, industry, and it seemed to be solidly based for all time. But in the early twentieth century, it began to have its foundations questioned. The challenge coming from physics was especially serious, because it was in this science that the modern mind was thought to have its firmest foundation. In particular, relativity theory, to a certain extent, and quantum theory, to a much greater extent, led to questioning the assumption of an intuitively imaginable and knowable order in the universe. The nature of the world began to fade out into something almost indescribable. For the most part, physicists began to give up the attempt to grasp the world as an intuitively comprehensible whole; they instead restricted their work mostly to developing a mathematical formalism with rules to apply in the laboratory and eventually in technology. Of course, a great deal of unity has emerged in this work, but it is almost entirely in the mathematical formalism. It has little or no imaginative or intuitive expression (whereas Newton's ideas were quite easily understandable by any reasonably educated person). . . .

The possibility of a postmodern physics, extended also to postmodern science in general, may be of crucial significance. A postmodern science should not separate matter and consciousness and should therefore not separate facts, meaning, and value. Science would then be inseparable from a kind of intrinsic morality, and truth and virtue would not be kept apart as they currently are in science. This separation is part of the reason we are in our present desperate situation.

Of course, this proposal runs entirely contrary to the prevailing view of what science should be, which is a morally neutral way of manipulating nature, either for good or for evil, according to the choices of the people who apply it. I hope in this essay to indicate how a very different approach to science is possible, one that is consistent and plausible and that fits better the actual development of modern physics than does the current approach.

MECHANISTIC PHYSICS

I begin by outlining briefly the mechanistic view in physics, which was characteristic of the modern view and which reached its highest point toward the end of the nineteenth century. This view remains the basis of the approach of [many] physicists and other scientists [to this day]. Although the more recent physics has dissolved the mechanistic view, not very many scientists and even fewer members of the general public are aware of this fact; therefore, the mechanistic view is still the dominant view as far as effectiveness is concerned. In discussing this mechanistic view, I start by listing the principal characteristics of mechanism in physics. To clarify this view, I contrast it with that of ancient times, which was organic rather than mechanistic.

The first point about mechanism is that the world is reduced as far as possible to a set of basic elements. Typically, these elements take the form of particles. They can be called atoms or sometimes these are broken into electrons, protons, and neutrons; now the most elementary particles are called quarks, maybe there will be a subquark. Whatever they may be called, the assumption is that a basic element exists which we either have or hope to have. To these elementary particles, various continuous fields, such as electromagnetic and gravitational fields, must be added.

Second, these elements are basically external to each other; not only are they separate in space, but even more important, the fundamental nature of each is independent of that of the other. Each particle just has its own nature; it may be somewhat affected by being pushed around by the others, but that is all. The elements do not grow organically as parts of a whole, but are rather more like parts of a machine whose forms are determined externally to the structure of the machine in which they are working. By contrast, organic parts, the parts of an organism, all grow together with the organism.

Third, because the elements only interact mechanically by sort of pushing each other around, the forces of interaction do not affect their inner natures. In an organism or a society, by contrast, the very nature of each part is profoundly affected by changes in the other parts, so that the parts are internally related. If a man comes into a group, the consciousness of the whole group may change, depending on what he does. He does not push people's consciousnesses around as if they were parts of a machine. In the mechanistic view, this sort of organismic behavior is admitted, but it is explained eventually by analyzing everything into still smaller particles out of which the organs of the body are made, such as DNA molecules, ordinary molecules, atoms, and so on. This view says that eventually everything is reducible to something mechanical.

The mechanistic program has been very successful and is still successful in certain areas, for example, in genetic engineering to control heredity by treating the molecules on which heredity depends. Advocates do admit that the program still has much to achieve, but this mechanistic reductionistic program assumes that there is nothing that cannot eventually be treated in this way—that if we just keep on going this way we will deal with anything that may arise.

The adherence to this program has been so successful as to threaten our very existence as well as to produce all sorts of other dangers, but, of course, such success does not prove its truth. To a certain extent the reductionistic picture is still an article of faith, and faith in the mechanistic reductionistic program still provides the motivation of most of the scientific enterprise, the faith that this approach can deal with everything. This is a counterpart of the religious faith that people had earlier which allowed them to do great things.

How far can this faith in mechanism be justified? People try endlessly to justify faith in their religions through theology, and much similar work has gone into justifying faith in mechanism through the philosophy of science. Of course, that the mechanism works in a very important domain is given, thereby bringing about a revolution in our life.

During the nineteenth century, the Newtonian worldview seemed so certain and complete that no serious scientist was able to doubt it. In fact, we may refer to Lord Kelvin, one of the leading theoretical physicists at the time. He expressed the opinion that physics was more or less finished, advising young people not to go into the field because further work was only a matter of minor refinements. He did point, however, to two small clouds on the horizon. One was the negative results of the Michelson-Morley experiment and the other was the difficulty in understanding black-body radiation. Now he certainly chose his clouds well: the first one led to the theory of relativity and the second to quantum theory. Those little clouds became tremendous storms; but the sky is not even as clear today as it was then—plenty of clouds are still around. The fact that relativity and quantum together overturned the Newtonian physics shows the danger of complacency about worldview, it shows that we constantly must look at our worldviews as provisional, as exploratory, and to inquire. We must have a worldview, but we must not make it an absolute thing that leaves no room for inquiry and change. We must avoid dogmatism.

THE BEGINNING OF NONMECHANISTIC PHYSICS: RELATIVITY THEORY

Relativity theory was the first important step away from the mechanistic vision. It introduced new concepts of space, time, and matter. Instead of having separate little particles as the constituents of matter, Einstein thought of a field spread through all space, which would have strong and weak regions. Some strong regions, which are stable, represent particles. If you watch a whirlpool or a vortex, you see the water going around and you see that the movement gets weaker the farther away it is from the center, but it never ends. Now the vortex does not actually exist; there is only the moving water. The vortex is a pattern and a form your mind abstracts from the sensations you have of moving water. If two vortices are put together, they will affect each other; a changing pattern will exist where they modify each other, but it will still be only one pattern. You can say that two exist, but this is only a convenient way of thinking. As they become even closer together, they may merge. When you have flowing water with patterns in them, none of those patterns actually has a separate existence. They are appearances or forms in the flowing movement, which the mind abstracts momentarily for the sake of convenience. The flowing pattern is the ultimate reality, at least at that level. Of course, all the nineteenth-century physicists knew this perfectly well, but they said that *really* water is made of little atoms, that neither the vortices nor the water are the reality: the reality is the little atoms out of which it is all made. So the problem did not bother them.

But with the theory of relativity, Einstein gave arguments showing that thinking of these separate atoms as existent would not be consistent. His solution was to think of a field not so different from the flowing water, a field that spreads through all space and time and in which every particle is a stable form of movement, just as the vortex or whirlpool is a temporarily stable form that can be thought of as an entity which can be given a name. We speak of a whirlpool, but one does not exist. In the same way, we can speak of a particle, but one does not exist: "particle" is a name for a certain form in the field of movement. If you bring two particles together, they will gradually modify each other and eventually become one. Consequently, this approach contradicted the assumption of separate, elementary, mechanical constituents of the universe. In doing so, it brought in a view which I call "unbroken wholeness or flowing wholeness": it has also been called "seamless wholeness." The universe is one seamless, unbroken whole, and all the forms we see in it are abstracted by our way of looking and thinking, which is convenient at times, helping us with our technology, for example.

Nonetheless, relativity theory retains certain essential features of mechanism, in that the fields at different points in space were thought to exist separately and not to be internally related. The separate existence of these basic elements was emphasized by the idea that they were only locally connected, that the field at one point could affect a field only infinitesimally nearby. There was no direct effect of a field here on something far away. This notion is now being called "locality" by physicists; it is the notion of no long-distance connection. This notion is essential to the kind of mechanistic materialism developing throughout the science of the modern era, the notion of separate elements not internally related and not connected to things far away. The animistic view of earlier times was that spirits were behind everything and that these spirits were not located anywhere. Therefore, things far away would tend to be related. This view was taken to be most natural by astrologers and alchemists. But that view had been turned completely around in the modern period, and the modem view seemed so fruitful and so powerful that there arose the utter conviction of its truth.

MORE FULLY NONMECHANISTIC PHYSICS: QUANTUM THEORY

With quantum theory, a much bigger change occurred. The main point is that all action or all motion is found in a discrete indivisible unit called a "quantum." In the early form of the theory, electrons had to jump from one orbit to the other without passing in between. The whole idea of the continuous motion of particles, an idea at the heart of mechanism, was thereby being questioned. The ordinary visible movement, like my hand moving, was thought to comprise a vast number of quantum movements, just as, if enough fine grains of sand are in the hourglass, the flow seems continuous. All movements were said to comprise very tiny, discrete movements that do not, as it were, go from one place to another by passing through the space in between. This was a very mysterious idea.

Second, matter and energy had a dual nature; they manifest either like a wave or like a particle, according to how they were treated in an experiment. An electron is ordinarily a particle, but it can also behave like waves, and light which ordinarily behaves like waves can also behave like particles; their behavior depends on the context in which they are treated. That is, the quality of the thing depends on the context. This idea is utterly opposed to mechanism, because in mechanism the particle is just what it is, no matter what the context. Of course, with complex things, this is a familiar fact; it is clear, for example, that organs depend very much on context, that the brain depends on the context, that the mind functions differently in a different context. The new suggestion of quantum theory is that this context dependence is true of the ultimate units of nature. They hence begin to look more like something organic than like something mechanical.

A third point of quantum theory was the property of nonlocal connection. In certain areas, things could apparently be connected with other things any distance away without any apparent force to carry the connection. This "nonlocality" was very opposed to what Einstein wanted and very opposed to mechanism.

A fourth new feature of quantum physics, which was against mechanism, was that the whole organizes the parts, even in ordinary matter. One can see it doing so in living matter, in organisms, where the state of the whole organizes the various parts in the organism. But something a bit similar happens in electrons, too, in various phenomena such as superconductivity. The whole of chemistry, in fact, depends on this idea.

In summary, according to quantum physics, ultimately no continuous motion exists; an internal relationship between the parts and the whole, among the various parts, and a context dependence, which is very much a part of the same thing, all do exist. An indivisible connection between elements also exists which cannot be further analyzed. All of that adds up to the notion that the world is one unbroken whole. Quantum physics thereby says what relativity theory said, but in a very different way.

These phenomena are evident only with highly refined modes of observation. At the ordinary order of refinement, which was available during the nineteenth century, there was no evidence that any of this was occurring. People formed the mechanistic philosophy on the basis of fairly crude observations, which demonstrates the danger of deciding a final philosophy on the basis of any particular observations; even our present observations may be too crude for something still deeper.

Now one may ask: if there has been such a disproof of mechanism, why is it that most scientists are still mechanistic? The first reason is that this disproof takes place only in a very esoteric part of modern physics, called "quantum mechanical field theories," which only a few people understand, and most of those only deal with it mathematically, being committed to the idea they could never understand it beyond that level. Second, most other physicists have only the vaguest idea of what quantum mechanical field theorists are doing, and scientists in other fields have still less knowledge about it. Science has become so specialized that people in one branch can apply another branch without really understanding what it means. In a way this is humorous, but it has some very serious consequences.

UNBROKEN WHOLENESS AND POSTMECHANISTIC PHYSICS

I propose a view that I have called "unbroken wholeness." Relativity and quantum physics agree in suggesting unbroken wholeness, although they disagree on everything else. That is, relativity requires strict continuity, strict determinism, and strict locality, while quantum mechanics requires just the opposite—discontinuity, indeterminism, and nonlocality. The two basic theories of physics have entirely contradictory concepts which have not been brought together; this is one of the problems that remains. They both agree, however, on the unbroken wholeness of the universe, although in different ways. So it has seemed to me that we could use this unbroken wholeness as our starting point for understanding the new situation.

The question is then how to understand this wholeness. The entire language of physics is now analytic. If we use this language, we are committed to analyzing into parts, even though our intention may be quite the opposite. Therefore, the task is quite difficult.

What I want to suggest is that one of the most important problems is that of *order*. Worldviews have always had views of order. The ancient Greeks had the view of Earth as the center of the universe and of various spheres in order of increasing perfection. In Newtonian physics, the order is that of the particles and the way they move. That is a mechanical order, and coordinates are used mathematically to express that order. What kind of order will enable us to consider unbroken wholeness?

What *is* order? That is a very deep question, because everything we say presupposes order. A few examples: There is the order of the numbers, the order of the words here, the order of the walls, the order in which the body works, the order in which thought works, the order in which language works. We cannot really define order, but we nevertheless understand order somewhat, because we cannot think, talk, or do anything without beginning from some kind of order.

The order physics has been using is the order of separation. Here the lens is the basic idea. If one takes a photograph, one point on the object corresponds to one point on the image. This fact has affected us very greatly, suggesting that everything is made of points. The camera was thereby a very important instrument for helping to strengthen the mechanistic philosophy. It gives an experience that allows everybody to see what is meant by the idea that the universe is nothing but separate parts.

Another instrument, the holograph, can also illustrate this point. The Greek word *holo* means "whole," and *graph* means "to write"; consequently, a holograph writes the whole. With the aid of a laser, which produces highly ordered light, the waves of light from everywhere can be brought to one spot, and just the waves, rather than the image of the object, can be photographed. What is remarkable is that in the resulting picture, each part of it can produce an image of the whole object. Unlike the picture produced by a camera, no point-to-point correspondence with the object obtains. Information about each object is enfolded in each part; an image is produced when this enfolded information is unfolded. The holograph hence suggests a new kind of knowledge and a new understanding of the universe in which information about the whole is enfolded in each part and in which the various objects of the world result from the unfolding of this information.

In my proposal of unbroken wholeness, I turn the mechanistic picture upside down. Whereas the mechanistic picture regarded discrete objects as the primary reality, and the enfolding and unfolding of organisms, including minds, as secondary phenomena, I suggest that the unbroken movements of enfolding and unfolding, which I call the "holomovement," is primary while the apparently discrete objects are secondary phenomena. They are related to the holomovement somewhat as the vortex, in the above example, is related to the unbroken flow of water. An essential part of this proposal is that the whole universe is actively enfolded to some degree in each of the parts. Because the whole is enfolded in each part, so are all the other parts, in some way and to some degree. Hence, the mechanistic picture, according to which the parts are only externally related to each other, is denied. That is, it is denied to be the primary truth; external relatedness is a secondary, derivative truth, applicable only to the secondary order of things, which I call the explicate or unfolded order. This is, of course, the order on which modern science has focused. The more fundamental truth is the truth of internal relatedness, because it is true of the more fundamental order, which I call the implicate order, because in this order the whole and hence all the other parts are enfolded in each part.

In my technical writings,[2] I have sought to show that the mathematical laws of quantum theory can be understood as describing the holomovement, in which the whole is enfolded in each region, and the region is unfolded into the whole. Whereas modern physics has tried to understand the whole reductively by beginning with the most elementary parts, I am proposing a postmodern physics which begins with the whole.

POSTMODERN SCIENCE AND QUESTIONS OF MEANING AND VALUE

We have seen that fragmentary thinking is giving rise to a reality that is constantly breaking up into disorderly, disharmonious, and destructive partial activities. Therefore, seriously exploring a mode of thinking that starts from the most encompassing possible whole and goes down to the parts (subwholes) in a way appropriate to the actual nature of things seems reasonable. This approach tends to bring about a different reality, one that is orderly, harmonious, and creative. For this actually to happen, however, a thoroughgoing end to fragmentation is necessary. . . .

Because we are enfolded inseparably in the world, with no ultimate division between matter and consciousness, *meaning and value are as much integral aspects of the world as they are of us*. If science is carried out with an amoral attitude, the world will ultimately respond to science in a destructive way. Postmodern science must therefore overcome the separation between truth and virtue, value and fact, ethics and practical necessity. To call for this nonseparation, is, of course, to ask for a tremendous revolution in our whole attitude to knowledge. But such a change is now necessary and indeed long overdue. Can humanity meet in time the challenge of what is required? The coming years will be crucial in revealing the answer to this question.

NOTES

1. See Huston Smith, *Beyond the Post-Modern Mind* (New York: Crossroad, 1982), esp. ch. 8, "Beyond the Modern Western Mind-Set."

2. See David Bohm, *Wholeness and the Implicate Order* (London: Routledge and Kegan Paul, 1980) and other references given therein.

34

PREDICTABILITY

Does the Flap of a Butterfly's Wings in Brazil Set Off a Tornado in Texas?

EDWARD LORENZ

L est I appear frivolous in even posing the title question, let alone suggesting that it might have an affirmative answer, let me try to place it in proper perspective by offering two propositions: (1) if a single flap of a butterfly's wings can be instrumental in generating a tornado, so also can all the previous and subsequent flaps of its wings, as can the flaps of the wings of millions of other butterflies, not to mention the activities of innumerable more powerful creatures, including our own species; and (2) if the flap of a butterfly's wings can be instrumental in generating a tornado, it can equally well be instrumental in preventing a tornado. More generally, I am proposing that over the years, minuscule disturbances neither increase nor decrease the frequency of occurrence of various weather events such as tornados; the most that they may do is to modify the sequence in which these events occur. The question which really interests us is whether they can do even this—whether, for example, two particular weather situations differing by as little as the immediate influence of a single butterfly will generally, after sufficient time, evolve into two situations differing by as much as the presence of a tornado. In more technical language: Is the behavior of the atmosphere unstable with respect to perturbations of small amplitude?

The connection between this question and our ability to predict the weather is evident. Since we do not know exactly how many butterflies there are, nor where they are all located, let alone which ones are flapping their wings at any

From: *The Essence of Chaos* (Seattle: University of Washington Press, 1993), pp. 181–84. Paper presented at the annual meeting of the American Association for the Advancement of Science, Section on Environmental Sciences: New Approaches to Global Weather: The Global Atmospheric Research Program, December 29, 1972.

instant, we cannot, if the answer to our question is affirmative, accurately predict the occurrence of tornados at a sufficiently distant future time. More significantly, our general failure to detect systems even as large as thunderstorms when they slip between weather stations may impair our ability to predict the general weather pattern even in the near future.

How can we determine whether the atmosphere is unstable? The atmosphere is not a controlled laboratory experiment; if we disturb it and then observe what happens, we shall never know what would have happened if we had not disturbed it. Any claim that we can learn what would have happened by referring to the weather forecast would imply that the question whose answer we seek has already been answered in the negative.

The bulk of our conclusions are based upon computer simulation of the atmosphere. The equations to be solved represent our best attempts to approximate the equations actually governing the atmosphere by equations which are compatible with present computer capabilities. Generally, two numerical solutions are compared. One of these is taken to simulate the actual weather, while the other simulates the weather which would have evolved from slightly different initial conditions, that is, the weather which would have been predicted with a perfect forecasting technique but imperfect observations. The difference between the solutions therefore simulates the error in forecasting. New simulations are continually being performed as more powerful computers and improved knowledge of atmospheric dynamics become available.

Although we cannot claim to have proven that the atmosphere is unstable, the evidence that it is so is overwhelming. The most significant results are the following:

1. Small errors in the coarser structure of the weather pattern—those features which are readily resolved by conventional observing networks—tend to double in about three days. As the errors become larger the growth rate subsides. This limitation alone would allow us to extend the range of acceptable prediction by three days every time we cut the observation error in half and would offer the hope of eventually making good forecasts several weeks in advance.

2. Small errors in the finer structure—for example, the positions of individual clouds—tend to grow much more rapidly, doubling in hours or less. This limitation alone would not seriously reduce our hopes for extended-range forecasting, since ordinarily we do not forecast the finer structure at all.

3. Errors in the finer structure, having attained appreciable size, tend to induce errors in the coarser structure. This result, which is less firmly established than the previous ones, implies that after a day or so there will be appreciable errors in the coarser structure, which will thereafter grow just as if they had been present initially. Cutting the observation error in the finer structure in half—a formidable task—would extend the range of acceptable prediction of even the coarser structure only by hours or less. The hopes for predicting two weeks or more in advance are thus greatly diminished.

4. Certain special quantities, such as weekly average temperatures and weekly total rainfall, may be predictable at a range at which entire weather patterns are not.

Regardless of what any theoretical study may imply, conclusive proof that good day-to-day forecasts can be made at a range of two weeks or more would be afforded by any valid demonstration that any particular forecasting scheme generally yields good results at that range. To the best of our knowledge, no such demonstration has ever been offered. Of course, even pure guesses will be correct a certain percentage of the time.

Returning now to the question as originally posed, we notice some additional points not yet considered. First of all, the influence of a single butterfly is not only a fine detail—it is confined to a small volume. Some of the numerical methods which seem to be well adapted for examining the intensification of errors are not suitable for studying the dispersion of errors from restricted to unrestricted regions. One hypothesis, unconfirmed, is that the influence of a butterfly's wings will spread in turbulent air, but not in calm air.

A second point is that Brazil and Texas lie in opposite hemispheres. The dynamical properties of the tropical atmosphere differ considerably from those of the atmosphere in temperate and polar latitudes. It is almost as if the tropical atmosphere were a different fluid. It seems entirely possible that an error might be able to spread many thousands of miles within the temperate latitudes of either hemisphere, while yet being unable to cross the equator.

We must therefore leave our original question unanswered for a few more years, even while affirming our faith in the instability of the atmosphere. Meanwhile, today's errors in weather forecasting cannot be blamed entirely, or even primarily, upon the finer structure of weather patterns. They arise mainly from our failure to observe even the coarser structure with near completeness, our somewhat incomplete knowledge of the governing physical principles, and the inevitable approximations which must be introduced in formulating these principles as procedures which the human brain or the computer can carry out. These shortcomings cannot be entirely eliminated, but they can be greatly reduced by an expanded observing system and intensive research. It is to the ultimate purpose of making not exact forecasts but the best forecasts which the atmosphere is willing to have us make that the Global Atmospheric Research Program is dedicated.

35

SCIENCE IN A WORLD
OF LIMITED PREDICTABILITY

ILYA PRIGOGINE

The twentieth century [was] a remarkable century in physics. It started with completely new theories and conceptions—quantum mechanics and relativity. Then in the second third of [the] century came some absolutely unexpected discoveries, which nobody could have predicted, such as the discovery that matter is unstable, and that elementary particles can transform into each other. The second discovery was that our universe has a history. Classically, the idea was that there could be no history of the universe because the universe contains everything. And the third discovery, I would say, was the discovery that non-equilibrium irreversibility can be a source of organization. These changes severely radicalize our views of space and time. And in essence, what we are trying to do now is to incorporate these unexpected discoveries into a more consistent picture. I want to emphasize that from the point of view of classical physics, there was a dichotomy—on the one hand, physics had the view of the universe as a giant automaton, at some stage we were satisfied with time-reversible and deterministic laws. On the other hand, when we see our own internal spiritual life, we see the importance of creativity, the fact that time is irreversible, and the fact that we have at least the feeling that we see, in a sense, order coming out of disorder—new ideas from fragments coming together.

Now, the view of classical physics is not an accident; in fact, it was elaborated in a very famous discussion between Gottfried Wilhelm Leibniz and Samuel Clarke (speaking in the name of Sir Isaac Newton). This is still a very

From: "The Rediscovery of Time: Science in a World of Limited Predictability," paper presented to the International Congress on "Geist & Natur" [Spirit and Nature], Hanover, Germany, May 21–27, 1988, excerpts.

interesting discussion to read today, . . . two centuries after it took place. Leibniz attacks Newton by saying that Newton has a poor idea of God because he believes that God is inferior to a good watchmaker. In the view of Newton, God had to repair the universe from time to time. In other words, in more modern language, one would say that Newton curiously was in favor of an evolutionary view of the universe. Clarke (Newton) said the idea of Leibniz, that everything was created in a single step, made God a *rex autiosus*, a *roi infini*, because once the creation of the universe had taken place, God had no more role. What is so important is that there is a theological component in this discussion in the way in which people understood the role of the scientist. It is clear that as people imagined God in the seventeenth century, he, of course, knew everything. For him there could be no difference between past, present, and future. And Leibniz's view became more and more important until Laplace said that for a sufficiently informed being nothing can happen which is not already in the present. For a scientist, who would be the representative of God on Earth, time would be an illusion, as Einstein said. In other words, one of the main components of classical science was to try to reach a knowledge of the universe which would be a knowledge we could imagine God would have. In physics, there have been a lot of demons—Maxwell's and Laplace's—the idea being that there may be a knowledge which would involve a kind of knowledge from above. One of the important things which we . . . learned [in the last] century is that this is impossible. In essence, we have only a window on nature, and out of the elements we have, we have to extrapolate. We are ourselves involved in the nature we describe.

The conflict between time as perceived by an evolutionary theory or by our existential experience and time of fundamental classical physics came to the forefront after the work of Ludwig Boltzmann. Boltzmann tried to give a mechanical interpretation of the second law of thermodynamics, of the increase of entropy, when he obtained a very interesting description of irreversibility. However, people were not convinced. They said, How can you even hope to obtain a description of irreversibility when the basic laws of nature are deterministic and reversible? That is impossible. Finally, Boltzmann had to retreat and to accept that on the basic level there is no time.

I was astonished that nobody, or very few people, protested. Nobody, or very few people at least, said, but there must be something difficult there, something which still has to be revised. How can we accept this dichotomy between a phenomenological description in which time exists and a fundamental description in which time would disappear? One of the main points in the evolution of [the twentieth] century is that now we understand better how we can go beyond this dichotomy; now we can understand irreversibility on a fundamental basis.

I would like to quote a sentence by the British mathematician James Lighthill, who wrote: "I have to speak on behalf of the broad global fraternity of researchers on mechanics. We collectively wish to apologize for having misled the general educated public by spreading ideas of all the determinisms of systems satisfying Newton's law of motion that after 1960 were to be proved incorrect." This statement is quite amazing. It helps us, of course, that an individual scientist apologizes for mistakes he may have made. That someone in the name of all the people working in classical dynamics would apologize for a mistake made for three centuries is something unlikely in the history of science. In other words, he says that the whole interpretation which was the basis of all our philosophy of space and time—the way in which Immanuel Kant, Henri Bergson, and others have seen classical science—was wrong.

I would like to explain briefly how these new aspects entered classical physics. And I will first speak briefly about nonequilibrium processes—thermodynamics—and then about classical mechanics, because I believe that it is there that we see the greatest change. Then I would like to speak about the way in which these ideas begin to appear in modern physics, that is, in quantum mechanics. Finally, I will say a few words about recent cosmology and the role of time and irreversibility in the creation of our universe.

First, thermodynamics is based, as you know, on the second law. The second law says that entropy for an isolated system will increase. If it is not isolated in any way, entropy will be produced by irreversible processes. Now, people had always tried to minimize this statement. They have said it is rather trivial. If I have two boxes, with a hundred particles in the left box and twenty particles in the right, the boxes bearing the same volume, of course, then after some time I will have fifty-fifty. That is rather trivial. Therefore, they said, don't worry about the second law. However, we now understand that this is completely wrong. Irreversible processes always create—or nearly always create—both order and disorder at the same time. Take again the example of the two boxes. If I have two boxes of hydrogen and nitrogen and I heat one and cool the other, then there will be more hydrogen on one side and more nitrogen on the other. In other words, I produce order. It is true that I have to pay a price. I have to pay a price because I have to dissipate energy through the heat flow. That this price leads to order in this and many other examples is now well understood. In fact, we are changing our paradigm. The classical paradigm was that the crystal is an ordered subject, an ordered object. However, we now know that even a nice crystal is far from being so ordered. The particles are moving, and this motion is described by incoherent thermal waves. On the other hand, what we believe to be disorder, turbulence, for example, appears today as highly ordered. In order to have turbulence, particles have to be correlated for an enormous distance, distances much larger than intermolecular distances, distances which are macroscopic. The coherence has be on a macroscopic scale. We see now that nonequilibrium processes contain much more order than most or even all equilibrium processes.

This completely changes our view of the relation between order and disorder, on the one hand, and equilibrium and nonequilibrium on the other. And . . . more and more of these kinds of phenomena have been discovered in hydrodynamics and chemistry. Not a month goes by without discovering another nonequilibrium type of structure. In essence, we see a kind of unexpected evolution of chemistry because chemistry is a science of irreversible processes. It is a science of nonlinear processes because we have collisions and interactions between particles. What we see now is that irreversibility and nonlinearity lead to self-organization. And we can follow this in many fields: the science of materials and many other examples. Another aspect I want to emphasize is that irreversibility leads to an enhancement of the role of fluctuation. We give up a deterministic description because nonlinearity leads through bifurcations to the possibility of different solutions. When you come to this type of bifurcation, the system can only be described in a probabilistic way. You cannot predict with certainty where the system will go. Quantum mechanics has become famous for introducing probabilistic concepts on the microscopic scale. Here we have a probabilistic concept on the macroscopic scale. So, as I said, nonequilibrium physics is in a period of exponential expansion. This is still phenomenology because it involves many particles. This involves systems which I cannot solve exactly. It is therefore interesting to understand how this irreversibility can be related to dynamics. And here we come to the revolution in dynamics to which Lighthill alluded.

What type of new dynamical systems have been discovered? The great names are Henri Poincaré at the end of the [nineteenth and the beginning of the twentieth century], and Andrei Nikolaevich Kolmogoroff, perhaps the greatest mathematician of [the twentieth] century. What Kolmogoroff discovered is that there are many dynamical systems which are so unstable that whatever the distance between two trajectories, the two trajectories will diverge exponentially in time. In other words, even if the distance between the two trajectories was initially very small, after some time it will become as large as you want. . . .

Now, how do unstable systems behave? How do they behave in contrast, let us say, with planetary motion? The important element is that unstable systems are not controllable. The classical view of nature, the classical view of the laws of nature, of our relation with nature, was domination. That we can control everything. If we change our initial conditions, the trajectories change slightly. If we change our initial wave function, the final wave function is changed slightly. We can, therefore, control things. And this remains true for many situations. When we send a *Sputnik* into space, fortunately we can control its trajectory. But that is not the general situation. If, for example, I take a few hundred particles and I give them initially all kinds of positions and velocities, whatever I do, I cannot prevent the system from reaching thermal equilibrium and a Maxwell-Boltzmann distribution after some time. In other words, the system escapes my control and behaves—if I can say so—like it wants to behave independently of the instructions which I gave it. You could say again that a system with a few hundred particles, a liquid or a gas, is a very complex system and it is not so astonishing that I cannot control it. But recently some of my colleagues performed numerical analyses, computer experiments, and other analytical calculations for the simplest problem in electrodynamics, the emission of radiation. In the emission of radiation, if I have a charge which is oscillating, it radiates according to classical electrodynamics and sends electromagnetic waves, that is, light, into space. In quantum mechanics, we would say it sends photons into space. Now what we have shown is that we can prepare the initial electron of the initial charge in many ways and we can even prevent the system from radiating for some time. But whatever we do, after some time it behaves in agreement with the laws of electrodynamics. In other words, it is escaping from our control. It is dissipating its vibrational energy into radiation, whatever we do in the preparation of the system.

The main point which I want to emphasize is that we see in nature the appearance of spontaneous processes which we cannot control in the strict sense in which it was imagined to be possible in classical mechanics. This is not giving up scientific rationality. After all, we have not chosen the world in which we are living. We scientists have to describe the world in which we are living. And the world in which we are living is highly unstable. What I want to emphasize, however is that this knowledge of instability may lead to other types of strategies, may lead to other types of interacting systems.

Well, thermodynamics, nonequilibrium processes, and classical dynamics are, of course, parts of classical physics. And it is very remarkable that parts of classical physics which were supposed to be in a final form or a nearly final form are precisely some of the fields which are changing so much in the present evolution of science. But, of course, we can expect that this evolution of classical physics will have very deep consequences in the more modern fields of quantum mechanics and relativity, because all the other fields of physics are, in essence, born inside classical physics. And they have used some of the concepts of stable dynamical systems, which were believed for such a long time to be the only type of dynamical systems. . . .

Quantum mechanics has this strange dual structure. On the one hand, you have the deterministic Schrödinger equation. Once you know the wave function, you can predict what will happen in the future just as you could calculate what happened in the past. Irreversibility and stochasticity come from our measurement. This is not such an absurd idea, because if I look, for example, at planetary motion, I know that it is reversible motion. But when I look at it, I use for the chemical reaction an irreversible process. Therefore, my observation of a time-reversible process introduces irreversibility. But quantum mechanics made an enormous statement when it said that all of irreversibility is due to my observation—in other words, to my measurement—and that was exactly the situation of classical physics before the discoveries of Kolmogoroff and Poincaré. What is very difficult to believe now is that in quantum physics as in classical physics, we have all the spontaneous processes: disintegration, emission of photons which again have nothing to do with our measurement, which comes from the intrinsic instability of dynamical systems. So I believe that the future revolution of quantum mechanics will be dealing with a better understanding of the classification of dynamical quantum systems.

It is quite remarkable to me that this change . . . has appeared at the very moment humanity is going through an age of transition, when instability, irreversibility, fluctuation, amplification are found in every human activity. The idea of chaos, of amplification, became very popular in the United States after October 19, 1987, the famous Black Monday of the crash of Wall Street. From this time on, everybody became interested in chaos and amplification. What is so interesting is that there is a kind of overall cultural atmosphere, be it in science, be it in human science, which is [still developing today].

Conclusion

PRINCIPLES OF ENVIRONMENTAL JUSTICE

THE FIRST NATIONAL PEOPLE OF COLOR ENVIRONMENTAL LEADERSHIP SUMMIT OCTOBER 24–27, 1991 WASHINGTON, DC

PREAMBLE

We, the people of color, gathered together at this multinational People of Color Environmental Leadership Summit, to begin to build a national and international movement of all peoples of color to fight the destruction and taking of our lands and communities, do hereby reestablish our spiritual interdependence to the sacredness of our Mother Earth; to respect and celebrate each of our cultures, languages, and beliefs about the natural world and our roles in healing ourselves; to ensure environmental justice; to promote economic alternatives which would contribute to the development of environmentally safe livelihoods; and to secure our political, economic, and cultural liberation that has been denied for over five hundred years of colonization and oppression, resulting in the poisoning of our communities and land and the genocide of our peoples, do affirm and adopt these Principles of Environmental Justice:

1. Environmental justice affirms the sacredness of Mother Earth, ecological unity, and the interdependence of all species, and the right to be free from ecological destruction.
2. Environmental justice demands that public policy be based on mutual respect and justice for all peoples, free from any form of discrimination or bias.
3. Environmental justice mandates the right to ethical, balanced, and responsible uses of land and renewable resources in the interest of a sustainable planet for humans and other living things.
4. Environmental justice calls for universal protection from nuclear

testing, extraction, production, and disposal of toxic/hazardous wastes and poisons and nuclear testing that threatens the fundamental right to clean air, land, water, and food.

5. Environmental justice affirms the fundamental right to political, economic, cultural, and environmental self-determination of all peoples.

6. Environmental justice demands the cessation of the production of all toxins, hazardous wastes, and radioactive materials, and that all past and current producers be held strictly accountable to the people for detoxification and the containment at the point of production.

7. Environmental justice demands the right to participate as equal partners at every level of decision making, including needs assessment, planning, implementation, enforcement, and evaluation.

8. Environmental justice affirms the right of all workers to a safe and healthy work environment, without being forced to choose between an unsafe livelihood and unemployment. It also affirms the right of those who work at home to be free from environmental hazards.

9. Environmental justice protects the right of victims of environmental injustice to receive full compensation and reparations for damages as well as quality healthcare.

10. Environmental justice considers governmental acts of environmental injustice a violation of international law, the Universal Declaration on Human Rights, and the United Nations Convention on Genocide.

11. Environmental justice must recognize a special legal and natural relationship of Native Peoples to the US government through treaties, agreements, compacts, and covenants affirming sovereignty and self-determination.

12. Environmental justice affirms the need for urban and rural ecological policies to clean up and rebuild our cities and rural areas in balance with nature, honoring the cultural integrity of all our communities, and providing fair access for all to the full range of resources.

13. Environmental justice calls for the strict enforcement of principles of informed consent, and a halt to the testing of experimental reproductive and medical procedures and vaccinations on people of color.

14. Environmental justice opposes the destructive operations of multinational corporations.

15. Environmental justice opposes military occupation, repression, and exploitation of lands, peoples, and cultures, and other life-forms.

16. Environmental justice calls for the education of present and future generations which emphasizes social and environmental issues, based on our experience and an appreciation of our diverse cultural perspectives.

17. Environmental justice requires that we, as individuals, make personal and consumer choices to consume as little of Mother Earth's resources and to produce as little waste as possible; and make the conscious deci-

sion to challenge and reprioritize our lifestyles to ensure the health of the natural world for present and future generations.

Adopted, October 27, 1991
The First National People of Color Environmental Leadership Summit
Washington, DC

INDEX

411